Introduction to Finite Element Analysis

WILEY SERIES IN COMPUTATIONAL MECHANICS

Series Advisors:

René de Borst
Perumal Nithiarasu
Tayfun E. Tezduyar
Genki Yagawa
Tarek Zohdi

Introduction to Finite Element Analysis

Formulation, Verification and Validation

Barna Szabó

Washington University in St. Louis, USA

Ivo Babuška

The University of Texas at Austin, USA

A John Wiley and Sons, Ltd., Publication

This edition first published 2011
© 2011 John Wiley & Sons, Ltd

Registered office
John Wiley & Sons Ltd, The Atrium, Southern Gate, Chichester, West Sussex, PO19 8SQ, United Kingdom

For details of our global editorial offices, for customer services and for information about how to apply for permission to reuse the copyright material in this book please see our website at www.wiley.com.

The right of the authors to be identified as the authors of this work has been asserted in accordance with the Copyright, Designs and Patents Act 1988.

All rights reserved. No part of this publication may be reproduced, stored in a retrieval system, or transmitted, in any form or by any means, electronic, mechanical, photocopying, recording or otherwise, except as permitted by the UK Copyright, Designs and Patents Act 1988, without the prior permission of the publisher.

Wiley also publishes its books in a variety of electronic formats. Some content that appears in print may not be available in electronic books.

Designations used by companies to distinguish their products are often claimed as trademarks. All brand names and product names used in this book are trade names, service marks, trademarks or registered trademarks of their respective owners. The publisher is not associated with any product or vendor mentioned in this book. This publication is designed to provide accurate and authoritative information in regard to the subject matter covered. It is sold on the understanding that the publisher is not engaged in rendering professional services. If professional advice or other expert assistance is required, the services of a competent professional should be sought.

MATLAB® is a trademark of The MathWorks, Inc. and is used with permission. The MathWorks does not warrant the accuracy of the text or exercises in this book. This book's use or discussion of MATLAB® software or related products does not constitute endorsement or sponsorship by The MathWorks of a particular pedagogical approach or particular use of the MATLAB® software.

Library of Congress Cataloguing-in-Publication Data

Szabó, B. A. (Barna Aladar), 1935-
  Introduction to finite element analysis : formulation, verification, and validation / Barna Szabó, Ivo Babuška.
    p. cm.
  Includes bibliographical references and index.
  ISBN 978-0-470-97728-6 (hardback)
1. Finite element method.   I. Babuška, Ivo.   II. Title.
  TA347.F5S979   2011
  620.001′51825–dc22

2010051233

A catalogue record for this book is available from the British Library.

Print ISBN: 9780470977286
ePDF ISBN: 9781119993827
oBook ISBN: 9781119993834
ePub ISBN: 9781119993483
Mobi ISBN: 9781119993490

Set in 10/12 Times by Aptara Inc., New Delhi, India.

This book is dedicated to our teachers and students.

If people do not believe that mathematics is simple, it is only because they do not realize how complicated life is.
—John von Neumann

Contents

About the Authors		xiii
Series Preface		xv
Preface		xvii

1 Introduction — 1
  1.1 Numerical simulation — 2
    1.1.1 Conceptualization — 2
    1.1.2 Validation — 5
    1.1.3 Discretization — 7
    1.1.4 Verification — 8
    1.1.5 Decision-making — 9
  1.2 Why is numerical accuracy important? — 11
    1.2.1 Application of design rules — 11
    1.2.2 Formulation of design rules — 12
  1.3 Chapter summary — 14

2 An outline of the finite element method — 17
  2.1 Mathematical models in one dimension — 17
    2.1.1 The elastic bar — 17
    2.1.2 Conceptualization — 24
    2.1.3 Validation — 27
    2.1.4 The scalar elliptic boundary value problem in one dimension — 28
  2.2 Approximate solution — 29
    2.2.1 Basis functions — 32
  2.3 Generalized formulation in one dimension — 33
    2.3.1 Essential boundary conditions — 35
    2.3.2 Neumann boundary conditions — 37
    2.3.3 Robin boundary conditions — 37
  2.4 Finite element approximations — 38
    2.4.1 Error measures and norms — 41
    2.4.2 The error of approximation in the energy norm — 43

CONTENTS

	2.5	FEM in one dimension		44
		2.5.1 The standard element		44
		2.5.2 The standard polynomial space		45
		2.5.3 Finite element spaces		47
		2.5.4 Computation of the coefficient matrices		49
		2.5.5 Computation of the right hand side vector		52
		2.5.6 Assembly		55
		2.5.7 Treatment of the essential boundary conditions		58
		2.5.8 Solution		61
		2.5.9 Post-solution operations		62
	2.6	Properties of the generalized formulation		67
		2.6.1 Uniqueness		67
		2.6.2 Potential energy		68
		2.6.3 Error in the energy norm		68
		2.6.4 Continuity		69
		2.6.5 Convergence in the energy norm		70
	2.7	Error estimation based on extrapolation		73
		2.7.1 The root-mean-square measure of stress		74
	2.8	Extraction methods		75
	2.9	Laboratory exercises		77
	2.10	Chapter summary		77
3	Formulation of mathematical models			79
	3.1	Notation		79
	3.2	Heat conduction		81
		3.2.1 The differential equation		83
		3.2.2 Boundary and initial conditions		83
		3.2.3 Symmetry, antisymmetry and periodicity		85
		3.2.4 Dimensional reduction		86
	3.3	The scalar elliptic boundary value problem		92
	3.4	Linear elasticity		93
		3.4.1 The Navier equations		97
		3.4.2 Boundary and initial conditions		97
		3.4.3 Symmetry, antisymmetry and periodicity		99
		3.4.4 Dimensional reduction		100
	3.5	Incompressible elastic materials		103
	3.6	Stokes' flow		105
	3.7	The hierarchic view of mathematical models		106
	3.8	Chapter summary		106
4	Generalized formulations			109
	4.1	The scalar elliptic problem		109
		4.1.1 Continuity		111
		4.1.2 Existence		112
		4.1.3 Approximation by the finite element method		112
	4.2	The principle of virtual work		115
	4.3	Elastostatic problems		117

		4.3.1	Uniqueness	119
		4.3.2	The principle of minimum potential energy	127
	4.4		Elastodynamic models	133
		4.4.1	Undamped free vibration	134
	4.5		Incompressible materials	140
		4.5.1	The saddle point problem	142
		4.5.2	Poisson's ratio locking	142
		4.5.3	Solvability	143
	4.6		Chapter summary	143
5	Finite element spaces			145
	5.1		Standard elements in two dimensions	145
	5.2		Standard polynomial spaces	146
		5.2.1	Trunk spaces	146
		5.2.2	Product spaces	147
	5.3		Shape functions	147
		5.3.1	Lagrange shape functions	148
		5.3.2	Hierarchic shape functions	150
	5.4		Mapping functions in two dimensions	152
		5.4.1	Isoparametric mapping	152
		5.4.2	Mapping by the blending function method	154
		5.4.3	Mapping of high-order elements	156
		5.4.4	Rigid body rotations	156
	5.5		Elements in three dimensions	157
	5.6		Integration and differentiation	158
		5.6.1	Volume and area integrals	159
		5.6.2	Surface and contour integrals	160
		5.6.3	Differentiation	161
	5.7		Stiffness matrices and load vectors	162
		5.7.1	Stiffness matrices	162
		5.7.2	Load vectors	164
	5.8		Chapter summary	164
6	Regularity and rates of convergence			167
	6.1		Regularity	167
	6.2		Classification	170
	6.3		The neighborhood of singular points	173
		6.3.1	The Laplace equation	173
		6.3.2	The Navier equations	175
		6.3.3	Material interfaces	183
		6.3.4	Forcing functions acting on boundaries	185
		6.3.5	Strong and weak singular points	192
	6.4		Rates of convergence	193
		6.4.1	The choice of finite element spaces	196
		6.4.2	Uses of a priori information	201
		6.4.3	A posteriori error estimation in the energy norm	209
		6.4.4	Adaptive and feedback methods	211
	6.5		Chapter summary	212

7	Computation and verification of data		215
	7.1	Computation of the solution and its first derivatives	215
	7.2	Nodal forces	217
		7.2.1 Nodal forces in the h-version	217
		7.2.2 Nodal forces in the p-version	220
		7.2.3 Nodal forces and stress resultants	221
	7.3	Verification of computed data	222
	7.4	Flux and stress intensity factors	228
		7.4.1 The Laplace equation	228
		7.4.2 Planar elasticity	232
	7.5	Chapter summary	235
8	What should be computed and why?		237
	8.1	Basic assumptions	238
	8.2	Conceptualization: drivers of damage accumulation	238
	8.3	Classical models of metal fatigue	240
		8.3.1 Models of damage accumulation	243
		8.3.2 Notch sensitivity	246
		8.3.3 The theory of critical distances	248
	8.4	Linear elastic fracture mechanics	250
	8.5	On the existence of a critical distance	252
	8.6	Driving forces for damage accumulation	253
	8.7	Cycle counting	254
	8.8	Validation	255
	8.9	Chapter summary	257
9	Beams, plates and shells		261
	9.1	Beams	261
		9.1.1 The Timoshenko beam	264
		9.1.2 The Bernoulli–Euler beam	269
	9.2	Plates	274
		9.2.1 The Reissner–Mindlin plate	276
		9.2.2 The Kirchhoff plate	280
		9.2.3 Enforcement of continuity: the HCT element	282
	9.3	Shells	283
		9.3.1 Hierarchic "thin-solid" models	286
	9.4	The Oak Ridge experiments	288
		9.4.1 Description	288
		9.4.2 Conceptualization	290
		9.4.3 Verification	291
		9.4.4 Validation: comparison of predicted and observed data	293
		9.4.5 Discussion	295
	9.5	Chapter summary	296
10	Nonlinear models		297
	10.1	Heat conduction	297
		10.1.1 Radiation	297
		10.1.2 Nonlinear material properties	298

	10.2	Solid mechanics	298
		10.2.1 Large strain and rotation	299
		10.2.2 Structural stability and stress stiffening	302
		10.2.3 Plasticity	306
		10.2.4 Mechanical contact	310
	10.3	Chapter summary	313
A	Definitions		315
	A.1	Norms and seminorms	315
	A.2	Normed linear spaces	316
	A.3	Linear functionals	316
	A.4	Bilinear forms	316
	A.5	Convergence	317
	A.6	Legendre polynomials	317
	A.7	Analytic functions	318
		A.7.1 Analytic functions in $\mathbb{R}^2$	318
		A.7.2 Analytic curves in $\mathbb{R}^2$	318
	A.8	The Schwarz inequality for integrals	319
B	Numerical quadrature		321
	B.1	Gaussian quadrature	322
	B.2	Gauss–Lobatto quadrature	323
C	Properties of the stress tensor		325
	C.1	The traction vector	325
	C.2	Principal stresses	326
	C.3	Transformation of vectors	327
	C.4	Transformation of stresses	328
D	Computation of stress intensity factors		331
	D.1	The contour integral method	331
	D.2	The energy release rate	333
		D.2.1 Symmetric (Mode I) loading	333
		D.2.2 Antisymmetric (Mode II) loading	334
		D.2.3 Combined (Mode I and Mode II) loading	335
		D.2.4 Computation by the stiffness derivative method	335
E	Saint-Venant's principle		337
	E.1	Green's function for the Laplace equation	337
	E.2	Model problem	338
F	Solutions for selected exercises		345

Bibliography 353

Index 359

About the Authors

Barna Szabó is co-founder and president of Engineering Software Research and Development, Inc. (ESRD), the company that produces the professional finite element analysis software StressCheck®. Prior to his retirement from the School of Engineering and Applied Science of Washington University in 2006 he served as the Albert P. and Blanche Y. Greensfelder Professor of Mechanics. His primary research interest is assurance of quality and reliability in the numerical simulation of structural and mechanical systems by the finite element method. He has published over 150 papers in refereed technical journals, several of them in collaboration with Professor Ivo Babuška, with whom he also published a book on finite element analysis (John Wiley & Sons, Inc., 1991). He is a founding member and Fellow of the US Association for Computational Mechanics. Among his honors are election to the Hungarian Academy of Sciences as External Member and an honorary doctorate.

Ivo Babuška's research has been concerned mainly with the reliability of computational analysis of mathematical problems and their applications, especially by the finite element method. He was the first to address a posteriori error estimation and adaptivity in finite element analysis. His research papers on these subjects published in the 1970s have been widely cited. His joint work with Barna Szabó on the p-version of the finite element method established the theoretical foundations and the algorithmic structure for this method. His recent work has been concerned with the mathematical formulation and treatment of uncertainties which are present in every mathematical model. In recognition of his numerous important contributions, Professor Babuška received may honors, which include honorary doctorates, medals and prizes and election to prestigious academies.

Series Preface

The series on *Computational Mechanics* will be a conveniently identifiable set of books covering interrelated subjects that have been receiving much attention in recent years and need to have a place in senior undergraduate and graduate school curricula, and in engineering practice. The subjects will cover applications and methods categories. They will range from biomechanics to fluid-structure interactions to multiscale mechanics and from computational geometry to meshfree techniques to parallel and iterative computing methods. Application areas will be across the board in a wide range of industries, including civil, mechanical, aerospace, automotive, environmental and biomedical engineering. Practicing engineers, researchers and software developers at universities, industry and government laboratories, and graduate students will find this book series to be an indispensible source for new engineering approaches, interdisciplinary research, and a comprehensive learning experience in computational mechanics.

This book, written by two well-recognized, leading experts on finite element analysis, gives an introduction to finite element analysis with an emphasis on validation – the process to ascertain that the mathematical/numerical model meets acceptance criteria – and verification – the process for acceptability of the approximate solution and computed data. The systematic treatment of formulation, verification and validation procedures is a distinguishing feature of this book and sets it apart from other texts on finite elements. It encapsulates contemporary research on proper model selection and control of modelling errors. Another unique feature of the book is that with a minimum of mathematical requisites it bridges the gap between engineering and mathematically-oriented introductory text books into finite elements.

Preface

Increasingly, engineering decisions are based on computed information with the expectation that the computed information will provide a reliable quantitative estimate of some attributes of a physical system or process. The question of how much reliance on computed information can be justified is being asked with increasing frequency and urgency. Assurance of the reliability of computed information has two key aspects: (a) selection of a suitable mathematical model and (b) approximation of the solution of the corresponding mathematical problem. The process by which it is ascertained that a mathematical model meets necessary criteria for acceptance (i.e., it is not unsuitable for purposes of analysis) is called *validation*. The process by which it is ascertained that the approximate solution, as well as the data computed from the approximate solution, meet necessary conditions for acceptance, given the goals of computation, is called *verification*. This book addresses the problems of verification and validation.

Obtaining approximate solutions for mathematical models with guaranteed accuracy is one of the principal goals of research in finite element analysis. An important result obtained in the mid-1980s was that exponential rates of convergence can be achieved through proper design of the finite element mesh and proper assignment of polynomial degrees for a large and important class of problems that includes elasticity, heat conduction and similar problems. This made it feasible to estimate and control the errors of discretization for many practical problems.

At present the problems of proper model selection and control of modeling errors are at the forefront of research. The concepts of hierarchic models and modeling strategies have been developed. Progress in this area makes many important practical applications possible.

The distinguishing feature of this book is that it presents a systematic treatment of formulation, verification and validation procedures, illustrated by examples. We believe that users of finite element analysis (FEA) software products must have a basic understanding of how mathematical models are constructed; what are the essential assumptions incorporated in a mathematical model; what is the algorithmic structure of the finite element method; how the discretization parameters affect the accuracy of the finite element solution; how the accuracy of the computed data can be assessed; and how to avoid common pitfalls and mistakes. Our primary objective in assembling the material presented in this book is to provide a basic working knowledge of the finite element method. A link to the student edition of a professional FEA software product called StressCheck® is provided in the companion website (www.wiley.com/go/szabo) to enable readers to perform computational experiments.[1]

---

[1] StressCheck® is a trademark of Engineering Software Research and Development, Inc., St. Louis, Missouri, USA.

Another important objective of this book is to prepare readers to follow and understand new developments in the field of FEA through continued self-study.

Engineering students typically take only one course in FEA, consisting of approximately 15 weeks of instruction (45 lecture hours). We have organized the material in this book so as to make efficient use of the available time. The book is written in such a way that the prerequisites are minimal. Junior standing in engineering with some background in potential flow and strength of materials are sufficient. For this reason the mathematical content is focused on the introduction of the essential concepts and terminology necessary for understanding applications of FEA in elasticity and heat conduction. Some key theorems are proven in a simple setting.

We would like to thank Dr. Norman F. Knight, Jr. and Dr. Sebastian Nervi for reviewing and commenting on the manuscript.

Barna Szabó
Washington University in St. Louis, USA

Ivo Babuška
The University of Texas at Austin, USA

1

Introduction

Engineering decision-making processes increasingly rely on information computed from approximate solutions of mathematical models. Engineering decisions have legal and ethical implications. The standard applied in legal proceedings in civil cases in the United States is to have opinions, recommendations and decisions "based upon a reasonable degree of engineering certainty." Codes of ethics of engineering societies impose higher standards. For example, the Code of Ethics of the Institute of Electrical and Electronics Engineers (IEEE) requires members "to accept responsibility in making engineering decisions consistent with the safety, health, and welfare of the public, and to disclose promptly factors that might endanger the public or the environment" and "to be honest and realistic in stating claims or estimates based on available data."

An important challenge facing the computational engineering community is to establish procedures for creating evidence that will show, with a high degree of certainty, that a mathematical model of some physical reality, formulated for a particular purpose, can in fact represent the physical reality in question with sufficient accuracy to make predictions based on mathematical models useful and justifiable for the purposes of engineering decision-making and the errors in the numerical approximation are sufficiently small. There is a large and rapidly growing body of work on this subject. See, for example, [38a], [68], [52], [51], [99]. The formulation and numerical treatment of mathematical models for use in support of engineering decision-making in the field of solid mechanics is addressed in a document issued by the American Society of Mechanical Engineers (ASME) and adopted by the American National Standards Institute (ANSI) [33]. The Simulation Interoperability Standards Organization (SISO) is another important source of information.

The considerations underlying the selection of mathematical models and methods for the estimation and control of modeling errors and the errors of discretization are the two main topics of this book. In this chapter a brief overview is presented and the basic terminology is introduced.

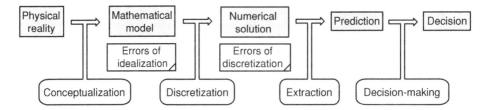

Figure 1.1 The main elements of numerical simulation and the associated errors.

1.1 Numerical simulation

The goal of numerical simulation is to make predictions concerning the response of physical systems to various kinds of excitation and, based on those predictions, make informed decisions. To achieve this goal, mathematical models are defined and the corresponding numerical solutions are computed. Mathematical models should be understood to be idealized representations of reality and should never be confused with the physical reality that they are supposed to represent.

The choice of a mathematical model depends on its intended use: What aspects of physical reality are of interest? What data must be predicted? What accuracy is required? The main elements of numerical simulation and the associated errors are indicated schematically in Figure 1.1.

Some errors are associated with the mathematical model and some errors are associated with its numerical solution. These are called errors of idealization and errors of discretization respectively. For the predictions to be reliable both kinds of errors have to be sufficiently small. The errors of idealization are also called modeling errors. Conceptualization is a process by which a mathematical model is formulated. Discretization is a process by which the exact solution of the mathematical model is approximated. Extraction is a process by which the data of interest are computed from the approximate solution. Some authors refer to the data of interest by the term "system response quantities" (SRQs).

1.1.1 Conceptualization

Mathematical models are operators that transform one set of data, the input, into another set, the output. In solid mechanics, for example, one is typically interested in predicting displacements, strains and stresses, stress intensity factors, limit loads, natural frequencies, etc., given a description of the solution domain, constitutive equations and boundary conditions (loading and constraints). Common to all models are the equations that represent the conservation of momentum (in static problems the equations of equilibrium), the strain–displacement relations and constitutive laws.

The end product of conceptualization is a mathematical model. The definition of a mathematical model involves specification of the following:

1. Theoretical formulation. The applicable physical laws, together with certain simplifications, are stated as a mathematical problem in the form of ordinary or partial differential equations, or extremum principles. For example, the classical differential equation for elastic beams is derived from the assumptions of the theory of elasticity supplemented by the assumption that the transverse variation of the longitudinal components of the

displacement vector can be approximated by a linear function without significantly affecting the data of interest, which are typically the displacements, bending moments, shear forces, natural frequencies, etc.

2. Specification of the input data. The input data comprise the following:

   (a) Data that characterize the solution domain. In engineering practice solution domains are usually constructed by means of computer-aided design (CAD) tools. CAD tools produce idealized representations of real objects. The details of idealization depend on the choice of the CAD tool and the skills and preferences of its operator.

   (b) Physical properties (elastic moduli, yield stress, coefficients of thermal expansion, thermal conductivities, etc.)

   (c) Boundary conditions (loads, constraints, prescribed temperatures, etc.)

   (d) Information or assumptions concerning the reference state and the initial conditions

   (e) Uncertainties. When some information needed in the formulation of a mathematical model is unknown then the uncertainty is said to be cognitive (also called epistemic). For example, the magnitude and distribution of residual stresses is usually unknown, some physical properties may be unknown, etc. Statistical uncertainties (also called aleatory uncertainties) are always present. Even when the average values of needed physical properties, loading and other data are known, there are statistical variations, possibly very substantial variations, in these data. Consideration of these uncertainties is necessary for proper interpretation of the computed information.

   Various methods are available for accounting for uncertainties. The choice of method depends on the quality and reliability of the available information. One such method, known as the Monte Carlo method, is to characterize input data as random variables and use repeated random sampling to compute their effects on the data of interest. If the probability density functions of the input data are sufficiently accurate and sufficiently large samples are taken then a reasonable estimate of the probability distribution of the data of interest can be obtained.

3. Statement of objectives. Definitions of the data of interest and the corresponding permissible error tolerances.

Conceptualization involves the application of expert knowledge, virtual experimentation and calibration.

Application of expert knowledge

Depending on the intended use of the model and the required accuracy of prediction, various simplifying assumptions are introduced. For example, the assumptions incorporated in the linear theory of elasticity, along with simplifying assumptions concerning the domain and the boundary conditions, are widely used in mechanical and structural engineering applications. In many applications further simplifications are introduced, resulting in beam, plate and shell models, planar models and axisymmetric models, each of which imposes additional restrictions on what boundary conditions can be specified and what data can be computed from the solution.

In the engineering literature the commonly used simplified models are grouped into separate model classes, called theories. For example, various beam, plate and shell theories have been developed. The formulation of these theories typically involves a statement on the assumed mode of deformation (e.g., plane sections remain plane and normal to the mid-surface of a deformed beam), the relationship between the functions that characterize the deformation and the strain tensor (e.g., the strain is proportional to the curvature and the distance from the neutral axis), application of Hooke's law, and statement of the equations of the equilibrium.

In undergraduate engineering curricula each model class is presented as a thing in itself and consequently there is a strong predisposition in the engineering community to view each model class as a separate entity. It is much more useful, however, to view any mathematical model as a special case of a more comprehensive model, rather than a member of a conventionally defined model class. For example, the usual beam, plate and shell models are special cases of a model based on the three-dimensional linear theory of elasticity, which in turn is a special case of a large family of models based on the equations of continuum mechanics that account for a variety of hyperelastic, elasto-plastic and other material laws, large deformation, contact, etc. This is the hierarchic view of mathematical models.

Given the rich variety of choices, model selection for particular applications is a non-trivial problem. The goal of conceptualization is to identify the simplest mathematical model that can provide predictions of the data of interest within a specified range of accuracy.

Conceptualization begins with the formulation of a tentative mathematical model based on expert knowledge. We will call this a working model. The term has the same connotation and meaning as the term working hypothesis. Since subjective judgment is involved, the formulation of the initial working model may differ from expert to expert. Nevertheless, assuming that software tools that allow systematic evaluation of mathematical models with respect to clearly defined objectives are available, it should be possible for experts to arrive at a close agreement on the definition of a mathematical model, given its intended use.

Virtual experimentation

Model selection involves systematic evaluation of the effects of various modeling assumptions on the data of interest and the sensitivity of the data of interest to uncertainties in the input data. This is done through a process called virtual experimentation.

For example, in solid mechanics one usually begins with a working model based on the linear theory of elasticity. The implied assumptions are that the strain is much smaller than unity, the stress is proportional to the strain, the displacements are so small that equilibrium equations written with respect to the undeformed configuration hold in the deformed configuration also, and the boundary conditions are independent of the displacement function. Once a verified solution is available, it is possible to examine the stress field and determine whether the stress exceeded the proportional limit of the material and whether this affects the data of interest significantly. Similarly, the effects of large deformation on the data of interest can be evaluated. Furthermore, it is possible to test the sensitivity of the data of interest to changes in boundary conditions. Virtual experimentation provides valuable information on the influence of various modeling assumptions on the data of interest.

Calibration

In the process of conceptualization there may be indications that the data of interest are sensitive functions to certain parameters that characterize material behavior or boundary conditions. If those parameters are not available then calibration experiments must be performed for the purpose of determining the needed parameters. In calibration the mathematical model is assumed to be correct and the parameters that characterize the model are selected such that the measured response matches the predicted response.

Example 1.1.1 If the goal of computation is to predict the number of load cycles that cause fatigue failure in a metal part then one or more empirical models must be chosen that require as input stress or strain amplitudes and material parameters. One of the widely used models for the prediction of fatigue life in low-cycle fatigue is the general strain–life model:

$$\epsilon_a = \frac{\bar{\sigma}_f}{E}(2N)^b + \bar{\epsilon}_f(2N)^c \tag{1.1}$$

where ϵ_a is the strain amplitude, N is the number of cycles to failure, E is the modulus of elasticity, $\bar{\sigma}_f$ is the fatigue strength coefficient, b is the fatigue strength exponent, $\bar{\epsilon}_f$ is the fatigue ductility coefficient and c is the fatigue ductility exponent. The parameters E, $\bar{\sigma}_f$, b, $\bar{\epsilon}_f$ and c are determined through calibration experiments. See, for example, [76]. Several variants of this model are in use. Standard procedures have been established for calibration experiments for metal fatigue.[1]

1.1.2 Validation

Validation is a process by which the predictive capabilities of a mathematical model are tested against experimental data. We will be concerned primarily with problems in solid mechanics for which the predictions can be tested through experiments especially designed for that purpose. This is a very large class of problems that includes all mathematical models designed for the prediction of the performance of mass-produced items. There are other important problems, such as the effects of earthquakes and other natural disasters, unique design problems, such as dams, siting of nuclear power plants and the like, for which the predictions based on mathematical models cannot be tested at full scale. In such cases the models are analyzed a posteriori and modified in the light of new information collected following an incident.

Associated with each mathematical model is a modeling error (illustrated schematically in Figure 1.1). Therefore it is necessary to have a process for testing the predictive capabilities of mathematical models. This process, called validation. is illustrated schematically in Figure 1.2.

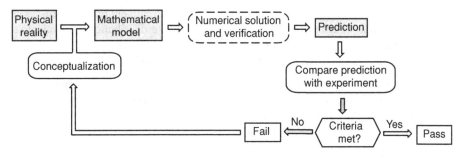

Figure 1.2 Validation.

---

[1] See, for example, the International Organization for Standardization ISO 12106:2003 and ISO 12107:2003.

For a validation experiment one or more metrics and the corresponding criteria are defined. If the predictions meet the criteria then the model is said to have passed the validation test, otherwise the model is rejected.

In large projects, such as the development of an aircraft, a series of validation experiments are performed starting with coupon tests for the determination of physical properties and failure criteria, then progressing to sub-components, components, parts, sub-assemblies and finally the entire assembly. The cost of experiments increases with complexity and hence the number of experiments decreases with complexity. The goal is to develop sufficiently reliable predictive capabilities such that the outcome of experiments involving sub-assemblies and assemblies will confirm the predictions. Finding problems late in the production cycle is generally very costly.

In evaluating the results of validation experiments it is important to bear in mind the limitations and uncertainties associated with the available information concerning the physical systems being modeled:

1. The solution domain is usually assumed to correspond to design specifications ("the blueprint"). In reality, parts, sub-assemblies and assemblies deviate from their specifications and the degree of deviation may not be known, or would be difficult to incorporate into a mathematical model.

2. For many materials the constitutive laws are known imperfectly and only in some average sense and within a narrow range of strain, strain rate, temperature and over a short time interval of loading.

3. The boundary conditions, other than stress-free boundary conditions, are not known with a high degree of precision, even under carefully controlled experimental conditions. The reason for this is that the loading and constraints typically involve mechanical contact which depends on the compliances of the structures that impose the load and constraints (e.g., testing machine, milling machine, assembly rig, etc.) and the physical properties of the contacting surfaces. In other words, the boundary conditions represent the influence of the environment on the object being modeled. The needed information is rarely available. Therefore subjective judgment of the analyst in the formulation of boundary conditions is usually unavoidable.

4. Due to the history of the material prior to manufacturing the parts that will be assembled into a machine or structure, such as casting, quenching, extrusion, rolling, forging, heat treatment, cold forming, machining and surface treatment residual stresses exist, the magnitude of which can be very substantial. The distribution of residual stress must satisfy the equations of equilibrium and the stress-free boundary conditions but otherwise it is generally unknown. See, for example, [47], [48].

5. Information concerning the probability distribution of the data that characterize the problem and their covariance functions is rarely available. In general, uncertainties increase with the complexity of models.

Remark 1.1.1 More than one mathematical model may have been proposed with identical objectives and it is possible that more than one mathematical model will meet the validation criteria. In that case the simpler model is.

Remark 1.1.2 Due to statistical variability in the data and errors in experimental observations, comparisons between prediction based on a mathematical model and the outcome of physical experiments must be understood in a statistical sense. The theoretical framework for model selection is based on Bayesian analysis.[2] Specifically, denoting a mathematical model by M, the newly acquired data by D and the background information by I, the probability that the model M is a predictor of the data D, given the background information I, can be written in terms of conditional probabilities:

$$\text{Prob}(M|D, I) \approx \text{Prob}(D|M, I) \times \text{Prob}(M|I). \tag{1.2}$$

In other words, Bayes' theorem relates the probability that a mathematical model is correct, given the measured data D and the background information I, to the probability that the measured data would have been observed if the model were functioning properly. See, for example, [74]. The term $\text{Prob}(M|I)$ is called prior probability. It represents expert opinion about the validity of M prior to coming into possession of some new data D. The term $\text{Prob}(D|M, I)$ is called the likelihood function. In this view competing mathematical models are assigned probabilities that represent the degree of belief in the reliability of each of the competing models, given the information available prior to acquiring additional information. In light of the new information, obtained by experiments, the prior probability is updated to obtain the term $\text{Prob}(M|D, I)$, called the posterior probability. An important and highly relevant aspect of Bayes' theorem is that it provides a framework for improvement of the probability estimate $\text{Prob}(M|D, I)$ based on new data.

1.1.3 Discretization

The finite element method (FEM) is one of the most powerful and widely used numerical methods for finding approximate solutions to mathematical problems formulated so as to simulate the responses of physical systems to various forms of excitation. It is used in various branches of engineering and science, such as elasticity, heat transfer, fluid dynamics, electromagnetism, acoustics, biomechanics, etc.

In the finite element method the solution domain is subdivided into elements of simple geometrical shape, such as triangles, squares, tetrahedra, hexahedra, and a set of basis functions are constructed such that each basis function is non-zero over a small number of elements only. This is called discretization. Details will be given in the following chapters. The set of all functions that can be written as linear combinations of the basis functions is called the finite element space. The accuracy of the data of interest depends on the finite element space and the method used for computing the data from the finite element solution. Associated with the finite element solution are errors of discretization, as indicated in Figure 1.1.

It is necessary to create finite element spaces such that the data of interest computed from the finite element solution are within acceptable error bounds with respect to their counterparts corresponding to the exact solution of the mathematical model.

The data of interest, such as the maximum displacement, temperature, stress, etc., are computed from the finite element solution u_{FE}. The data of interest will be denoted by $\Phi_i(u_{FE})$, $i = 1, \ldots, n$, in the following. The objective is to compute $\Phi_i(u_{FE})$ and to ensure

---

[2] Thomas Bayes (1702–1761).

8   INTRODUCTION

that the relative errors are within prescribed tolerances:

$$\frac{|\Phi_i(u_{EX}) - \Phi_i(u_{FE})|}{|\Phi_i(u_{EX})|} \leq \tau_i \tag{1.3}$$

where u_{EX} is the exact solution. Of course u_{EX} is not known in general, but it is known that $\Phi_i(u_{EX})$ is independent of the finite element space. The error in $\Phi_i(u_{FE})$ depends on the finite element space and the method used for computing $\Phi_i(u_{FE})$. The errors of discretization are controlled through suitable enlargement of the finite element spaces, and by various procedures used for computing $\Phi_i(u_{FE})$.

1.1.4 Verification

Verification is concerned with verifying that (a) the input data are correct, (b) the computer code is functioning properly and (c) the errors in the data of interest meet necessary conditions to be within permissible tolerances.

Common errors in input are incorrectly entered data, such as mixed units and errors in data entry. Such errors are easily found in a careful review of the input data.

The primary responsibility for ensuring that the code is functioning properly rests with the code developers. However, computer codes tend to have programming errors, especially in their less frequently traversed branches, and the user shares in the responsibility of verifying that the code is functioning properly.

In verification accuracy is understood to be with respect to the exact solution of the mathematical model, not with respect to physical reality. The process of verification of the numerical solution is illustrated schematically in Figure 1.3. The term extraction refers to methods used for computing $\Phi_i(u_{FE})$. Details are presented in the following chapters.

Remark 1.1.3 Verification and validation are possible only when the mathematical model is properly formulated with respect to the goals of computation. For example, in linear elasticity the solution domain must not have sharp re-entrant corners or edges if the goal of computation is to determine the maximum stress. Point constraints and point forces can be used only when certain criteria are met etc. Details are given in the following chapters. Unfortunately, using mathematical models without regard to their limitations is a commonly occurring conceptual error.

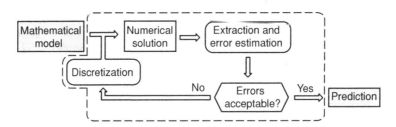

Figure 1.3 Verification of the numerical solution.

Remark 1.1.4 The process illustrated schematically in Figure 1.1 is often referred to as finite element modeling. This term is unfortunate because it mixes two conceptually different aspects of numerical simulation: the definition of a mathematical model and its numerical solution by the finite element method.

1.1.5 Decision-making

The goal of numerical simulation is to support various engineering decision-making processes. There is an implied expectation of reliability: one cannot reasonably base decisions on computed information without believing that the information is sufficiently reliable to support those decisions. Demonstration of the reliability of mathematical models used in support of engineering decision-making is an essential part of any modeling effort. In fact, the role of physical testing is to calibrate and validate mathematical models so that a variety of load cases and design alternatives can be evaluated.

In the following we illustrate the importance of the reliability of numerical simulation processes through brief descriptions of four well-documented examples of the consequences of large errors in prediction either because improper mathematical models were used or because large errors occurred in the numerical solution. Additional examples can be found in [61], [62]. Undoubtedly, there are many undocumented instances of substantial loss attributable to errors in predictions based on mathematical models.

Example 1.1.2 The Tacoma Narrows Bridge, the first suspension bridge across Puget Sound (Washington State, USA), collapsed on November 7, 1940, four months after its opening. Wind blowing at 68 km/h caused sufficiently large oscillations in the 853 m main span to collapse the span.

Until that time bridges were designed on the basis of equivalent static forces. The possibility that relatively small periodic aerodynamic forces (the effects of Kármán vortices)[3] may become significant was not considered. The Kármán vortices were first analyzed in 1911 and the results were presented in the Göttingen Academy in the same year.[4] The designers were either unaware of those results or did not see their relevance to the Tacoma Narrows Bridge, the failure of which was caused by insufficient torsional stiffness to resist the periodic excitation induced by Kármán vortices.

Example 1.1.3 The roof of the Hartford Civic Center Arena collapsed on January 18, 1978. The roof structure, measuring 91.4 by 109.7 m (300 by 360 ft), was a space frame, an innovative design at that time. It was analyzed using a mathematical model that accounted for linear response only. Furthermore, the connection details were greatly simplified in the model. In linear elastostatic analysis it is assumed that the deformation of a structure is negligibly small and hence it is sufficient to satisfy the equations of equilibrium in the undeformed configuration.

The roof frame was assembled on the ground. Once the roof was lifted into its final position, its deflection was measured to be twice of what was predicted by the mathematical model:

---

[3] Theodore von Kármán (1881–1963).
[4] Von Kármán, Th. and Edson, L., *The Wind and Beyond: Theodore von Kármán, Pioneer in Aviation and Pathfinder in Space*, Little, Brown, Boston, MA, 1967, pp. 211–215.

10   INTRODUCTION

When notified of this condition, the engineers expressed no concern, explaining that such discrepancies had to be expected in view of the simplifying assumptions of the theoretical calculation.[5]

Subsequent investigation identified that reliance on an oversimplified model that did not represent the connection details properly and failed to account for geometric nonlinearities was the primary cause of failure.

Example 1.1.4 The Vaiont Dam, one of the highest dams in the world (262 m), was completed in the Dolomite Region of the Italian Alps, 100 km north of Venice, in 1961. On October 9, 1963, after heavy rains, a massive landslide into the reservoir caused a large wave that overtopped the dam by up to 245 m and swept into the valley below, resulting in the loss of an estimated 2000 lives.[6] The courts found that, due to the *predictability* of the landslide, three engineers were criminally responsible for the disaster. The dam withstood the overload caused by the wave. This incident serves as an example of a full scale test of a major structure caused by an unexpected event.

Example 1.1.5 The consequences of large errors of discretization are exemplified by the Sleipner accident. The gravity base structure (GBS) of the Sleipner A offshore platform, made of reinforced concrete, sank during ballast test operations in Gandsfjorden, south of Stavenger, Norway, on August 23, 1991. The economic loss was estimated to be 700 million dollars.

The main function of the GBS was to support a platform weighing 56 000 tons. The GBS consisted of 24 caisson cells with a base area of 16 000 m^2. Four cells were elongated to form shafts designed to support the platform. The total concrete volume of the GBS was 75 000 m^3. The accident occurred as the GBS was being lowered to a depth of approximately 99 m. Failure first occurred in two triangular cells, called tri-cells, next to one of the shafts. When the GBS hit the sea bed, seismic events measuring 3 on the Richter scale were recorded in the Stavenger area.[7]

There is general agreement among the investigators that the accident was caused by large errors in the finite element analysis, the goal of which was to estimate the requirements for reinforcement of the concrete cells by steel bars:

> The global finite element analysis gave a 47% underestimation of the shear forces in the tri-cell walls. This error was caused by the use of a coarse finite element mesh with some skewed elements used for analysis of the tri-cell walls.[8]

---

[5] Levy, M. and Salvadori, M., *Why Buildings Fall Down: How Structures Fail*, W. W. Norton, New York, 2002.

[6] See, for example, Hendron, A. J. and Patten, F. D., The Vaiont Slide. US Corps of Engineers Technical Report GL-85-8 (1985).

[7] Jacobsen, B., The loss of the Sleipner A Platform, *Proceedings of the 2nd International Offshore and Polar Engineering Conference*, International Society of Offshore and Polar Engineers, Vol. 1, 1992.

[8] Rettedal, W. K., Gudmestad, O. T. and Aarum, T., Design of concrete platforms after Sleipner A-1 sinking, *Proceedings of the 12th International Conference on Offshore Mechanics and Arctic Engineering*, Vol. 1, Offshore Technology, pp. 309–310, ASME, 1993.

A check of the global response analysis revealed serious inaccuracies in the interpretation of results from finite element analyses giving a shear force in a critical section of the cell wall that was less than 60% of the correct value.[9]

1.2 Why is numerical accuracy important?

A number of difficulties associated with accurate representation of a real physical system by mathematical means were noted in Section 1.1.2. Given these difficulties, it may seem reasonable to ask: "If we do not know the input data with sufficient accuracy, then why should we be concerned with the accuracy of the numerical solution?" In answering this question we consider two important areas of application of mathematical models: the application of design rules and the formulation of design rules. It is shown in the following that both require estimation and control of the numerical accuracy.

1.2.1 Application of design rules

Design and design certification involve application of existing design rules, established by various codes, regulations and guidelines. The design rules are typically stated in the form of required minimum factors of safety:

$$FS := \frac{\Phi_{\lim}}{\Phi_{\max}(u_{EX})} \geq (FS)_{\text{design}} \tag{1.4}$$

where FS is the realized factor of safety, $\Phi_{\lim} > 0$ is the limiting (not to exceed) value of some entity (such as maximum bending moment, maximum stress, etc.), $\Phi_{\max}(u_{EX}) > 0$ is the exact value of the same entity corresponding to the exact solution of the mathematical model and $(FS)_{\text{design}}$ is the minimum value of the factor of safety specified by the applicable design rules. It is the designer's responsibility to ensure that the applicable design rules are followed.

We will denote by $\Phi_{\max}(u_{FE})$ the value of $\Phi_{\max}$ computed from the finite element solution. Let us suppose that, due to numerical errors, it is possible to guarantee only that the relative error is not greater than τ:

$$\frac{|\Phi_{\max}(u_{EX}) - \Phi_{\max}(u_{FE})|}{\Phi_{\max}(u_{EX})} \leq \tau, \quad 0 \leq \tau < 1; \tag{1.5}$$

in other words, $\Phi_{\max}(u_{FE})$ may underestimate $\Phi_{\max}(u_{EX})$ by $100\tau\%$. Therefore we have

$$\Phi_{\max}(u_{EX}) \leq \frac{1}{(1-\tau)} \Phi_{\max}(u_{FE}). \tag{1.6}$$

On substituting this expression into Equation (1.4), we obtain

$$\frac{\Phi_{\lim}}{\Phi_{\max}(u_{FE})} \geq \frac{(FS)_{\text{design}}}{1-\tau}. \tag{1.7}$$

---

[9] Holand, I., The Sleipner accident, in *From Finite Elements to the Troll Platform - Ivar Holand 70th Anniversary*, K. Bell, editor, The Norwegian Institute of Technology, Trondheim, pp. 157–168, 1994.

On comparing Equation (1.7) with Equation (1.4) it is seen that, to compensate for numerical errors in the computation of $\Phi_{\max}(u_{FE})$, it is necessary to increase the required factor of safety to $(FS)_{\text{design}}/(1-\tau)$. For example, if the accuracy of $\Phi_{\max}(u_{FE})$ can be guaranteed to 20% (i.e., $\tau = 0.20$) then $(FS)_{\text{design}}$ must be increased by 25%. Since $(FS)_{\text{design}}$ was chosen conservatively to account for the uncertainties, the economic penalties associated with using an increased factor of safety generally far outweigh the costs associated with guaranteeing the accuracy of the data of interest to within a small relative error (say 5%).

Application of design rules is a task of verification, that is, verification that the correct data are used, the computer code is functioning properly and the tolerance τ in Equation (1.5) is sufficiently small.

Remark 1.2.1 In aerospace engineering the design requirements are stated in terms of minimum acceptable margins of safety (MS). By definition, $MS = FS - 1$.

Remark 1.2.2 The Federal Aviation Regulations (FAR), issued by the Federal Aviation Administration of the US Department of Transportation, state in Sec. 25.303 that "a factor of safety of 1.5 must be applied to the prescribed limit load."

Economic considerations dictate that the realized factor of safety should not be much larger than $(FS)_{\text{design}}$. This is especially true in aerospace engineering where avoidance of weight penalties makes it necessary to ensure that the realized factors of safety are reasonably close to the mandated factors of safety.

Example 1.2.1 The yield strength in shear of hot rolled 0.2% carbon steel is 165 MPa and the usual factor of safety for static loads is 1.65 (so that the allowable maximum shear stress is 100 MPa).[10] If the numerical computations could underestimate the maximum shear stress by as much as 20% then the factor of safety would have to be increased to 2.06, that is, the allowable maximum value would be reduced to 80 MPa.

1.2.2 Formulation of design rules

The responsibility for formulating design rules rests with committees of experts appointed by professional societies and regulatory bodies. Some of the design rules have been evolving for a long time while others are still in their early stages of development. For example, the ASME Boiler and Pressure Vessel Code dates back to 1914. It has been adopted in part or in its entirety by all 50 states of the United States. The code is being updated by the Boiler and Pressure Committee which meets regularly to consider requests for interpretations, revision, and to develop new rules. On the other hand, design rules for structural components of aircraft made of laminated composites are in their early stages of development

The formulation and revision of design rules are applications of the process of validation discussed in Section 1.1.2 and schematically illustrated in Figure 1.2 in the following sense. The object of design is the physical reality. The mathematical model is the collection of physical laws that experts consider relevant to decision-making. The formulas and rules are, respectively, the metrics and criteria used in the prediction that the object of design will

---

[10] See, for example, Popov, E. P., *Engineering Mechanics of Solids*, 2nd edition, Prentice Hall, Upper Saddle River, NJ, 1998.

function as intended when the stated criteria are satisfied. In light of new experience, the model, metrics and criteria are revised.

Formulation of design rules involves definition of certain entities Φ_k ($k = 1, 2, \ldots$), such as the maximum principal stress, some specific combinations of stress and strain components, etc., that characterize failure and the corresponding limiting values. In the following the subscript k will be dropped and the discussion will be concerned with a generic design rule, that is, the determination of $\Phi_{\lim}$ and evaluation of the associated uncertainties. The factor of safety is determined on the basis of assessment of uncertainties and consideration of the consequences of failure.

Suppose that a mathematical model predicts that failure will occur when Φ reaches its critical value $\Phi_{\lim}$. First, a set of calibration experiments have to be performed with the objective to determine $\Phi_{\lim}$. Second, another set of experiments have to be conducted to test whether failure can be predicted on the basis of $\Phi_{\lim}$. These are validation experiments. In general Φ cannot be observed directly, therefore it must be inferred from correlations between computed data and experimental observations.

Let Y_{ij} be the ith ideal observation of the jth experiment and let $\phi_i(u_{EX}^{(j)})$ be the corresponding functional[11] computed from the exact solution $u_{EX}^{(j)}$ so that if there were no experimental errors and the mathematical model were correct then we would have

$$Y_{ij} - \phi_i(u_{EX}^{(j)}) = 0.$$

Due to experimental errors we actually observe y_{ij} and compare it with $\phi_i(u_{FE}^{(j)})$, the finite element approximation to $\phi_i(u_{EX}^{(j)})$. Let us write

$$Y_{ij} = y_{ij} \pm e_{ij}^{\exp}$$

and

$$\phi_i(u_{EX}^{(j)}) = \phi_i(u_{FE}^{(j)}) \pm e_{ij}^{\text{fea}}$$

where $e_{ij}^{\exp}$ (resp. e_{ij}^{fea}) is the experimental (resp. approximation) error. Then

$$y_{ij} - \phi_i(u_{FE}^{(j)}) = Y_{ij} \mp e_{ij}^{\exp} - \phi_i(u_{EX}^{(j)}) \pm e_{ij}^{\text{fea}}.$$

Using the triangle inequality, we have

$$\underbrace{|y_{ij} - \phi_i(u_{FE}^{(j)})|}_{\text{apparent error}} \leq \underbrace{|Y_{ij} - \phi_i(u_{EX}^{(j)})|}_{\text{true error}} + |e_{ij}^{\exp}| + |e_{ij}^{\text{fea}}|. \tag{1.8}$$

This result shows that in testing a mathematical model it is essential to have both the experimental errors and the errors of discretization under control, otherwise it will not be possible to know whether the apparent error is due to an error in the hypothesis, errors in the numerical approximation, or errors in the experiment. Furthermore, means for the estimation

---

[11] A functional is a real number defined on a space of functions. In the present context a functional is a real number computed from the exact solution or the finite element solution.

and control of discretization errors, in terms of the data of interest, must be provided by the computer code.

The aim of experiments needs to include the development of reliable statistical information on the basis of which the factor of safety is established.

1.3 Chapter summary

The principal aim of this book is to present the theoretical and practical considerations relevant to (a) the validation of mathematical models and (b) verification of the data of interest computed from finite element solutions. Some fundamental concepts and basic terminology were introduced, as follows.

Mathematical model

A mathematical representation of a physical system or process intended for predicting some set of responses is called a mathematical model. It transforms one set of data, the input, into another set, the output.

Conceptualization

Conceptualization is a process by which a mathematical model is defined, for a particular application. Conceptualization involves (a) application of expert knowledge, (b) virtual experimentation and (c) calibration.

Discretization

Discretization is a process by which a mathematical problem is formulated that can be solved on digital computers. The solution approximates the exact solution of a given mathematical model.

Validation

Validation is a process by which the predictive capabilities of mathematical models are tested. Ideally, experiments are conducted especially to test whether a mathematical model meets necessary conditions for acceptance from the perspective of its intended use. If the predictions based on the model are not sufficiently close to the outcome of physical experiments then the model is rejected. The quality of predictions is evaluated on the basis of one or more metrics and the corresponding criteria, formulated prior to the execution of validation experiments. Of course, the comparisons must take into account uncertainties in the model parameters, the accuracy of numerical solutions and errors in the experiments.

It is generally feasible to perform full scale validation experiments for mass-produced items. For one-of-a-kind objects, such as dams, bridges, buildings, etc., it is not feasible to perform full scale validation experiments. In such cases the model is analyzed a posteriori in light of new information collected following an incident.

Verification

Verification is a process by which it is ascertained that the data of interest computed from the approximate solution meet the necessary conditions for acceptance. Verification is understood in relation to the exact solution of a mathematical model, not in relation to the physical reality that the mathematical model is supposed to represent.

Uncertainties

It is important to distinguish between statistical (aleatory) and cognitive (epistemic) uncertainties. In the process of calibration the mathematical model is assumed to be correct and the parameters that characterize the model are selected such that the measured response matches the predicted response. Consequently aleatory and epistemic uncertainties are unavoidably mixed. Epistemic uncertainties can be reduced when the predictive capabilities of alternative models are systematically tested and revised as new experimental information becomes available. This involves reinterpretation of the available experimental data with reference to the revised model.

2

An outline of the finite element method

In this chapter an outline of the finite element method (FEM) is presented in a one-dimensional setting. It will be generalized to two and three dimensions in subsequent chapters.

Throughout the book the units of physical data will be identified in terms of the standard SI[1] notation. Any consistent set of units may be used, however.

2.1 Mathematical models in one dimension

The formulation of mathematical models will be discussed in Chapter 3. Here a simple mathematical model that will serve as the basis for the discussion of the conceptual and algorithmic aspects of the FEM is formulated.

2.1.1 The elastic bar

The elastostatic response of an elastic body is characterized by the displacement vector

$$\vec{u} = u_x(\vec{r})\,\vec{e}_x + u_y(\vec{r})\,\vec{e}_y + u_z(\vec{r})\,\vec{e}_z \qquad (2.1)$$

where $\vec{r}$ is the position vector

$$\vec{r} = x\vec{e}_x + y\vec{e}_y + z\vec{e}_z \qquad (2.2)$$

where $\vec{e}_x, \vec{e}_y, \vec{e}_z$ are orthonormal basis vectors. The displacement $\vec{u}$ and the position vector $\vec{r}$ are measured in units of meters (m). We will be concerned with a highly simplified

---

[1] Système International d'Unités (International System of Units).

18   AN OUTLINE OF THE FINITE ELEMENT METHOD

model of a three-dimensional elastic body, the elastic bar. For the elastic bar we assume that $\vec{u} = u_x(x\vec{e}_x)\vec{e}_x$, that is, only one displacement component is non-zero and $x\vec{e}_x$ is the only independent variable. In order to simplify notation we will write $u(x)$ instead of $\vec{u} = u_x(x\vec{e}_x)\vec{e}_x$.

We assume that the centroidal axis of the bar is coincident with the x-axis. We denote the length of the bar by ℓ (m). The mathematical model of an elastic bar is based on equations that represent the strain–displacement relationship, the stress–strain relationship and equilibrium:

1. **The strain–displacement relationship**. The total strain is

$$\epsilon_x \equiv \epsilon = \frac{du}{dx} \equiv u'. \tag{2.3}$$

The total strain ϵ is the sum of the mechanical strain ϵ_m and the thermal strain $\epsilon_t = \alpha T_\Delta$ where $\alpha = \alpha(x) \geq 0$ is the coefficient of thermal expansion ($1/K$ units). Therefore the mechanical strain is

$$\epsilon_m = \epsilon - \epsilon_t = u' - \alpha T_\Delta. \tag{2.4}$$

Note that strain is dimensionless.

2. **The stress–strain relationship**. In one dimension Hooke's law states that the stress is proportional to the mechanical strain:

$$\sigma_x \equiv \sigma = E\epsilon_m = E(u' - \alpha T_\Delta) \tag{2.5}$$

where $E = E(x) > 0$ is the modulus of elasticity (MPa units). Since strain is dimensionless, stress is also in MPa units. Positive stress is called tensile stress, negative stress is called compressive stress.

3. **Equilibrium**. It is assumed that the stress is constant over the cross-sectional area. The equilibrium equations are written in terms of the bar force F_b defined by

$$F_b := \int_A \sigma \, dy dz = \sigma A = AE(u' - \alpha T_\Delta) \tag{2.6}$$

where $A = A(x) > 0$ (m^2) is the area of the cross-section. The bar may be subjected to distributed forces and/or volume forces T_b in N/m units and tractions exerted by elastic springs:

$$T_s := c(d - u) \tag{2.7}$$

where $d = d(x)$ (m) is displacement imposed on the distributed spring and $c = c(x) \geq 0$ is the spring rate in N/m^2 units. When the force T_b accounts for volume forces then it is understood that the volume forces have been integrated over the cross-sectional area and therefore T_b is in N/m units. Distributed forces are also called traction forces or traction loads.

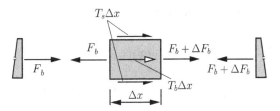

Figure 2.1 Bar element.

Referring to Figure 2.1, and considering the equilibrium of an isolated part of the bar of length Δx, we write

$$\Delta F_b + T_b \Delta x + T_s \Delta x = 0.$$

Note that F_b, T_b and T_s are vectors. F_b is positive in the direction of the positive (outward) normal to the cross-section. This is consistent with the action–reaction principle which states that if a force acts upon a body, then an equal and opposite force must act upon another body. The bar element in Figure 2.1 was "cut" from the bar and therefore equal and opposite forces act on the mating cross-sections. The bar element is said to be in tension (resp. compression) when F_b is positive (resp. negative).

Assuming that F_b is a continuous and differentiable function,

$$\Delta F_b = \frac{dF_b}{dx}\Delta x + O(\Delta x).$$

Letting $\Delta x \to 0$, we have the equilibrium equation:

$$\frac{dF_b}{dx} + T_b + T_s = 0. \tag{2.8}$$

On combining Equations (2.6), (2.7) and (2.8), the ordinary differential equation that models the mechanical response of elastic bars to applied traction forces is obtained:

$$-\frac{d}{dx}\left(AE\frac{du}{dx}\right) + cu = T_b + cd - \frac{d}{dx}(AE\alpha T_\Delta) \quad \text{on } x \in I \tag{2.9}$$

where I represents the set of points x that lie in the interval $0 < x < \ell$. In the following we will write $I = (0, \ell)$.

Dimensionless form

Mathematical problems can always be written in dimensionless form. For example, referring to Equation (2.9), we introduce $X \equiv x/\ell$, $U \equiv u/\ell$, $AE \equiv A_0 E_0 f(X)$, $c \equiv c_0 g(X)$, $T_b \equiv T_0 b(X)$, $d \equiv d_0 h(X)$, $\alpha T_\Delta \equiv t(X)$ where $f(X), g(X), b(X), h(X)$ and $t(X)$ are dimensionless functions, and their multiplying coefficients are constants that have the appropriate dimensions of length, force, etc. We can write Equation (2.9) as follows:

$$-\frac{d}{dX}\left(\frac{A_0 E_0}{\ell}f(X)\frac{dU}{dX}\right) + c_0 \ell g(X)U =$$

$$T_0 b(X) + c_0 d_0 g(X)h(X) - \frac{d}{dX}\left(\frac{A_0 E_0}{\ell}f(X)t(X)\right) \quad \text{on } X \in (0, 1).$$

On dividing by $A_0 E_0/\ell$ we have the dimensionless form of Equation (2.9):

$$-\frac{d}{dX}\left(f(X)\frac{dU}{dX}\right) + \frac{c_0 \ell^2}{A_0 E_0}g(X)U =$$

$$\frac{T_0 \ell}{A_0 E_0}b(X) + \frac{c_0 d_0 \ell}{A_0 E_0}g(X)h(X) - \frac{d}{dX}(f(X)t(X)), \quad X \in (0, 1). \quad (2.10)$$

The choice of the length scale is arbitrary; here we have chosen ℓ. When formulating a mathematical problem it is sound practice to check that the data are given in consistent units and therefore the dimensions of the coefficients cancel. This is part of the verification process.

Remark 2.1.1 The solution of a mathematical problem, whether by classical or numerical means, is independent of the physical dimensions associated with the variables and coefficients. The physical dimensions enter only in the interpretation of the solution and data computed from the solution.

Boundary conditions

We will be considering linear boundary conditions associated with Equation (2.9). These are shown schematically in Figure 2.2. A brief description follows.

1. Displacement boundary conditions, also called kinematic boundary conditions, are shown in Figure 2.2(a). The given displacement is denoted by $\hat{u}_0 \vec{e}_x$ at $x = 0$ (resp. $\hat{u}_\ell \vec{e}_x$ at $x = \ell$).

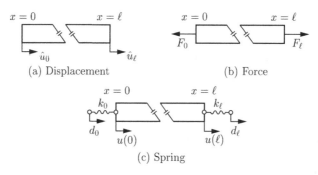

Figure 2.2 Elastic bar: linear boundary conditions.

2. The bar forces $F_b(0)$ and $F_b(\ell)$ applied on the boundaries, denoted respectively by F_0 and F_ℓ, may be prescribed at one or both ends of the bar, see Figure 2.2(b). By definition,

$$F_0 := [AE(u' - \alpha T_\Delta)]_{x=0} \qquad F_\ell := [AE(u' - \alpha T_\Delta)]_{x=\ell}. \qquad (2.11)$$

Recall that bar forces are positive when tensile.

3. Spring boundary conditions are linear relationships between the bar forces F_0 and F_ℓ and the corresponding displacements at the boundary points, as indicated in Figure 2.2(c):

$$F_0 = [AE(u' - \alpha T_\Delta)]_{x=0} = k_0(u(0) - d_0) \qquad (2.12)$$

$$F_\ell = [AE(u' - \alpha T_\Delta)]_{x=\ell} = k_\ell(d_\ell - u(\ell)) \qquad (2.13)$$

where $k_0 > 0$ (resp. $k_\ell > 0$) is the spring constant (in N/m units) at $x = 0$ (resp. $x = \ell$) and d_0 (resp. d_ℓ) is a displacement imposed on the spring at $x = 0$ (resp. $x = \ell$).

Of course, the displacement, force and spring boundary conditions may occur in any combination.

Symmetry, antisymmetry and periodicity

Under special conditions the mathematical problem can be formulated on a smaller domain and extended by symmetry, antisymmetry or periodicity to the full domain.

The line of symmetry is a line that passes through the mid-point of the interval ℓ and is perpendicular to the x-axis. By symmetry we will understand mirror image symmetry.

A scalar function defined on I is said to be symmetric if in symmetrically located points with respect to the line of symmetry the function has the same absolute value and the same sign. A scalar function is said to be antisymmetric if in symmetrically located points with respect to the line of symmetry the function has the same absolute value but opposite sign.

Vector functions are symmetric if in symmetrically located points the absolute values are equal and the basis vector components normal to the line of symmetry have opposite sense; the basis vector components parallel to the line of symmetry have the same sense. For example, assuming that the y-axis is coincident with the line of symmetry, in Figure 2.3(a) $\vec{v}$ is the symmetric image of $\vec{u}$.

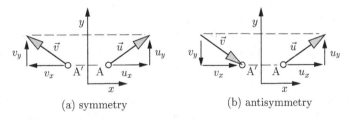

(a) symmetry   (b) antisymmetry

Figure 2.3 Symmetry and antisymmetry of a vector function in two dimensions. Points A and A′ are equidistant from the y-axis.

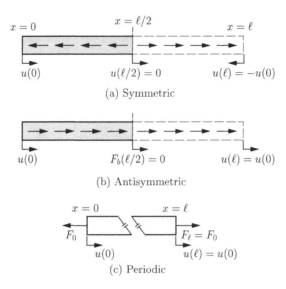

Figure 2.4 Elastic bar: symmetric, antisymmetric and periodic loading and boundary conditions.

Vector functions are antisymmetric if in symmetrically located points the absolute values are equal and the basis vector components normal to the line of symmetry $\vec{e}_x$ have the same sense; the basis vector components parallel to the line of symmetry have opposite sense. For example, assuming that the y-axis is coincident with the line of symmetry, in Figure 2.3(b) $\vec{v}$ is the antisymmetric image of $\vec{u}$.

In many instances the coefficients $AE(x)$ and $c(x)$ are symmetric functions with respect to the line of symmetry. If in such cases $T_b(x)$, $d(x)$, $\alpha T_\Delta(x)$ and the boundary conditions are also symmetric (resp. antisymmetric) then the solution is a symmetric (resp. antisymmetric) function with respect to the line of symmetry. When the solution is a symmetric or antisymmetric function then the problem can be solved on half of the interval and extended to the entire interval by symmetry or antisymmetry.

The displacement, force and traction are vector functions. In the case of the elastic bar these vector functions have only one non-zero component, which is perpendicular to the line of symmetry. Examples of symmetric and antisymmetric loading and constraints are shown in Figures 2.4(a) and 2.4(b). In the symmetric case the boundary condition is $u(\ell/2) = 0$. In the antisymmetric case the boundary condition is $F_b(\ell/2) = 0$.

When $AE(x)$, $c(x)$, $T_b(x)$, $d(x)$ and $\alpha T_\Delta(x)$ are periodic functions, the length of the period being ℓ, that is, $(AE)_{x=0} = (AE)_{x=\ell}$, $c(0) = c(\ell)$, $T_b(0) = T_b(\ell)$, $d(0) = d(\ell)$, $\alpha T_\Delta(0) = \alpha T_\Delta(\ell)$, $u(0) = u(\ell)$ and $F(0) = F(\ell)$, then the solution is a periodic function and the boundary conditions are said to be periodic. The solution obtained for the interval $(0, \ell)$ is extended to $-\infty < x < \infty$. Periodic boundary conditions are illustrated in Figure 2.4(c).

Remark 2.1.2 Considering the mathematical model for the elastic bar, it was stated that when the solution is symmetric or antisymmetric then the corresponding mathematical problem can be solved on half of the domain. In the symmetric case the boundary condition is $u(\ell/2) = 0$. In the antisymmetric case the boundary condition is $F(\ell/2) = 0$, which is equivalent to $u'(\ell/2) = 0$. In Section 2.1.4 we will consider an identical mathematical problem where $u(x)$

is a scalar function. In that case the symmetry boundary condition is $u'(\ell/2) = 0$ and the antisymmetry boundary condition is $u(\ell/2) = 0$. This can be easily visualized by considering mirror image symmetry.

Remark 2.1.3 Equation (2.9), together with specific boundary conditions, is a mathematical model of an elastic bar. This problem is solved with the goal to obtain some desired information, called data of interest, such as the displacement $u(x)$, the axial strain $u'(x)$, or the axial force $AE(u'(x) - \alpha T_\Delta)$ at all or specific points, or at points where their maxima occur. Incorporated in the model are the assumptions that $|\epsilon_m| \ll 1$, $|\epsilon_t| \ll 1$ and $|\sigma| \leq \sigma_{\text{pl}}$ where σ_{pl} is the proportional limit of the material.

Example 2.1.1 Consider the problem

$$-(AEu')' + cu = T_b, \qquad u(0) = 0, \qquad F_b(\ell) = F_\ell$$

where AE and c are positive constants and $T_b = b_0 + b_1 x/\ell$, where b_0, b_1 are given constants (in N/mm units). To solve this problem we define $\lambda^2 := c/(AE)$. The general solution can be found in standard texts on ordinary differential equations and engineering mathematics:

$$u = C_1 \cosh \lambda x + C_2 \sinh \lambda x + \frac{b_0}{c} + \frac{b_1}{c}\frac{x}{\ell}. \qquad (2.14)$$

From the boundary conditions we find

$$C_1 = -\frac{b_0}{c}$$

$$C_2 = \frac{1}{\lambda \cosh \lambda \ell} \left[\frac{F_\ell}{AE} + \frac{b_0}{c} \lambda \sinh \lambda \ell - \frac{b_1}{c\ell} \right].$$

Example 2.1.2 Consider the problem of Example 2.1.1 with periodic boundary conditions

$$-(AEu')' + cu = T_b, \qquad u(0) = u(\ell), \qquad F_b(0) = F_b(\ell)$$

where, as in Example 2.1.1, AE and c are positive constants and $T_b = b_0 + b_1 x/\ell$, with b_0, b_1 as given constants. The general solution is given by (2.14). On applying the periodic boundary conditions we find

$$u = \frac{b_1}{2c} \cosh \lambda x + \frac{b_1}{2c} \frac{\sinh \lambda \ell}{1 - \cosh \lambda \ell} \sinh \lambda x + \frac{b_0}{c} + \frac{b_1}{c}\frac{x}{\ell}. \qquad (2.15)$$

Exercise 2.1.1 Consider an elastic bar constrained by a distributed spring of stiffness c. Assume that AE, c and the coefficient of thermal expansion α are constants, $T_b = 0$, $d = 0$. The boundary conditions are $u(0) = 0$, $F_b(\ell) = 0$. The bar is subjected to a temperature change $T_\Delta(x) = b_0$ (constant). Write down the solution for this problem.

Exercise 2.1.2 Consider an elastic bar constrained by a distributed spring of stiffness c. Assume that AE, $c > 0$ are constants and $T_b = 0$, $d = 0$, $T_\Delta = 0$. Let $u(0) = \hat{u}_0$, $F_b(\ell) = k_\ell(d_\ell - u(\ell))$. Write down the solution for this problem.

2.1.2 Conceptualization

We have formulated mathematical models suitable for predicting static responses of elastic bars. We tacitly assumed that the physical properties and boundary conditions were given. In many practical applications not all of the needed information is available. Therefore it is necessary to perform and interpret calibration experiments. The procedure is illustrated by the following example.

Example 2.1.3 One of the methods used for ensuring that the foundations of a large building are sufficiently stiff to resist the dead and live loads without undergoing excessive settlement is to drive large elastic bars, called piles, into the soil. Suppose that two experts were consulted on the question of how to estimate the stiffness of a pile and both experts agreed that the mathematical model should be based on the following differential equation:

$$-AEu'' + cu = 0, \quad AEu'(0) = F_0, \quad u'(\ell) = 0 \qquad (2.16)$$

where c represents the action of the soil on the pile. The goal is to predict the displacement u_0 at the top of the pile as a function of the applied axial force. The notation is shown in Figure 2.5.

Both experts recommended using the nominal value for the modulus of elasticity of steel $E = 200\,\text{GPa}$; however, one expert recommended that c should be treated as a constant and the other expert recommended that $c = kx$, where k is a constant, should be assumed. In other words, different mathematical models were proposed for the same problem. In the following we refer to these as Model A and Model B, respectively. In order to determine c,

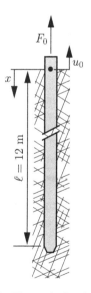

Figure 2.5 Example 2.1.3: notation.

an HP305 × 110 test pile[2] was driven into the soil to the depth of 12.0 m. The cross-sectional area is 1.402×10^{-2} m^2.

The pull test yielded the following results. When the applied force F_0 is 200 kN then the measured upward displacement u_0 is 9.0 mm; at $F_0 = 300$ kN, $u_0 = 13.5$ mm; at $F_0 = 400$ kN, $u_0 = 18.0$ mm. In other words, the experimental measurements yielded $F_0/u_0 = 22.22$ kN/mm. The observation that the ratio F_0/u_0 is constant is consistent with the assumption, made by both experts, that the linear differential equation (2.16) is a reasonable mathematical model of the pile driven into the soil. If the ratio F_0/u_0 changed substantially with the applied load then this model would have to be rejected on the basis of the outcome of calibration experiments.

Calibration of Model A

Equation (2.16) can be rewritten as

$$-u'' + \lambda^2 u = 0, \quad u'(0) = F_0/AE, \ u'(\ell) = 0 \quad \text{where } \lambda := \sqrt{\frac{c}{AE}} \tag{2.17}$$

the solution of which is

$$u(x) = \frac{F_0}{\lambda AE}\left(\sinh \lambda x - \frac{\cosh \lambda \ell}{\sinh \lambda \ell} \cosh \lambda x\right).$$

Note that with the sign convention adopted in Section 3.4.4 and illustrated in Figure 2.2, the upward displacement is negative, that is, $u(0) = -u_0$. Therefore the force–displacement relationship is

$$F_0 = AEu'(0) = AE\frac{\lambda \sinh \lambda \ell}{\cosh \lambda \ell} u_0 \tag{2.18}$$

which can be written as

$$G(\lambda) := \frac{F_0}{u_0} - AE\frac{\lambda \sinh \lambda \ell}{\cosh \lambda \ell} = 0. \tag{2.19}$$

For the three data pairs $F_0/u_0 = 22.22$ kN/mm was measured. We need to find λ such that $G(\lambda) = 0$. Various root finding methods are available. One of the most commonly used methods is the Newton–Raphson method.[3] In this method we select a trial value for λ, denoted by λ_1, and compute the corresponding $G(\lambda_1)$ and $G'(\lambda_1) := (dG/d\lambda)_{\lambda=\lambda_1}$. The choice of λ_1 must be such that $(dG/d\lambda)_{\lambda=\lambda_1} \neq 0$. We then compute λ_{k+1} from

$$\lambda_{k+1} = \lambda_k - \frac{G(\lambda_k)}{G'(\lambda_k)} \quad \text{for } k = 2, 3, \ldots$$

---

[2] This designation indicates that the cross-section is H-shaped, the nominal depth of the cross-section is 305 mm and the mass is approximately 110 kg/m.

[3] Sir Isaac Newton (1642–1727), Joseph Raphson (1642–1727).

26 AN OUTLINE OF THE FINITE ELEMENT METHOD

Table 2.1 The derivatives of $u(x)$.

n	$D^n u(x)$	$D^n u(0)$
0	u	$-u_0$
1	Du	F_0/AE
2	$\lambda^2 xu$	0
3	$\lambda^2 u + \lambda^2 x Du$	$-\lambda^2 u_0$
4	$2\lambda^2 Du + \lambda^2 x D^2 u$	$2\lambda^2 F_0/AE$
5	$3\lambda^2 D^2 u + \lambda^2 x D^3 u$	0
6	$4\lambda^2 D^3 u + \lambda^2 x D^4 u$	$-4\lambda^4 u_0$
7	$5\lambda^2 D^4 u + \lambda^2 x D^5 u$	$10\lambda^4 F_0/AE$

and continue the process until $\lambda_{k+1} - \lambda_k$ is sufficiently small. By this method we find $\lambda = 2.6112 \times 10^{-2}\,\text{m}^{-1}$ and from the definition of λ given in Equation (2.17) we have $c = 1912\,\text{kN/m}^2$. This completes the calibration step for Model A.

Calibration of Model B

Calibration of Model B involves solution of the problem

$$-AEu'' + kxu = 0, \quad AEu'(0) = F_0, \; u'(\ell) = 0 \qquad (2.20)$$

which will be written as

$$-u'' + \lambda^2 xu = 0, \quad u'(0) = F_0/AE, \; u'(\ell) = 0 \quad \text{where } \lambda := \sqrt{\frac{k}{AE}}. \qquad (2.21)$$

This is known as the Airy equation,[4] see for example [64]. We will use a Taylor series[5] expansion to find an approximate solution. We denote the nth derivative of u by $D^n u$. The derivatives for $n = 0, 1, \ldots, 7$ are shown in Table 2.1.

We see that for $k \geq 3$

$$D^k u = (k-2)\lambda^2 D^{k-3} u + \lambda^2 x D^{k-2} u.$$

The Taylor series expansion of $u(x)$ is

$$u(x) = -u_0 + \frac{F_0}{AE} x - \frac{\lambda^2}{3!} u_0 x^3 + \frac{2\lambda^2}{4!} \frac{F_0}{AE} x^4 - \frac{4\lambda^4}{6!} u_0 x^6 + \frac{10\lambda^4}{7!} \frac{F_0}{AE} x^7 - \cdots .$$

Letting $u'(\ell) = 0$ we get

$$0 = \frac{F_0}{AE} - \frac{\lambda^2}{2} u_0 \ell^2 + \frac{\lambda^2}{3} \frac{F_0}{AE} \ell^3 - \frac{\lambda^4}{30} u_0 \ell^5 + \frac{\lambda^4}{72} \frac{F_0}{AE} \ell^6 - \cdots . \qquad (2.22)$$

---

[4] Sir George Biddell Airy (1801–1892).
[5] Sir Brook Taylor (1685–1731).

Therefore we need to find λ such that

$$G(\lambda) \approx \frac{F_0}{u_0}\left(\frac{1}{AE} + \frac{\lambda^2 \ell^3}{3AE} + \frac{\lambda^4 \ell^6}{72AE}\right) - \frac{\lambda^2 \ell^2}{2} - \frac{\lambda^4 \ell^5}{30} = 0. \quad (2.23)$$

Using the experimental result $F_0/u_0 = 22.22$ kN/mm, we find $\lambda \approx 1.0767 \times 10^{-2}$ m$^{-3/2}$ and hence $k \approx 325.1$ kN/m^3.

In this example the conceptual development of a mathematical model was illustrated in a simple setting. Model A and Model B differ by the definition of the constant c. The characterizing parameters c and k were determined by calibration. Calibration is part of the conceptualization process because definition of the mathematical model depends on information obtained by calibration experiments.

Exercise 2.1.3 Determine whether using 4 significant figures in the estimate $k \approx 325.1$ kN/m^3 in Example 2.1.3 is justified.

2.1.3 Validation

Making a prediction based on a mathematical model concerning the outcome of a physical experiment, then testing to see whether the prediction is correct, is called validation. Validation involves one or more metrics and the corresponding criteria. The metrics and criteria depend on the intended use of the model. For testing the model described in Example 2.1.3 we define the metric to be the ratio F_0/u_0 and the criterion is the corresponding tolerance. Validation is illustrated by the following example.

Example 2.1.4 On examining the pull test data in Example 2.1.3, we see that each 100 kN increment in the applied force resulted in a 4.5 mm increment in displacement. Therefore the assumption that the pile is supported by a distributed linear spring is consistent with the available observations. However, it is not possible to determine from these observations how the spring coefficient c varies with x.

Let us assume that a second pile, driven to 8.5 m depth (i.e., $\ell = 8.5$ in Figure 2.5), is to be tested. Based on Model A and Model B we predict the test results shown in the second and third columns of Table 2.2 and we state our criterion as follows: a model will be rejected if the difference between the predicted and observed values of F_0/u_0 exceeds the tolerance of 5%.

Table 2.2 Predicted and observed data.

F_0 Applied (kN)	u_0 Model A (mm)	u_0 Model B (mm)	u_0 Experiment (mm)
200	12.5	17.4	16.5
300	18.8	26.0	25.1
400	25.0	34.7	33.7

28   AN OUTLINE OF THE FINITE ELEMENT METHOD

Let us suppose that we observe the set of displacements shown in the fourth column of Table 2.2. Since the ratio predicted by Model A, $(F_0/u_0)^A_{\text{pred}} = 16.0$ kN/mm, and the observed ratio, $(F_0/u_0)_{\text{obs}} = 11.9$ kN/mm, differ by more than 5%, Model A is rejected. On the other hand, the ratio predicted by Model B, $(F_0/u_0)^B_{\text{pred}} = 11.5$ kN/mm, and the observed ratio differ by less than 5%. Therefore Model B passes the validation test.

Remark 2.1.4   Example 2.1.4 illustrates some of the difficulties associated with validation of mathematical models. Typically only a very limited number of experimental observations are available. The information being sought, in this case $c(x)$, is not observable directly but must be inferred from some observable information. If force–displacement data were available for one depth only then it would not be possible to decide whether c is constant or not. Based on two pile tests of differing lengths, it was possible to reject the hypothesis that c is a constant and establish that the available information is consistent with linear variation of the form $c = kx$, but it was not possible to establish with certainty that c varies linearly.

The probability that a model adequately represents physical reality increases with the number of successful predictions of the outcomes of independent experiments, but the inherent epistemic uncertainty cannot be removed completely by any number of experiments [38a], [53]. In fact, it is possible to construct several models that match a given set of observations. In engineering and scientific applications the simplest model is preferred.

Remark 2.1.5   In order to focus on the main points of calibration in Example 2.1.3 and validation in Example 2.1.4, the input data and physical observations were treated without consideration of their statistical aspects. Since there are uncertainties in model parameters, comparing predictions with the outcome of experiments should be understood in a statistical sense.

Let us assume that, having considered statistical uncertainties in the input data, we predict a log-normal probability density function (pdf) for the material constant k in Example 2.1.4. Let us assume further that the criterion for rejection was set at the 95% confidence interval. We make an experimental observation and compute k_{exp}. Let us assume that k_{exp} falls within the 95% confidence interval. This shows that the outcome of the experiment is consistent with the prediction based on the model at the 95% confidence level. This should not be interpreted to mean that we are 95% confident that the model is valid. What this means is that the chance that a valid model would be rejected is 5%. The chance of rejecting a valid model would be reduced by setting the confidence interval at (say) 99%; however, the chance of not rejecting an invalid model would be increased.

Exercise 2.1.4   Using the calibration results for Model B in Example 2.1.3, predict the F_0/u_0 ratio for a pile driven to a depth of 17.5 m.

2.1.4   The scalar elliptic boundary value problem in one dimension

Equation (2.9) is a second-order elliptic ordinary differential equation (ODE). In Chapter 3 it will be shown that the mathematical model of steady state heat conduction in a bar will result in a second-order elliptic ODE also. Although the physical meanings of the unknown functions and the coefficients differ, the mathematical problem is essentially the same. For this reason we will focus on the mathematical problem

$$-(\kappa u')' + cu = f(x) \quad \text{on} \quad 0 \leq x \leq \ell \tag{2.24}$$

where $\kappa(x) \geq \kappa_0 > 0$, $c(x) \geq 0$ and $f(x)$ are bounded functions subject to the restriction that the indicated operations are defined.

The boundary conditions are analogous to those described in Section 2.1.1, but in the mathematical literature they are known by different names. The displacement boundary condition is called the essential or Dirichlet boundary condition.[6] The force boundary condition is called the Neumann boundary condition.[7] The spring boundary condition is called the mixed or Robin boundary condition.[8] The Neumann and Robin boundary conditions are also called natural boundary conditions.

Although Equation (2.24) may be understood to represent an elastic bar, where u is the displacement vector, or heat conduction in a bar, where u is the temperature, a scalar function, symmetry and antisymmetry are treated differently: when u is a scalar function then the symmetry boundary condition is $u'(\ell/2) = 0$ and the antisymmetry condition is $u(\ell/2) = 0$. See Remark 2.1.2.

2.2 Approximate solution

A brief introduction to approximation based on minimizing the error of an integral expression is presented in the following.

Consider the problem given by Equation (2.24) with the boundary conditions $u(0) = u(\ell) = 0$ and let us seek to approximate u by u_n, defined as follows:

$$u_n := \sum_{j=1}^{n} a_j \varphi_j(x) \qquad \varphi_j(x) := x^j(\ell - x) \tag{2.25}$$

such that the integral

$$\mathcal{I} := \frac{1}{2} \int_0^\ell \left(\kappa(u' - u_n')^2 + c(u - u_n)^2\right) dx \tag{2.26}$$

is minimum. It will be shown in the following that minimization of the error in the sense of this integral will allow us to find an approximation to the exact solution u without knowing u.

The function u_n is called a *trial function*. The functions $\varphi_j(x)$ are called basis functions. Selection of the type and number of basis functions will, of course, influence the error of approximation $u - u_n$. Discussion of this point is postponed in order to keep our focus on the basic algorithmic structure of the method.

---

[6] Johann Peter Gustav Lejeune Dirichlet (1805–1859).
[7] Franz Ernst Neumann (1798–1895).
[8] Victor Gustave Robin (1855–1897).

30   AN OUTLINE OF THE FINITE ELEMENT METHOD

Note that $\varphi_j(0) = \varphi_j(\ell) = 0$, hence u_n satisfies the prescribed boundary conditions for any choice of the coefficients a_i. From the minimum condition we have

$$\frac{\partial \mathcal{I}}{\partial a_i} = 0: \quad \int_0^\ell \left(\kappa(u' - u'_n)\varphi'_i + c(u - u_n)\varphi_i \right) dx = 0, \quad i = 1, 2, \ldots, n. \tag{2.27}$$

Recalling the product rule, we write

$$\kappa u' \varphi'_i = \left(\kappa u' \varphi_i \right)' - (\kappa u')' \varphi_i$$

and substitute this expression into Equation (2.27) to obtain

$$\underbrace{\left(\kappa u' \varphi_i \right)_{x=\ell} - \left(\kappa u' \varphi_i \right)_{x=0}}_{0} + \int_0^\ell \underbrace{\left(-(\kappa u')' + cu \right)}_{f(x)} \varphi_i \, dx - \int_0^\ell \left(\kappa u'_n \varphi'_i + c u_n \varphi_i \right) dx = 0$$

where the first two terms are zero on account of the boundary conditions. This equation can be written as

$$\int_0^\ell \left(\kappa u'_n \varphi'_i + c u_n \varphi_i \right) dx = \int_0^\ell f(x) \varphi_i \, dx, \quad i = 1, 2, \ldots, n. \tag{2.28}$$

Observe that Equation (2.28) represents n algebraic equations in the n unknowns a_i. Therefore we are able to compute an approximation to $u(x)$ without knowing $u(x)$, since only the given function $f(x)$ is needed. Specifically, Equation (2.28) is equivalent to

$$[K]\{a\} = \{r\} \tag{2.29}$$

where $\{a\} := \{a_1 \ a_2 \ \ldots \ a_n\}^T$ and the elements of $[K]$ and $\{r\}$ are, respectively,

$$k_{ij} := \int_0^\ell \left(\kappa(x)\varphi'_i(x)\varphi'_j(x) + c(x)\varphi_i(x)\varphi_j(x) \right) dx \tag{2.30}$$

$$r_i := \int_0^\ell f(x)\varphi_i(x) \, dx. \tag{2.31}$$

Example 2.2.1 Consider the problem on $I = (0, \ell)$,

$$-u'' + u = x \qquad u(0) = u(\ell) = 0,$$

and assume that the goal is to determine $u'(0)$. Let $\ell = 1$. The exact solution of this problem is

$$u = -\frac{2e}{e^2 - 1} \sinh x + x \quad \text{and therefore} \quad u'(0) = 1 - \frac{2e}{e^2 - 1} \approx 0.149\,08$$

where e is the base of the natural logarithm. We will seek to approximate u using the basis functions $\varphi_j(x)$ given in Equation (2.25) with $n = 2$. Therefore

$$k_{11} = \int_0^1 [(\varphi_1')^2 + \varphi_1^2]\,dx = \int_0^1 \left[(1-2x)^2 + x^2(1-x)^2\right] dx = \frac{11}{30}$$

$$k_{12} = k_{21} = \int_0^1 [\varphi_1'\varphi_2' + \varphi_1\varphi_2]\,dx = \int_0^1 \left[(1-2x)(2x-3x^2) + x^3(1-x)^2\right] dx = \frac{11}{60}$$

$$k_{22} = \int_0^1 [(\varphi_2')^2 + \varphi_2^2]\,dx = \int_0^1 \left[(2x-3x^2)^2 + x^4(1-x)^2\right] dx = \frac{1}{7}$$

and

$$r_1 = \int_0^1 x\varphi_1\,dx = \int_0^1 x^2(1-x)\,dx = \frac{1}{12}$$

$$r_2 = \int_0^1 x\varphi_2\,dx = \int_0^1 x^3(1-x)\,dx = \frac{1}{20}.$$

The problem is then to solve the system of linear equations:

$$\begin{bmatrix} 11/30 & 11/60 \\ 11/60 & 1/7 \end{bmatrix} \begin{Bmatrix} a_1 \\ a_2 \end{Bmatrix} = \begin{Bmatrix} 1/12 \\ 1/20 \end{Bmatrix}.$$

On solving we obtain $a_1 = 0.145\,88$, and $a_2 = 0.162\,79$, therefore the approximate solution is

$$u_n = u_2 = 0.145\,88\,x(1-x) + 0.162\,79\,x^2(1-x)$$

and hence $u_n'(0) = 0.145\,88$ and the relative error is

$$\frac{|u'(0) - u_n'(0)|}{|u'(0)|} = 0.021 \quad (2.1\%).$$

In this example the exact solution was known and hence the relative error in the data of interest could be computed. In general the exact solution is not known, therefore the relative error in the data of interest has to be estimated. Methods of error estimation will be discussed in subsequent chapters.

Exercise 2.2.1 Determine the relative error of $(u_n')_{x=\ell}$ for the problem solved in Example 2.2.1.

Remark 2.2.1 In engineering computations the goal is to determine some data of interest. The data of interest are typically numbers or functions that depend on the solution $u(x)$ and/or its first derivative. For example, if Equation (2.24) is understood to represent an elastic bar then we may be interested in computing the reaction force at $x = 0$, defined by $F_0 = (\kappa u')_{x=0}$. If,

32   AN OUTLINE OF THE FINITE ELEMENT METHOD

on the other hand, Equation (2.24) is understood to represent heat conduction then we may be interested in the rate of heat flow exiting the bar at $x = 0$, which is defined by $q_0 = -(\kappa u')_{x=0}$.

We will be interested in finding an approximate solution, computing the data of interest from the approximate solution, as illustrated by Example 2.2.1, and estimating the relative error in the data of interest.

Remark 2.2.2 Observe that Equation (2.29) can be obtained by minimizing the quadratic expression

$$\pi(u_n) := \frac{1}{2}\int_0^\ell \left(\kappa(u'_n)^2 + c(u_n)^2\right) dx - \int_0^\ell f u_n \, dx$$

$$= \frac{1}{2}\{a\}^T[K]\{a\} - \{a\}^T\{r\} \tag{2.32}$$

with respect to a_i. Therefore the function that minimizes $\pi(u_n)$ is also closest to the exact solution u in the sense that the error defined by the integral expression of Equation (2.26) is minimized. This method is known as the Rayleigh–Ritz method[9] or simply as the Ritz method. The functional $\pi(u)$ is called potential energy.

Exercise 2.2.2 Compute the coefficients a_1 and a_2 of Example 2.2.1 by minimizing $\pi(u_n)$ with respect to a_1 and a_2.

2.2.1   Basis functions

We defined the polynomial basis functions $\varphi_j(x) := x^j(\ell - x)$, $j = 1, 2, \ldots, n$, in Equation (2.25) and sought to minimize Equation (2.26) with respect to the coefficients a_j of these basis functions. This led to the definition of n algebraic equations in n unknowns, represented by Equation (2.29). The solution of Equation (2.29) is unique, provided that $[K]$ is a non-singular matrix.

In order to ensure that $[K]$ is non-singular, the basis functions must be linearly independent. By definition, a set of functions $\varphi_j(x)$ ($j = 1, 2, \ldots, n$) is linearly independent if

$$\sum_{j=1}^n a_j \varphi_j(x) = 0$$

implies that $a_j = 0$ for $j = 1, 2, \ldots, n$. It is left to the reader to show that $\varphi_j(x)$ ($j = 1, 2, \ldots, n$) are linearly independent.

Given a set of linearly independent functions $\varphi_j(x)$ ($j = 1, 2, \ldots, n$), the set of functions S defined by

$$S := \left\{ u_n \mid u_n = \sum_{j=1}^n a_j \varphi_j(x) \right\}$$

is called the span and $\varphi_j(x)$ are basis functions of S.

---

[9] Lord Rayleigh (John William Strutt; 1842–1919), Walter Ritz (1878–1909).

We could have defined other polynomial basis functions, for example,

$$u_n := \sum_{i=1}^{n} c_i \psi_i(x) \qquad \psi_i(x) := x(\ell - x)^i. \tag{2.33}$$

When one set of basis functions $\{\varphi\} := \{\varphi_1 \; \varphi_2 \; \ldots \; \varphi_n\}^T$ can be written in terms of another set $\{\psi\} := \{\psi_1 \; \psi_2 \; \ldots \; \psi_n\}^T$ in the form

$$\{\psi\} = [B]\{\varphi\} \tag{2.34}$$

where $[B]$ is an invertible matrix of constant coefficients, then both sets of basis functions are said to have the same span. The following exercise demonstrates that the approximate solution depends on the span of the basis functions, not on the basis functions.

Exercise 2.2.3 Solve the problem of Example 2.2.1 using the basis functions

$$\psi_1(x) = x(1-x), \qquad \psi_2(x) = x(1-x)(1-2x)$$

and show that the solution $u_2 = b_1 \psi_1(x) + b_2 \psi_2(x)$ is the same as the solution in Example 2.2.1. In this exercise the span is the set of polynomials of degree 3, subject to the restriction that they vanish at the boundary points.

Exercise 2.2.4 Let $\varphi_i(x) = x^i(\ell - x)$, $\psi_i(x) = x(\ell - x)^i$ and

$$u_n = \sum_{i=1}^{3} a_i \varphi_i(x) = \sum_{i=1}^{3} c_i \psi_i(x).$$

Determine the matrix $[B]$ as defined in Equation (2.34) and, assuming that the values of a_i are given, find an expression for c_i in terms of a_i ($i = 1, 2, 3$) and $[B]$.

2.3 Generalized formulation in one dimension

We have seen in Section 2.2 that it was possible to obtain an approximate solution to a differential equation without knowing the exact solution. This depended on a seemingly fortuitous choice of the integral expression $\mathcal{I}$ and zero boundary conditions, allowing us to replace the unknown exact solution with the known function f following integration by parts. In this section the reasons for the choice of I are explained in a general setting, without restriction on the boundary conditions.

Once again our starting point is Equation (2.24)

$$-\left(\kappa u'\right)' + cu = f(x)$$

34 AN OUTLINE OF THE FINITE ELEMENT METHOD

subject to boundary conditions to be discussed later. Let us multiply this equation by an arbitrary function $v(x)$ defined on $I = (0, \ell)$ and integrate:

$$\int_0^\ell \left(-(\kappa u')' + cu\right) v \, dx = \int_0^\ell fv \, dx. \tag{2.35}$$

Clearly, if u is the solution of Equation (2.24) then this equation will be satisfied for all v for which the indicated operations are defined. Integrating the first term by parts,

$$-\int_0^\ell (\kappa u')' v \, dx = -\int_0^\ell \left[(\kappa u'v)' - \kappa u'v'\right] dx$$

$$= -\left[\kappa u'v\right]_{x=\ell} + \left[\kappa u'v\right]_{x=0} + \int_0^\ell \kappa u'v' \, dx$$

we have

$$\int_0^\ell (\kappa u'v' + cuv) \, dx = \int_0^\ell fv \, dx + \left[\kappa u'v\right]_{x=\ell} - \left[\kappa u'v\right]_{x=0}. \tag{2.36}$$

Note that the integrand $(\kappa u')'v$ became $\kappa u'v'$ plus two boundary terms. This equation will be the starting point for our discussion of the generalized formulation. The specific statement of the generalized formulation for a particular problem depends on the boundary conditions. Some useful definitions and notation are introduced in the following.

Definition 2.3.1 The function u (resp. v) in Equation (2.36) is called the trial (resp. test) function.

Definition 2.3.2 In one dimension the strain energy is defined by

$$U(u) = \frac{1}{2} \int_0^\ell \left(\kappa (u')^2 + cu^2\right) dx \tag{2.37}$$

where $\kappa \geq \kappa_0 > 0$ and $c \geq 0$. Therefore $U(u) \geq 0$. When $c \neq 0$ then $U(u) = 0$ only when $u = 0$ in the interval $I = (0, \ell)$. When $c = 0$ then $U(u) = 0$ when $u(x) = C$ where C is an arbitrary constant.

Definition 2.3.3 The energy space, denoted by $E(I)$, is the set of functions defined on the interval I that satisfy the inequality

$$E(I) := \left\{ u \mid U(u) \leq C < \infty \right\} \tag{2.38}$$

where C is some positive number. For any $u \in E(I)$ and $v \in E(I)$ the integral expressions in Equation (2.36) are defined. This follows from the Schwarz inequality,[10] see Appendix A.

---
[10] Hermann Amandus Schwarz (1843–1921).

Definition 2.3.4 When $u(0) = \hat{u}_0$ and/or $u(\ell) = \hat{u}_\ell$ are specified on the boundaries then the boundary condition is called an *essential* or *Dirichlet* boundary condition. The functions in $E(I)$ that satisfy the essential boundary conditions are called *admissible* functions. The set of all admissible functions is called the *trial space* and is denoted by $\tilde{E}(I)$. This notation should be understood as follows:

(a) If essential boundary conditions are specified at $x = 0$ and $x = \ell$ then

$$\tilde{E}(I) := \{u \mid u \in E(I), \ u(0) = \hat{u}_0, \ u(\ell) = \hat{u}_\ell\}. \tag{2.39}$$

Corresponding to $\tilde{E}(I)$ is the *test space* $E^0(I)$ defined as follows:

$$E^0(I) := \{u \mid u \in E(I), \ u(0) = 0, \ u(\ell) = 0\}. \tag{2.40}$$

(b) If an essential boundary condition is specified only at $x = 0$ then

$$\tilde{E}(I) := \{u \mid u \in E(I), \ u(0) = \hat{u}_0\} \tag{2.41}$$

$$E^0(I) := \{u \mid u \in E(I), \ u(0) = 0\}. \tag{2.42}$$

(c) If an essential boundary condition is specified only at $x = \ell$ then

$$\tilde{E}(I) := \{u \mid u \in E(I), \ u(\ell) = \hat{u}_\ell\} \tag{2.43}$$

$$E^0(I) := \{u \mid u \in E(I), \ u(\ell) = 0\}. \tag{2.44}$$

(d) If the essential boundary conditions are homogeneous, that is, $\hat{u}_0 = 0$, $\hat{u}_\ell = 0$, then $\tilde{E}(I) = E^0(I)$.

(e) If essential boundary conditions are not prescribed on either boundary then $\tilde{E}(I) = E^0(I) = E(I)$.

(f) If periodic boundary conditions are prescribed then both the trial and test spaces are

$$\hat{E}(I) = \{u \mid u \in E(I), \ u(0) = u(\ell)\}. \tag{2.45}$$

Remark 2.3.1 Note that $\tilde{E}(I)$ is not a linear space. Refer to Appendix A, Section A.2. It is seen that $\tilde{E}(I)$ does not satisfy condition 1 whereas $E^0(I)$ and $\hat{E}(I)$ satisfy all of the conditions of Section A.2.

2.3.1 Essential boundary conditions

Essential boundary conditions are enforced by restriction. This was done in the special case discussed in Section 2.2 where the homogeneous essential boundary conditions $\hat{u}_0 = \hat{u}_\ell = 0$ were used and the basis functions were defined in Equation (2.25) so that the boundary conditions prescribed on u were satisfied for an arbitrary choice of the coefficients a_i.

The known boundary conditions are imposed on the trial functions u and the test function v is set to zero on the boundary points where essential boundary conditions were prescribed. In this way the boundary terms (the terms in the square brackets in Equation (2.36)) vanish and the generalized formulation is stated as follows: "Find $u \in \tilde{E}(I)$ such that

$$\underbrace{\int_0^\ell \left(\kappa u'v' + cuv\right) dx}_{B(u,v)} = \underbrace{\int_0^\ell fv\,dx}_{F(v)} \quad \text{for all } v \in E^0(I)." \tag{2.46}$$

We will use the shorthand notation $B(u, v)$ for the left hand side and $F(v)$ for the right hand side, as indicated in Equation (2.46). $B(u, v)$ is a symmetric *bilinear form*, that is, it is linear with respect to each of its arguments and $B(u, v) = B(v, u)$, and $F(v)$ is a *linear functional*. The properties of bilinear forms and linear functionals are given in Appendix A.

Alternatively we can select an arbitrary function $u^\star$ from $\tilde{E}(I)$ and write

$$u = \bar{u} + u^\star \tag{2.47}$$

where $\bar{u} \in E^0(I)$. Clearly, the prescribed boundary conditions are satisfied for any choice $\bar{u} \in E^0(I)$. Substituting Equation (2.47) into Equation (2.36), the generalized formulation can be stated as follows: "Find $\bar{u} \in E^0(I)$ such that

$$\underbrace{\int_0^\ell \left(\kappa \bar{u}'v' + c\bar{u}v\right) dx}_{B(\bar{u},v)} = \underbrace{\int_0^\ell fv\,dx - \int_0^\ell \left(\kappa(u^\star)'v' + cu^\star v\right) dx}_{F(v)} \tag{2.48}$$

for all $v \in E^0(I)$."

Example 2.3.1 Let us state the generalized formulation for the following problem:

$$-u'' = (2+x)e^x \qquad u(0) = 1, \ u(2) = -1.$$

In this case $\tilde{E}(I) = \{u \mid u \in E(I),\ u(0) = 1,\ u(2) = -1\}$. Let us select $u^\star = 1 - x$ and substitute $u = \bar{u} + u^\star$ into Equation (2.36). The statement of the generalized formulation is now: "Find $\bar{u} \in E^0(I)$ such that $B(\bar{u}, v) = F(v)$ for all $v \in E^0(I)$" where

$$B(\bar{u}, v) := \int_0^2 \bar{u}'v'\,dx, \qquad F(v) := \int_0^2 (2+x)e^x v\,dx - \int_0^2 (-1)v'\,dx.$$

Example 2.3.2 In this example it is shown that Equation (2.36) leads to the same equations as those obtained in Section 2.2. To obtain an approximation to the solution of Equation (2.24), we substitute u_n from Equation (2.25) for u and similarly substitute v_n for v:

$$v_n := \sum_{i=1}^n b_i \varphi_i(x), \qquad \varphi_i(x) := x^i(\ell - x)$$

where b_i, $i = 1, 2, \ldots, n$, are a set of arbitrary numbers. Since $v_n(0) = v_n(\ell) = 0$, the terms in the square brackets in Equation (2.36) vanish and we have

$$\{b\}^T [K]\{a\} = \{b\}^T \{r\}$$

where $\{b\} := \{b_1 \; b_2 \; \ldots \; b_n\}^T$ and the definitions of $[K]$ and $\{r\}$ are the same as in Equations (2.30) and (2.31). Equivalently,

$$\{b\}^T \left([K]\{a\} - \{r\}\right) = 0.$$

Since this relationship must hold for any choice of $\{b\}$, we must have $[K]\{a\} = \{r\}$, which is exactly the same as the result obtained in Section 2.2 with k_{ij} (resp. r_i) defined by Equation (2.30) (resp. Equation (2.31)).

Exercise 2.3.1 Show that

(a) $B(u_1 + u_2, v) = B(u_1, v) + B(u_2, v)$

(b) $B(u + v, u + v) = B(u, u) + 2B(u, v) + B(v, v)$.

2.3.2 Neumann boundary conditions

When u' or more commonly $F = \kappa u'$ is prescribed on a boundary then the boundary condition is called a Neumann boundary condition. The treatment of Neumann boundary conditions is straightforward. Let $F_0 = (\kappa u')_{x=0}$ and $F_\ell = (\kappa u')_{x=\ell}$ be given. Substituting F_ℓ and F_0 into Equation (2.36), the generalized formulation is stated as follows: "Find $u \in E(I)$ such that $B(u, v) = F(v)$ for all $v \in E(I)$" where

$$B(u, v) = \int_0^\ell \left(\kappa u' v' + cuv\right) dx, \quad F(v) = \int_0^\ell fv \, dx + F_\ell v(\ell) - F_0 v(0). \quad (2.49)$$

Note that there are no restrictions on u or v at the boundary points.

Remark 2.3.2 When $c = 0$ and Neumann boundary conditions are prescribed then, since Equation (2.49) must hold for all choices of $v \in E(I)$, it must hold for $v = C$ where C is a constant. Therefore we must have

$$\int_0^\ell f \, dx + F_\ell - F_0 = 0. \quad (2.50)$$

This means that f, F_0 and F_ℓ cannot be assigned arbitrarily. The tractions acting on the bar and the bar forces acting on the boundary points must be in equilibrium.

2.3.3 Robin boundary conditions

A linear combination of u' and u is given at the boundary:

$$(\kappa u')_{x=0} = \beta_0(u(0) - U_0)$$

$$(\kappa u')_{x=\ell} = \beta_\ell(U_\ell - u(\ell))$$

where $\beta_0 > 0$, $\beta_\ell > 0$, U_0, U_ℓ are input data analogous to the spring rates k_0, k_ℓ and spring displacements d_0, d_ℓ defined in Section 2.1.1. Substituting these expressions into Equation (2.36), the generalized formulation is once again stated as follows: "Find $u \in E(I)$ such that $B(u, v) = F(v)$ for all $v \in E(I)$" where

$$B(u, v) := \int_0^\ell \left(\kappa u'v' + cuv\right) dx + \beta_0 u(0)v(0) + \beta_\ell u(\ell)v(\ell)$$

$$F(v) := \int_0^\ell fv\, dx + \beta_0 U_0 v(0) + \beta_\ell U_\ell v(\ell).$$

There are no restrictions on u or v at the boundary points. The spring boundary condition described in Section 2.1.1 is a Robin boundary condition.

Example 2.3.3 Any combination of Dirichlet, Neumann and Robin boundary conditions may be prescribed. For example, let us consider the problem

$$-\left(\kappa u'\right)' + cu = f(x) \qquad u(0) = \hat{u}_0, \quad (\kappa u')_{x=\ell} = \beta_\ell(U_\ell - u(\ell)).$$

In this case $\tilde{E}(I)$ is defined by Equation (2.41), $E^0(I)$ is defined by Equation (2.42) and

$$B(\bar{u}, v) := \int_0^\ell \left(\kappa \bar{u}'v' + c\bar{u}v\right) dx + \beta_\ell u(\ell)v(\ell)$$

$$F(v) := \int_0^\ell fv\, dx + \beta_\ell U_\ell v(\ell) - \int_0^\ell \left(\kappa (u^\star)'v' + cu^\star v\right) dx$$

where $u^\star$ is an arbitrary fixed function from $\tilde{E}(I)$. For example, we may select $u^\star = \hat{u}_0(1 - x/\ell)$ or simply $u^\star = \hat{u}_0$.

The generalized formulation of this problem is stated as follows: "Find $\bar{u} \in E^0(I)$ such that $B(\bar{u}, v) = F(v)$ for all $v \in E^0(I)$." The exact solution is then $u = \bar{u} + u^\star$.

Exercise 2.3.2 State the generalized formulation for the following problem:

$$-\left(\kappa u'\right)' + cu = f(x) \qquad (\kappa u')_{x=0} = -\hat{q}_0, \quad u(\ell) = 0.$$

Exercise 2.3.3 State the generalized formulation for the following problem:

$$-\left(\kappa u'\right)' + cu = f(x) \qquad (\kappa u')_{x=0} = \beta_0(u(0) - U_0), \quad u(\ell) = \hat{u}_\ell.$$

Exercise 2.3.4 State the generalized formulation for the following problem:

$$-\left(\kappa u'\right)' + cu = f(x) \qquad (\kappa u')_{x=0} = -\hat{q}_0, \quad (\kappa u')_{x=\ell} = \beta_\ell(U_\ell - u(\ell)).$$

2.4 Finite element approximations

We have recast a differential equation in the form of a generalized formulation which reads: "Find $u \in \tilde{E}(I)$ such that $B(u, v) = F(v)$ for all $v \in E^0(I)$." It may appear that nothing has been gained. This problem is more difficult to solve than the differential equation was, since there are an infinite number of trial functions for u that must be tested against an infinite number of test functions v.

One of the main advantages of the generalized formulation is that it serves as a framework for obtaining approximate solutions. To obtain an approximate solution we construct a finite-dimensional subspace of $E(I)$, as we have done in Section 2.2, where we selected

$$u_n = \sum_{j=1}^{n} a_j \varphi_j(x)$$

with $n = 2$. The family of functions that can be written in this way will be denoted by $S(I)$. The functions $\varphi_j(x)$, called basis functions, will be defined such that $S(I) \subset E(I)$. The number n is the dimension of $S(I)$. We will use the notation $\tilde{S}(I) := S(I) \cap \tilde{E}(I)$; $S^0(I) = S(I) \cap E^0(I)$. The dimension of space $S^0(I)$, that is, the number of linearly independent functions in $S^0(I)$, is called the number of degrees of freedom and denoted by N.

In the FEM the space S is constructed by partitioning the solution domain into elements and defining polynomial basis functions on the elements. Approximation spaces so constructed are called *finite element spaces*. A particular partition is called a finite element mesh and will be denoted by Δ. The number of elements will be denoted by $M(\Delta)$. A simple illustration is given in the following example.

Example 2.4.1 Typical finite element basis functions in one dimension are illustrated in Figure 2.6 where the domain $I = (0, \ell)$ is partitioned into three intervals (i.e., $M(\Delta) = 3$), called finite elements, and the polynomial degrees $p_1 = 2$, $p_2 = 1$ and $p_3 = 3$ are assigned. The length of the elements is denoted by ℓ_k, $k = 1, 2, 3$. There are four node points, labeled x_i, $i = 1, 2, 3, 4$. There are seven basis functions, labeled $\varphi_1(x), \ldots, \varphi_7(x)$. The numbering of the basis functions is arbitrary; however, it is good practice to number them by polynomial degree. The first four basis functions are piecewise linear. For example,

$$\varphi_2(x) = \begin{cases} \dfrac{x - x_1}{x_2 - x_1} & \text{if } x_1 \leq x \leq x_2 \\ \dfrac{x_3 - x}{x_3 - x_2} & \text{if } x_2 < x \leq x_3 \\ 0 & \text{otherwise.} \end{cases}$$

The basis functions $\varphi_5(x)$, $\varphi_6(x)$ are quadratic functions. For example,

$$\varphi_6(x) = \begin{cases} (x - x_3)(x - x_4) & \text{if } x_3 \leq x \leq x_4 \\ 0 & \text{otherwise.} \end{cases}$$

The basis function $\varphi_7(x)$ is a cubic polynomial on element 3 and is zero outside of element 3. The finite element space, characterized by the mesh and the polynomial degree of elements shown in Figure 2.6, is the set of all functions that can be written in the form

$$u = \sum_{j=1}^{7} a_j \varphi_j(x). \qquad (2.51)$$

40 AN OUTLINE OF THE FINITE ELEMENT METHOD

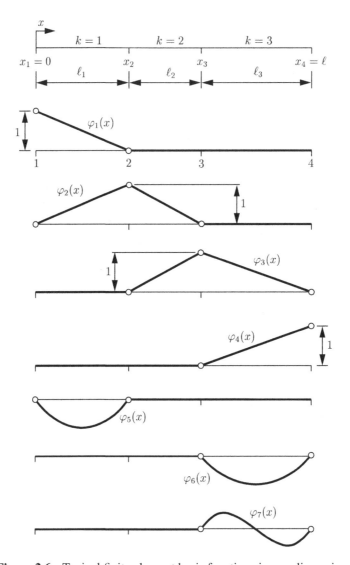

Figure 2.6 Typical finite element basis functions in one dimension.

Exercise 2.4.1 Refer to Figure 2.6. Write down the basis function $\varphi_7(x)$.

Exercise 2.4.2 Show that the set of basis functions

$$\psi_1(x) := \varphi_1(x) + \varphi_2(x), \quad \psi_2(x) := \varphi_2(x) - \varphi_1(x),$$
$$\psi_j(x) = \varphi_j(x) \quad j = 3, 4, \ldots, 7 \qquad (2.52)$$

has the same span as the set $\varphi_j(x)$, $j = 1, 2, \ldots, 7$, defined in Example 2.4.1.

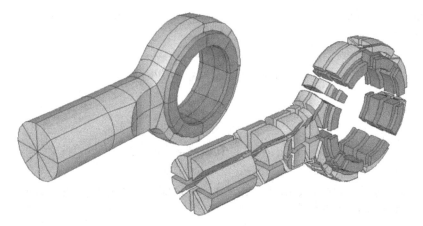

Figure 2.7 Example of a three-dimensional finite element mesh.

Remark 2.4.1 The key difference between the original form of the Rayleigh–Ritz method (see, for example, [42]) and the FEM is that in the Rayleigh–Ritz method the basis functions are analytic functions defined on the entire solution domain whereas in the FEM the basis functions are characterized by piecewise polynomials that are non-zero over a few elements only. In one-dimensional applications, for example, they are non-zero over at most two elements, as seen in Figure 2.6. This makes it possible to construct algorithms suitable for handling a great variety of problems very efficiently.

The partition of the domain into finite elements makes it possible to construct basis functions analogous to those discussed in Example 2.4.1 on complicated domains such as shown in Figure 2.7.

In the FEM the approximating functions are piecewise polynomials. There are two reasons for this: Piecewise polynomials are advantageous from the point of view of implementation and they have favorable approximation properties.

2.4.1 Error measures and norms

Since we will be solving various problems approximately, we need to ask: "What is the error of approximation?" Various quantitative measures of error are used in connection with finite element analyses. The most useful measure is the relative error in terms of the data of interest.

There are other measures, called norms, useful for measuring the quality of the approximate solution. Three norms are defined for functions of a single variable in the following. Their generalization to two and three dimensions is straightforward. Norms are analogous to the length of a vector in Euclidean space.[11] The definitive properties of norms are listed in Appendix A, Section A.1.

The energy norm

It is natural to use the energy norm in connection with the formulation discussed in Section 2.2 (page 29) because the coefficients a_j in Equation (2.25) are determined in such a way that

---

[11] Euclid of Alexandria (c.325–265 BC).

the error measured in the energy norm is minimal. This is Theorem 2.4.2 (see page 43). By definition,

$$\|u\|_E := \sqrt{U(u)} \equiv \sqrt{\frac{1}{2}B(u,u)}. \qquad (2.53)$$

In Example 2.2.1

$$B(u,u) = \int_0^\ell \left(\kappa(u')^2 + cu^2\right) dx$$

and $\|u - u_n\|_E = 1.01606 \times 10^{-3}$ and $\|u\|_E = 0.10074$. Therefore the relative error measured in the energy norm is

$$(e_r)_E := \frac{\|u - u_n\|_E}{\|u\|_E} = 0.0101 \ (1.01\%).$$

The maximum norm

The maximum norm of a continuous function $u(x)$ defined on the interval $\bar{I}$ is

$$\|u\|_{\max} := \max_{x \in \bar{I}} |u(x)| \qquad (2.54)$$

and the relative error in maximum norm is defined by

$$(e_r)_{\max} := \frac{\|u - u_n\|_{\max}}{\|u\|_{\max}}. \qquad (2.55)$$

Often the percent relative error is given. The maximum norm is usually approximated by computing the maximum on some fine grid. The abscissa at which $\|u - u_n\|_{\max}$ is computed may be different from the abscissa at which the reference value $\|u\|_{\max}$ was computed.

In Example 2.2.1, using 100 equally spaced grid points, $\max |u - u_n| = 2.3 \times 10^{-4}$ at $x = 0.51$; $\max |u| = 5.83 \times 10^{-2}$ at $x = 0.60$. Therefore the relative error is $(e_r)_{\max} = 0.39\%$.

The L_2 norm

The L_2 norm of a function u defined on the interval $I = (0, \ell)$ is

$$\|u\|_{L_2} := \sqrt{\int_0^\ell u^2 \, dx}. \qquad (2.56)$$

The definition of relative error in the L_2 norm is analogous to Equation (2.55). In Example 2.2.1, $\|u - u_n\|_{L_2} = 1.487 \times 10^{-4}$ and $\|u\|_{L_2} = 4.183 \times 10^{-2}$. Therefore the percent relative error measured in the L_2 norm is

$$(e_r)_{L_2} := \frac{\|u - u_n\|_{L_2}}{\|u\|_{L_2}} = 0.0036 \ (0.36\%).$$

Remark 2.4.2 The error depends on the norm in which it is measured. The choice of the norm depends on the purpose of computation.

Exercise 2.4.3 Obtain an approximate solution for the problem

$$-u'' + u = 1, \qquad u(0) = u(1) = 0$$

by minimizing π given by Equation (2.32). Use $n = 2$ and the same basis functions as in Example 2.2.1. Determine the exact solution and compute the relative error in the maximum norm and energy norm.

2.4.2 The error of approximation in the energy norm

In Section 2.2 we minimized the integral expression (2.26) to obtain an approximate solution to the problem (2.24). We are now in a position to generalize this to any combination of the three kinds of boundary conditions discussed in Section 2.3.

In the following we will denote the exact solution by u_{EX} and the finite element solution by u_{FE}. The approximation problem is stated as follows: "Find $u_{FE} \in \tilde{S}(I)$ such that $B(u_{FE}, v) = F(v)$ for all $v \in S^0(I)$."

Theorem 2.4.1 The error of approximation $e := u_{EX} - u_{FE}$ is orthogonal to all test functions in $S^0(I)$ in the following sense:

$$B(e, v) = 0 \quad \text{for all } v \in S^0(I). \tag{2.57}$$

This is a basic property of the error of approximation, known as the Galerkin orthogonality.[12]

Proof: Since $S^0(I) \subset E^0(I)$,

$$B(u_{EX}, v) = F(v) \quad \text{for all } v \in S^0(I)$$
$$B(u_{FE}, v) = F(v) \quad \text{for all } v \in S^0(I).$$

On subtracting the second equation from the first, Equation (2.57) is obtained. □

An important theorem is proven in the following. This theorem establishes that the FEM will select the coefficients of the basis functions in such a way that the energy norm of the error $\|e\|_E$ will be minimum.

Theorem 2.4.2 The finite element solution minimizes the error in the energy norm on the space $\tilde{S}(I)$:

$$\|u_{EX} - u_{FE}\|_E = \min_{u \in \tilde{S}(I)} \|u_{EX} - u\|_E. \tag{2.58}$$

---

[12] Boris Grigorievich Galerkin (1871–1945).

Proof: We have seen a direct application of this theorem in Section 2.2. Once again we write $e := u_{EX} - u_{FE}$. For an arbitrary $v \in S^0(I)$, $\|v\|_E \neq 0$, we have

$$\|e+v\|_E^2 = \frac{1}{2}B(e+v, e+v) = \frac{1}{2}B(e,e) + \underbrace{B(e,v)}_{0} + \underbrace{\frac{1}{2}B(v,v)}_{\|v\|_E^2 > 0}.$$

By Equation (2.57), $B(e, v) = 0$, therefore for any $\|v\|_E \neq 0$ we have $\|e + v\|_E^2 > \|e\|_E^2$ which was to be proven. $\square$

This theorem shows that the selection $S(I)$ is of crucial importance, since the error of approximation is determined by $S(I)$. This theorem also shows that if u_{EX} happens to lie in $S(I)$ then $u_{FE} = u_{EX}$. Furthermore, the theorem shows that if we construct a sequence of finite element spaces $S_1 \subset S_2 \subset \cdots \subset S_m$ and compute the corresponding finite element solutions $u_{FE}^{(1)}, u_{FE}^{(2)}, \ldots, u_{FE}^{(m)}$, then the error measured in the energy norm will decrease monotonically with respect to increasing m.

2.5 FEM in one dimension

In this section the key algorithmic procedures common to all finite element computer programs are outlined in the simplest setting. Although the discussion covers the one-dimensional case only, analogous procedures apply to two- and three-dimensional problems.

2.5.1 The standard element

In order make computation of the coefficient matrices and load vectors suitable for implementation in a computer program, the computations are performed element by element. In one dimension the kth element is characterized by the node points x_k and x_{k+1} of the mesh Δ. The mesh is the set of elements $I_k := \{x \mid x_k < x < x_{k+1}\}$, $k = 1, 2, \ldots, M(\Delta)$, and $\ell_k := x_{k+1} - x_k$ is the size of element k.

In order to standardize the element-level computations, a standard element I_{st} is defined:

$$I_{\text{st}} := \{\xi \mid -1 < \xi < +1\}. \tag{2.59}$$

The standard element is mapped into the kth element by the mapping function:

$$x = Q_k(\xi) := \frac{1-\xi}{2} x_k + \frac{1+\xi}{2} x_{k+1} \qquad \xi \in I_{\text{st}}. \tag{2.60}$$

The inverse map is

$$\xi = Q_k^{-1}(x) := \frac{2x - x_k - x_{k+1}}{x_{k+1} - x_k} \qquad x \in I_k. \tag{2.61}$$

Remark 2.5.1 The mapping of the standard element I_{st} onto the "real" element I_k is not unique. For example, the mapping

$$x = \frac{1}{2}\xi(\xi - 1)x_k + \frac{1}{2}\xi(1 + \xi)x_{k+1} \qquad \xi \in I_{st}$$

would serve the same purpose as Equation (2.60). It will be shown, however, that in general the simplest mapping functions are preferable. In special cases, however, there are some exceptions.

2.5.2 The standard polynomial space

The polynomials of degree p defined on the standard element I_{st} will be denoted by $S^p(I_{st})$. The basis functions that span $S^p(I_{st})$ are usually called *shape functions*. We will discuss two types of shape functions used in finite element analysis (FEA) programs: Lagrange[13] and *hierarchic* shape functions that are scaled integrals of Legendre[14] polynomials. We will use the notation $N_i(\xi)$ ($i = 1, 2, \ldots, p+1$) for both types of shape functions. For $S^1(I_{st})$ the shape functions for both sets are

$$N_1(\xi) := \frac{1-\xi}{2} \tag{2.62}$$

$$N_2(\xi) := \frac{1+\xi}{2}. \tag{2.63}$$

Lagrange shape functions

The standard domain is partitioned into p sub-intervals. The lengths of the sub-intervals may vary. The node points are $\xi_1 = -1$, $\xi_2 = 1$ and $-1 < \xi_3 < \xi_4 < \cdots < \xi_{p+1} < 1$. The shape functions for $S^p(I_{st})$ are

$$N_i(\xi) := \prod_{\substack{k=1 \\ k \neq i}}^{p+1} \frac{\xi - \xi_k}{\xi_i - \xi_k} \qquad i = 1, 2, \ldots, p+1, \qquad \xi \in I_{st}. \tag{2.64}$$

These shape functions have the following important properties:

$$N_i(\xi_j) = \begin{cases} 1 & \text{if } i = j \\ 0 & \text{if } i \neq j \end{cases} \quad \text{and} \quad \sum_{i=1}^{p+1} N_i(\xi) = 1. \tag{2.65}$$

For example, for $p = 2$ the equally spaced node points are $\xi_1 = -1$, $\xi_2 = 1$, $\xi_3 = 0$ and the three Lagrange shape functions as illustrated in Figure 2.8.

Exercise 2.5.1 Write down the Lagrange shape functions for $S^3(I_{st})$ using equally spaced node points.

---

[13] Joseph-Louis Lagrange (1736–1813).
[14] Adrien-Marie Legendre (1752–1833).

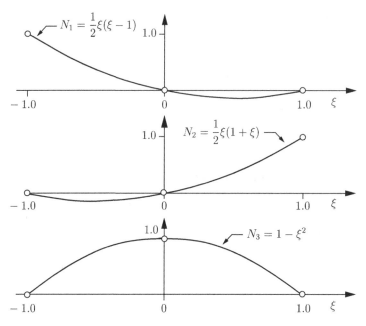

Figure 2.8 Lagrange shape functions in one dimension, $p = 2$.

Hierarchic shape functions based on Legendre polynomials

It is advantageous to retain the definitions (2.62) and (2.63) for $p = 1$ and then for $p \geq 2$ define the shape functions as follows:

$$N_i(\xi) = \sqrt{\frac{2i-3}{2}} \int_{-1}^{\xi} P_{i-2}(t)\, dt \qquad i = 3, 4, \ldots, p+1 \qquad (2.66)$$

where $P_i(t)$ are the Legendre polynomials defined in Appendix A. These shape functions have the following important properties:

(a) Orthogonality. For $i, j \geq 3$,

$$\int_{-1}^{+1} \frac{dN_i}{d\xi} \frac{dN_j}{d\xi}\, d\xi = \begin{cases} 1 & \text{if } i = j \\ 0 & \text{if } i \neq j. \end{cases} \qquad (2.67)$$

This property follows directly from the orthogonality of Legendre polynomials, see Equation (A.13) in Appendix A.

(b) The shape functions of $S^{p-1}(I_{st})$ are a subset of the shape functions of $S^p(I_{st})$. Shape functions that have this property are called hierarchic shape functions.

(c) These shape functions vanish at the endpoints of I_{st}: $N_i(-1) = N_i(+1) = 0$ for $i \geq 3$.

The first five hierarchic shape functions are shown in Figure 2.9. Observe that all roots lie in I_{st}.

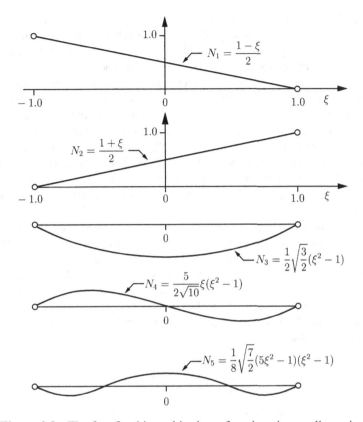

Figure 2.9 The first five hierarchic shape functions in one dimension.

Exercise 2.5.2 Show that for the hierarchic shape functions, defined by Equation (2.66), $N_i(-1) = N_i(+1) = 0$ for $i \geq 3$.

Exercise 2.5.3 Show that the hierarchic shape functions defined by Equation (2.66) can be written in the form

$$N_i(\xi) = \frac{1}{\sqrt{2(2i-3)}} (P_{i-1}(\xi) - P_{i-3}(\xi)) \qquad i = 3, 4, \ldots. \tag{2.68}$$

Hint: Use Equation (A.12) in Appendix A.

2.5.3 Finite element spaces

We are now in a position to provide a precise definition of finite element spaces in one dimension: a finite element space S is a set of continuous functions characterized by the finite element mesh Δ, the assigned polynomial degrees p_k and the mapping functions Q_k, $k = 1, 2, \ldots, M(\Delta)$. Specifically,

$$S = S(\Delta, \mathbf{p}, \mathbf{Q}) = \{u \mid u \in E(I), \ u(Q_k(\xi)) \in \mathcal{S}^{p_k}(I_{\text{st}}), \ k = 1, 2, \ldots, M(\Delta)\} \tag{2.69}$$

where **p** and **Q** represent, respectively, the arrays of the assigned polynomial degrees and the mapping functions. The expression $u(Q_k(\xi)) \in S^{p_k}(I_{st})$ indicates that, on element I_k, $u(x)$ is mapped from the standard polynomial space $S^{p_k}(I_{st})$. When linear mapping is used then the finite element space is composed of piecewise polynomial functions.

It is shown in Section 2.6.4 that if $u \in E(I)$ then u must be a continuous function. Of course, it would be possible to enforce levels of continuity higher than the minimum required. For example, we could require the first derivatives of the basis functions to be continuous also. However, enforcement of higher levels of continuity than the minimum required is a restriction on the space of admissible functions which, in view of Theorem 2.4.2, is detrimental to the generality and good overall approximation properties of the method. For this reason we will be concerned with finite element spaces that are exactly and minimally continuous. In other words, no more than the minimal continuity required for satisfying the condition that $u \in E(I)$ will be enforced.

We will be concerned with approximations-based sequences of finite element spaces created by systematic mesh refinement or increase of the polynomial degree of elements, or a combination of both. Systematic enlargement of finite element spaces by mesh refinement, increase of the polynomial degree(s) of elements or a combination of both is called, respectively, *h*-extension, *p*-extension and *hp*-extension. When a sequence of finite element spaces has the property that $S_1 \subset S_2 \subset S_3 \subset \cdots \subset S_n$ then the sequence is called a hierarchic sequence.

A family of meshes Δ_k is said to be quasiuniform if there exist positive constants C_1, C_2, independent of k, such that

$$C_1 \leq \frac{h_{\max}^{(k)}}{h_{\min}^{(k)}} \leq C_2, \quad k = 1, 2, \ldots \tag{2.70}$$

where $h_{\max}^{(k)}$ (resp. $h_{\min}^{(k)}$) is the diameter of the largest (resp. smallest) element in mesh Δ_k.

A mesh is geometrically graded toward the point $x = 0$ on the interval $0 < x < \ell$ if the node points are located as follows:

$$x_k = \begin{cases} 0 & \text{for } k = 1 \\ q^{M(\Delta)+1-k}\ell & \text{for } k = 2, 3, \ldots, M(\Delta) + 1 \end{cases}$$

where $0 < q < 1$ is called the grading factor or common factor. Such meshes are called geometric meshes.

A mesh is said to be a radical mesh if, on the interval $0 < x < \ell$, the node points are located by

$$x_k = \left(\frac{k-1}{M(\Lambda)}\right)^\theta \ell \quad \theta > 1, \quad k = 1, 2, \ldots, M(\Delta) + 1.$$

It will be demonstrated in Chapter 6 that for a large and important class of problems, the ideal meshes are geometric meshes when the mesh is fixed and the polynomial degree of elements is increased, and the ideal meshes are radical meshes when the polynomial degree is fixed and the number of elements is increased [70].

Exercise 2.5.4 Consider the family of mapping functions

$$x = Q_k(\alpha, \xi) := \frac{1}{2}\xi(\xi - 1)x_k + (1 - \xi^2)(\alpha x_k + (1 - \alpha)x_{k+1}) + \frac{1}{2}\xi(1 + \xi)x_{k+1}$$

where $1/4 < \alpha < 3/4$. Show that by letting $\alpha = 1/2$ the shape function N_2 is mapped into $(x - x_k)/\ell_k$ and all mapped shape functions are polynomials. Show also that by letting $\alpha = 3/4$ the shape function N_2 is mapped into $\sqrt{(x - x_k)/\ell_k}$. In one dimension this mapping is not admissible because the mapped shape function does not lie in the energy space. Similar mappings, however, are admissible in two and three dimensions where elements mapped by analogous functions are called quarter-point elements.

This exercise illustrates that when the mapping is linear ($\alpha = 1/2$) the mapped shape functions are polynomials.

2.5.4 Computation of the coefficient matrices

In the FEM the coefficient matrices are computed element by element. These computations produce element-level matrices that are "assembled" in a separate step. The procedure is outlined in the following.

Computation of the stiffness matrix

The first term of the bilinear form $B(u_n, v_n)$ is computed as a sum of integrals over the elements

$$\int_0^\ell \kappa(x) u_n' v_n' \, dx = \sum_{k=1}^{M(\Delta)} \int_{x_k}^{x_{k+1}} \kappa(x) u_n' v_n' \, dx$$

where n is the number of basis functions that span the finite element space, see Section 2.4. We will be concerned with the evaluation of the integral

$$\int_{x_k}^{x_{k+1}} \kappa(x) u_n' v_n' \, dx = \int_{x_k}^{x_{k+1}} \kappa(x) \left(\sum_{j=1}^{p_k+1} a_j \frac{dN_j}{dx} \right) \left(\sum_{i=1}^{p_k+1} b_i \frac{dN_i}{dx} \right) dx.$$

The shape functions N_i are defined on the standard domain I_{st}. Therefore the indicated differential operations cannot be performed directly. Using

$$\frac{d}{dx} = \frac{d}{d\xi} \frac{d\xi}{dx} = \frac{2}{x_{k+1} - x_k} \frac{d}{d\xi} \equiv \frac{2}{\ell_k} \frac{d}{d\xi} \quad \text{and} \quad dx = \frac{x_{k+1} - x_k}{2} d\xi \equiv \frac{\ell_k}{2} d\xi,$$

where $\ell_k := x_{k+1} - x_k$ is the length of the kth element, we have

$$\int_{x_k}^{x_{k+1}} \kappa(x) u_n' v_n' \, dx = \frac{2}{\ell_k} \int_{-1}^{+1} \kappa(Q_k(\xi)) \left(\sum_{j=1}^{p_k+1} a_j \frac{dN_j}{d\xi} \right) \left(\sum_{i=1}^{p_k+1} b_i \frac{dN_i}{d\xi} \right) d\xi.$$

Defining

$$k_{ij}^{(k)} = \frac{2}{\ell_k} \int_{-1}^{+1} \kappa(Q_k(\xi)) \frac{dN_i}{d\xi} \frac{dN_j}{d\xi} \, d\xi, \tag{2.71}$$

the following expression is obtained:

$$\int_{x_k}^{x_{k+1}} \kappa(x) u'_n v'_n \, dx = \sum_{i=1}^{p_k+1} \sum_{j=1}^{p_k+1} k_{ij}^{(k)} a_j b_i \equiv \{b\}^T [K^{(k)}] \{a\}. \tag{2.72}$$

The terms of the coefficient matrix $k_{ij}^{(k)}$ are computable from the mapping, the definition of the shape functions and the function $\kappa(x)$. The matrix $[K^{(k)}]$ is called the element stiffness matrix. Observe that $k_{ij}^{(k)} = k_{ji}^{(k)}$, hence $[K^{(k)}]$ is symmetric. This follows directly from the symmetry of $B(u, v)$ and the fact that $u_n \in S$, $v_n \in S$ and the same basis functions are used for u_n and v_n.

In the FEM the integrals are evaluated by numerical methods. Numerical integration is discussed in Appendix B. The number of integration points must be sufficiently large so as to ensure that the error of approximation is not influenced significantly by the errors in numerical integration. In the important special case where $\kappa(x) = \kappa_k$ is constant on I_k, it is possible to compute $[K^{(k)}]$ once and for all. This is illustrated by the following example.

Example 2.5.1 When $\kappa(x) = \kappa_k$ is constant on I_k and the hierarchic shape functions defined in Section 2.5.2 are used, then, except for the first two rows and columns, the elemental stiffness matrix is perfectly diagonal:

$$[K^{(k)}] = \frac{2\kappa_k}{\ell_k} \begin{bmatrix} 1/2 & -1/2 & 0 & 0 & \cdots & 0 \\ & 1/2 & 0 & 0 & & 0 \\ & & 1 & 0 & & 0 \\ & & & 1 & & 0 \\ & (\text{sym.}) & & & \ddots & \vdots \\ & & & & & 1 \end{bmatrix}. \tag{2.73}$$

Exercise 2.5.5 Assume that $\kappa(x) = \kappa_k$ is constant on I_k. Using the Lagrange polynomials defined in Section 2.5.2 for $p = 2$, compute $k_{11}^{(k)}$ and $k_{13}^{(k)}$ in terms of κ_k and ℓ_k.

Computation of the Gram matrix

The second term of the bilinear form is also computed as a sum of integrals over the elements:

$$\int_0^\ell c(x) u_n v_n \, dx = \sum_{k=1}^{M(\Delta)} \int_{x_k}^{x_{k+1}} c(x) u_n v_n \, dx.$$

We will be concerned with evaluation of the integral

$$\int_{x_k}^{x_{k+1}} c(x) u_n v_n \, dx = \int_{x_k}^{x_{k+1}} c(x) \left(\sum_{j=1}^{p_k+1} a_j N_j \right) \left(\sum_{i=1}^{p_k+1} b_i N_i \right) dx$$

$$= \frac{\ell_k}{2} \int_{-1}^{+1} c(Q_k(\xi)) \left(\sum_{j=1}^{p_k+1} a_j N_j \right) \left(\sum_{i=1}^{p_k+1} b_i N_i \right) d\xi.$$

Defining

$$m_{ij}^{(k)} := \frac{\ell_k}{2} \int_{-1}^{+1} c(Q_k(\xi))N_i N_j \, d\xi, \qquad (2.74)$$

the following expression is obtained:

$$\int_{x_k}^{x_{k+1}} c(x) u_n v_n \, dx = \sum_{i=1}^{p_k+1} \sum_{j=1}^{p_k+1} m_{ij}^{(k)} a_j b_i = \{b\}^T [M^{(k)}]\{a\} \qquad (2.75)$$

where $\{a\} := \{a_1 \ a_2 \ \ldots \ a_{p_k+1}\}^T$, $\{b\}^T := \{b_1 \ b_2 \ \ldots \ b_{p_k+1}\}$ and

$$[M^{(k)}] := \begin{bmatrix} m_{11}^{(k)} & m_{12}^{(k)} & \cdots & m_{1,p_k+1} \\ m_{21}^{(k)} & m_{22}^{(k)} & \cdots & m_{2,p_k+1} \\ \vdots & & \ddots & \vdots \\ m_{p_k+1,1}^{(k)} & m_{p_k+1,2}^{(k)} & \cdots & m_{p_k+1,p_k+1} \end{bmatrix}.$$

The terms of the coefficient matrix $m_{ij}^{(k)}$ are computable from the mapping, the definition of the shape functions and the function $c(x)$. The matrix $[M^{(k)}]$ is called the elemental Gram matrix[15] or the elemental mass matrix. Observe that $[M^{(k)}]$ is symmetric. In the important special case where $c(x) = c_k$ is constant on I_k, it is possible to compute $[M^{(k)}]$ once and for all. This is illustrated by the following example.

Example 2.5.2 When $c(x) = c_k$ is constant on I_k and the hierarchic shape functions defined in Section 2.5.2 are used then the elemental Gram matrix is strongly diagonal. For example, for $p_k = 5$ the elemental Gram matrix is

$$[M^{(k)}] = \frac{c_k \ell_k}{2} \begin{bmatrix} 2/3 & 1/3 & -1/\sqrt{6} & 1/3\sqrt{10} & 0 & 0 \\ & 2/3 & -1/\sqrt{6} & -1/3\sqrt{10} & 0 & 0 \\ & & 2/5 & 0 & -1/5\sqrt{21} & 0 \\ & & & 2/21 & 0 & -1/7\sqrt{45} \\ & \text{(sym.)} & & & 2/45 & 0 \\ & & & & & 2/77 \end{bmatrix}. \qquad (2.76)$$

---

[15] Jörgen Pedersen Gram (1752–1833).

Remark 2.5.2 For $p_k \geq 2$ a simple closed form expression can be obtained for the diagonal terms and the off-diagonal terms. Using Equation (2.68) it can be shown that

$$m_{ii}^{(k)} = \frac{c_k \ell_k}{2} \frac{1}{2(2i-3)} \int_{-1}^{+1} (P_{i-1}(\xi) - P_{i-3}(\xi))^2 \, d\xi$$

$$= \frac{c_k \ell_k}{2} \frac{2}{(2i-1)(2i-5)}, \quad i \geq 3 \qquad (2.77)$$

and all off-diagonal terms are zero for $i \geq 3$, with the exceptions

$$m_{i,i+2}^{(k)} = m_{i+2,i}^{(k)} = -\frac{c_k \ell_k}{2} \frac{1}{(2i-1)\sqrt{(2i-3)(2i+1)}}, \quad i \geq 3. \qquad (2.78)$$

Exercise 2.5.6 Assume that $c(x) = c_k$ is constant on I_k. Using the Lagrange shape functions defined in Section 2.5.2 for $p = 2$, compute $m_{11}^{(k)}$ and $m_{13}^{(k)}$ in terms of c_k and ℓ_k.

2.5.5 Computation of the right hand side vector

Computation of the right hand side vector involves numerical evaluation of the functional $F(v)$, given that $v \in S^0$. In particular, we write

$$F(v_n) = \int_0^\ell f(x) v_n \, dx = \sum_{k=1}^{M(\Delta)} \int_{x_k}^{x_{k+1}} f(x) v_n \, dx.$$

The element-level integral is computed from the definition of v_n on I_k:

$$\int_{x_k}^{x_{k+1}} f(x) v_n \, dx = \frac{\ell_k}{2} \int_{-1}^{+1} f(Q_k(\xi)) \left(\sum_{i=1}^{p_k+1} b_i^{(k)} N_i \right) d\xi = \sum_{i=1}^{p_k+1} b_i^{(k)} r_i^{(k)} \qquad (2.79)$$

where

$$r_i^{(k)} := \frac{\ell_k}{2} \int_{-1}^{+1} f(Q_k(\xi)) N_i(\xi) \, d\xi \qquad (2.80)$$

which can be computed from the mapping, the given function $f(x)$ and the definition of the shape functions.

Example 2.5.3 Let us assume that $f(x)$ is a linear function on I_k. In this case $f(x)$ can be written as

$$f(x) = \frac{1-\xi}{2} f(x_k) + \frac{1+\xi}{2} f(x_{k+1}) = f(x_k) N_1(\xi) + f(x_{k+1}) N_2(\xi)$$

and, assuming that the hierarchic shape functions defined in Section 2.5.2 are used,

$$r_1^{(k)} = f(x_k)\frac{\ell_k}{2}\int_{-1}^{+1} N_1^2\, d\xi + f(x_{k+1})\frac{\ell_k}{2}\int_{-1}^{+1} N_1 N_2\, d\xi = \frac{\ell_k}{6}(2f(x_k) + f(x_{k+1}))$$

$$r_2^{(k)} = f(x_k)\frac{\ell_k}{2}\int_{-1}^{+1} N_1 N_2\, d\xi + f(x_{k+1})\frac{\ell_k}{2}\int_{-1}^{+1} N_2^2\, d\xi = \frac{\ell_k}{6}(f(x_k) + 2f(x_{k+1}))$$

$$r_3^{(k)} = f(x_k)\frac{\ell_k}{2}\int_{-1}^{+1} N_1 N_3\, d\xi + f(x_{k+1})\frac{\ell_k}{2}\int_{-1}^{+1} N_2 N_3\, d\xi$$

$$= -\frac{\ell_k}{6}\sqrt{\frac{3}{2}}(f(x_k) + f(x_{k+1})).$$

Exercise 2.5.7 Assume that $f(x)$ is a linear function on I_k and the hierarchic shape functions defined in Section 2.5.2 are used. Compute $r_4^{(k)}$ and show that $r_i^{(k)} = 0$ for $i > 4$. Hint: Make use of Equation (2.68).

Exercise 2.5.8 Let

$$f(x) = f_k \sin\frac{x - x_k}{\ell_k}\pi, \quad x \in I_k$$

where f_k is a constant. Compute $r_5^{(k)}$ numerically in terms of f_k and ℓ_k using three, four and five Gauss points. See Appendix B. Use the hierarchic basis functions defined in Section 2.5.2.

Exercise 2.5.9 Assuming that $f(x)$ is a linear function on I and the Lagrange shape functions defined in Section 2.5.2 for $p = 2$ are used, compute $r_1^{(k)}$.

Loading by a concentrated force

A concentrated force F_0 acting on an elastic bar at $x = x_0$ is understood as a surface loading $T(x)$ defined by

$$T(x) = \begin{cases} 0 & \text{if } 0 < x < x_0 - \Delta x/2 \\ F_0/\Delta x & \text{if } x_0 - \Delta x/2 \le x \le x_0 + \Delta x/2 \\ 0 & \text{if } x_0 + \Delta x/2 < x < \ell \end{cases}$$

where $\Delta x \to 0$. In the following we make use of the result, obtained in Section 2.6.4, that, in one dimension, if $v \in E(I)$ then v is a continuous function. Writing

$$\int_0^\ell T(x)v\, dx = \int_{x_0-\Delta x/2}^{x_0+\Delta x/2} \frac{F_0}{\Delta x}v\, dx$$

54   AN OUTLINE OF THE FINITE ELEMENT METHOD

and since $v \in E(I)$ is continuous (see proof in Section 2.6.4), we have

$$\lim_{\Delta x \to 0} \int_{x_0 - \Delta x/2}^{x_0 + \Delta x/2} \frac{F_0}{\Delta x} v \, dx = F_0 v(x_0). \tag{2.81}$$

The computation of the right hand side terms corresponding to a concentrated force F_0 involves identification of the element I_k in which x_0 lies. It is then necessary to find $\xi_0 = Q_k^{-1}(x_0)$ and compute

$$r_i^{(k)} = F_0 N_i(\xi_0), \qquad i = 1, 2, \ldots, p_k + 1. \tag{2.82}$$

If x_0 is a node point then either element sharing that node point may be chosen. The reason for this is discussed in Section 2.5.6.

Thermal loading

When an elastic bar is subjected to a temperature change $T_\Delta := T - T_0$ then $F(v)$ includes the term

$$\int_0^\ell A E \alpha T_\Delta \frac{dv}{dx} \, dx = \sum_{k=1}^{M(\Delta)} \int_{x_k}^{x_{k+1}} A E \alpha T_\Delta \frac{dv}{dx} \, dx.$$

On the kth element

$$\int_{x_k}^{x_{k+1}} A E \alpha T_\Delta \frac{dv}{dx} \, dx = \int_{-1}^{+1} A E \alpha T_\Delta \frac{dv}{d\xi} \, d\xi = \sum_{i=1}^{p_k+1} b_i \tilde{r}_i^{(k)}$$

where

$$\tilde{r}_i^{(k)} := \int_{-1}^{+1} \underbrace{A E \alpha T_\Delta}_{x \to Q_k(\xi)} \frac{dN_i}{d\xi} \, d\xi, \qquad i = 1, 2, \ldots, p_k + 1 \tag{2.83}$$

is the element-level right hand side vector corresponding to the temperature change.

Remark 2.5.3 When $AE\alpha T_\Delta = \tau_k$ is constant on I_k and the hierarchic basis functions defined in Section 2.5.2 are used then $\tilde{r}_1^{(k)} = -\tau_k$, $\tilde{r}_2^{(k)} = \tau_k$ and $\tilde{r}_i^{(k)} = 0$ for $i > 3$.

Example 2.5.4 Assuming that $AE\alpha T_\Delta = \tau$ is a linear function on I_k and the hierarchic basis functions defined in Section 2.5.2 are used, then

$$\tilde{r}_1^{(k)} = \int_{-1}^{+1} \left(\frac{1-\xi}{2} \tau(x_k) + \frac{1+\xi}{2} \tau(x_{k+1}) \right) \left(-\frac{1}{2} \right) d\xi = -\frac{1}{2}(\tau(x_k) + \tau(x_{k+1})).$$

Exercise 2.5.10 Assuming that $AE\alpha T_\Delta = \tau$ is a linear function on I_k and the hierarchic basis functions defined in Section 2.5.2 are used, determine $\tilde{r}_3^{(k)}$.

2.5.6 Assembly

The bilinear form was computed element by element, resulting in the expressions shown in Equations (2.72), (2.75). These element-level expressions must be summed to obtain the coefficient matrix for the entire problem:

$$B(u_n, v_n) = \sum_{k=1}^{M(\Delta)} \{b^k\}^T \left([K^{(k)}] + [M^{(k)}]\right) \{a^k\}.$$

Similarly, the linear functional was computed element by element, resulting in the expressions shown in Equation (2.79), which may be augmented by the terms corresponding to concentrated forces applied at nodes, thermal loads and essential boundary conditions, the treatment of which is discussed in Section 2.5.7:

$$F(v_n) = \sum_{k=1}^{M(\Delta)} \{b^k\}^T \{r^{(k)}\}.$$

The indicated summations are performed in the assembly process.

Prior to the summation a unique identifying number must be assigned to each basis function and the corresponding coefficients. At the element level the shape functions and their coefficients are numbered from 1 to p_{k+1}. At the domain level the basis functions and their coefficients are numbered from 1 to n, where n is the dimension of the finite element space $S(I, \Delta, \mathbf{p}, \mathbf{Q})$. The numbering of the basis functions at the domain level is arbitrary; however, the numbering influences the structure of the assembled coefficient matrix.

Since the basis functions must be continuous on the domain I, the re-numbering must be consistent with the requirement of continuity. Typically the basis functions associated with nodes (i.e., the basis functions which are non-zero at the nodes) are numbered first. The node number is assigned to the basis function. For example, as shown in Figure 2.10, node k is shared by element I_{k-1} and element I_k. The basis function $\varphi_k(x)$ is composed of two linear segments. One of the segments is mapped from the shape function N_2, the other from the

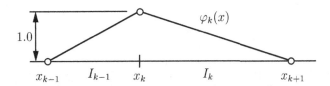

Figure 2.10 Typical nodal basis function.

shape function N_1:

$$\varphi_k(x) = \begin{cases} N_2(Q_{k-1}^{-1}(x)) & \text{for } x \in I_{k-1} \\ N_1(Q_k^{-1}(x)) & \text{for } x \in I_k. \end{cases}$$

Therefore the local (i.e., element-level) variables a_2, b_2 associated with element I_{k-1} and the local variables a_1, b_1 associated with element I_k must be assigned the same "global" number k. The basis functions mapped from the internal shape functions are zero at the node points, hence automatically satisfy the condition of continuity on I.

Example 2.5.5 Consider the three-element mesh shown in Figure 2.6. The p-distribution is $\{2, 1, 3\}$. The basis functions, mapped from the hierarchic shape functions, are illustrated in Figure 2.6. Note that the basis functions are numbered first by nodes then by p-level. Assigning the global numbering to the coefficients $a_i, b_i, i = 1, 2, \ldots, 7$, but retaining the local numbering for the stiffness coefficients, with the element numbers indicated by superscripts, we have

$$B(u_n, v_n) = \{b_1 \ b_2 \ b_5\} \begin{bmatrix} c_{11}^{(1)} & c_{12}^{(1)} & c_{13}^{(1)} \\ c_{21}^{(1)} & c_{22}^{(1)} & c_{23}^{(1)} \\ c_{31}^{(1)} & c_{32}^{(1)} & c_{33}^{(1)} \end{bmatrix} \begin{Bmatrix} a_1 \\ a_2 \\ a_5 \end{Bmatrix}$$

$$+ \{b_2 \ b_3\} \begin{bmatrix} c_{11}^{(2)} & c_{12}^{(2)} \\ c_{21}^{(2)} & c_{22}^{(2)} \end{bmatrix} \begin{Bmatrix} a_2 \\ a_3 \end{Bmatrix}$$

$$+ \{b_3 \ b_4 \ b_6 \ b_7\} \begin{bmatrix} c_{11}^{(3)} & c_{12}^{(3)} & c_{13}^{(3)} & c_{14}^{(3)} \\ c_{21}^{(3)} & c_{22}^{(3)} & c_{23}^{(3)} & c_{24}^{(3)} \\ c_{31}^{(3)} & c_{32}^{(3)} & c_{33}^{(3)} & c_{34}^{(3)} \\ c_{41}^{(3)} & c_{42}^{(3)} & c_{43}^{(3)} & c_{44}^{(3)} \end{bmatrix} \begin{Bmatrix} a_3 \\ a_4 \\ a_6 \\ a_7 \end{Bmatrix}$$

where $c_{ij}^{(k)} := k_{ij}^{(k)} + m_{ij}^{(k)}$. Of course, the element-level matrices are symmetric. The full matrices are displayed for the purpose of clarity only. The 7×7 coefficient matrix is obtained by performing the summation:

$$B(u_n, v_n) = \sum_{j=1}^{7} \sum_{i=1}^{7} c_{ij} a_j b_i = \{b_1 \ b_2 \ \cdots \ b_7\} \begin{bmatrix} c_{11} & c_{12} & \cdots & c_{17} \\ c_{21} & c_{22} & & c_{27} \\ \vdots & & & \vdots \\ c_{71} & c_{72} & \cdots & c_{77} \end{bmatrix} \begin{Bmatrix} a_1 \\ a_2 \\ \vdots \\ a_7 \end{Bmatrix}$$

$$\equiv \{b\}^T [C] \{a\}$$

where (for example) $c_{11} = c_{11}^{(1)}$, $c_{22} = c_{22}^{(1)} + c_{11}^{(2)}$, $c_{36} = c_{13}^{(3)}$, etc. The matrix $[C]$ is called the unconstrained coefficient matrix or unconstrained global stiffness matrix. This matrix needs to be modified in order to account for the restrictions associated with the essential boundary conditions. This point will be discussed in Section 2.5.7.

The assembly of the right hand side vector from the element-level right hand side vectors is analogous to the procedure just described:

$$F(v_n) = \{b_1\ b_2\ b_5\} \begin{Bmatrix} r_1^{(1)} \\ r_2^{(1)} \\ r_3^{(1)} \end{Bmatrix} + \{b_2\ b_3\} \begin{Bmatrix} r_1^{(2)} \\ r_2^{(2)} \end{Bmatrix} + \{b_3\ b_4\ b_6\ b_7\} \begin{Bmatrix} r_1^{(3)} \\ r_2^{(3)} \\ r_3^{(3)} \\ r_4^{(3)} \end{Bmatrix}$$

$$= \{b_1\ b_2\ \cdots\ b_7\} \begin{Bmatrix} r_1 \\ r_2 \\ \vdots \\ r_7 \end{Bmatrix}$$

where (for example) $r_2 = r_2^{(1)} + r_1^{(2)}$.

Exercise 2.5.11 Assume that $\kappa(x) = \kappa_k$ is constant on each of the three elements shown in Figure 2.6 and $c(x) = 0$ and hierarchic shape functions are used. Write down the assembled coefficient matrix in terms of κ_k, ℓ_k. Let $\mathbf{p} = \{2\ 1\ 3\}$.

Exercise 2.5.12 An elastic bar is constrained by a distributed spring characterized by coefficient c. The axial stiffness of the bar is AE and the coefficient of thermal expansion is α. Assume that AE, c and the temperature are constants. The boundary condition at $x = 0$ is $AEu' = k_0(u(0) - d_0)$. The boundary condition at $x = \ell$ is the prescribed displacement $u(\ell) = \hat{u}_\ell$. Using two elements of equal length and $p = 1$, write down the assembled coefficient matrix and the right hand side vector. The notation is shown in Figure 2.11.

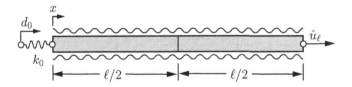

Figure 2.11 Notation for Exercise 2.5.12.

Exercise 2.5.13 Let

$$f(x) = \begin{cases} C_1 + C_2 x & \text{for } x \in I_1 \\ C_3 & \text{for } x \in I_2 \\ 0 & \text{for } x \in I_3 \end{cases}$$

where C_1, C_2, C_3 are constants. Using the basis functions defined in Example 2.5.5, determine elements r_2 and r_3 of the assembled right hand side vector.

58 AN OUTLINE OF THE FINITE ELEMENT METHOD

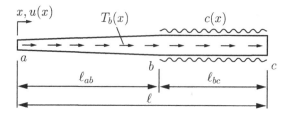

Figure 2.12 Elastic bar: notation.

Exercise 2.5.14 If a concentrated force is acting on a node point then either element sharing that node point may be chosen for computing the corresponding element-level load vector term. Explain why.

Exercise 2.5.15 The elastic bar shown in Figure 2.12 consists of a tapered and a prismatic section. The cross-sectional area of the tapered section varies linearly from A_a to A_b. The modulus of elasticity E and the spring coefficient c are constants. The boundary conditions are $F_a = F_c = 0$. Let $T_b(x) = T_b$ be a constant. Using two elements and $p = 1$ on both elements, write down each term of the assembled coefficient matrix and load vector for this problem.

2.5.7 Treatment of the essential boundary conditions

When essential boundary conditions are specified then it is necessary to modify the assembled coefficient matrix $[C]$ and the right hand side vector $\{r\}$ using the procedure described in Section 2.3.1. For example, when using the hierarchic shape functions, we select $u^\star$ as follows:

$$u^\star = \begin{cases} \hat{u}_0 N_1(Q_1^{-1}(x)) & \text{for } x \in I_1 \\ \hat{u}_\ell N_2(Q_M^{-1}(x)) & \text{for } x \in I_M \\ 0 & \text{elsewhere} \end{cases}$$

where $M = M(\Delta)$ is the number of elements. Such a choice is illustrated in Figure 2.13.

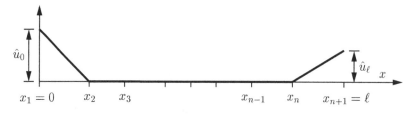

Figure 2.13 Typical choice of the function $u^\star$ in one dimension.

FEM IN ONE DIMENSION   59

It is necessary to compute the integral

$$\int_0^\ell \left(\kappa(x)(u^\star)'v' + c(x)u^\star v\right) dx = \hat{u}_0 \int_{x_1}^{x_2} \left(\kappa(x)N_1'(Q_1^{-1}(x))v' + c(x)N_1(Q_1^{-1}(x))v\right) dx$$

$$+ \hat{u}_\ell \int_{x_n}^{x_{n+1}} \left(\kappa(x)N_2'(Q_M^{-1}(x))v' + c(x)N_2(Q_M^{-1}(x))v\right) dx.$$

Noting that

$$\int_{x_1}^{x_2} \kappa(x)(u^\star)'v' \, dx = \hat{u}_0 \frac{2}{\ell_1} \int_{-1}^{+1} \kappa(Q_1(\xi)) \frac{dN_1}{d\xi} \sum_{i=1}^{p_1+1} b_i^{(1)} \frac{dN_i}{d\xi} d\xi$$

$$= \hat{u}_0 \sum_{i=1}^{p_1+1} k_{i1}^{(1)} b_i^{(1)}$$

and upon treating the other terms similarly, we have

$$\int_0^\ell \left(\kappa(x)(u^\star)'v' + c(x)u^\star v\right) dx = \hat{u}_0 \sum_{i=1}^{p_1+1} \left(k_{i1}^{(1)} + m_{i1}^{(1)}\right) b_i^{(1)}$$

$$+ \hat{u}_\ell \sum_{i=1}^{p_M+1} \left(k_{i2}^{(M)} + m_{i2}^{(M)}\right) b_i^{(M)}.$$

These expressions represent the first columns of the stiffness and Gram matrices of element 1 multiplied by $\hat{u}_0$ and the second columns of the stiffness and Gram matrices of element M multiplied by $\hat{u}_\ell$. Therefore enforcement of non-zero essential boundary conditions involves multiplication of the first columns of the stiffness and Gram matrices of element 1 (resp. element $M(\Delta)$) by $\hat{u}_0$ (resp. $\hat{u}_\ell$) and subtracting these columns from the right hand side vector. Furthermore, since $b_1^{(1)} = b_2^{(M)} = 0$, the corresponding rows in the assembled coefficient matrix and right hand side vector are deleted. Therefore the number of equations is reduced by the number of essential boundary conditions specified. The number of equations following enforcement of the essential boundary conditions is the number of degrees of freedom N.

Example 2.5.6 Let us consider the following problem on $I = (0, 2)$:

$$-u'' + 12u = 0 \qquad u(0) = 1.25, \ u(2) = 5.50 \qquad (2.84)$$

and let us use a uniform mesh with four elements, $\mathbf{p} = \{1\ 1\ 1\ 1\}$. Using Equations (2.73) and (2.76) the unconstrained coefficient matrix $[C]$ is readily assembled:

$$[C] = \begin{bmatrix} 4 & -1 & 0 & 0 & 0 \\ -1 & 8 & -1 & 0 & 0 \\ 0 & -1 & 8 & -1 & 0 \\ 0 & 0 & -1 & 8 & -1 \\ 0 & 0 & 0 & -1 & 4 \end{bmatrix}.$$

60 AN OUTLINE OF THE FINITE ELEMENT METHOD

Multiplying the first column by $a_1 = 1.25$, the fifth column by $a_5 = 5.50$ and subtracting these columns from the right hand side vector (which is the zero vector in this case), we obtain

$$\begin{bmatrix} 8 & -1 & 0 \\ -1 & 8 & -1 \\ 0 & -1 & 8 \end{bmatrix} \begin{Bmatrix} a_2 \\ a_3 \\ a_4 \end{Bmatrix} = \begin{Bmatrix} 1.25 \\ 0 \\ 5.50 \end{Bmatrix}.$$

The number of degrees of freedom is 3. On solving we find $a_2 = 0.169\,859$, $a_3 = 0.108\,871$, $a_4 = 0.701\,109$.

Exercise 2.5.16 Using one element, $p = 4$ and hierarchic basis functions, write down the system of equations following enforcement of the essential boundary conditions for the problem in Example 2.5.6.

Exercise 2.5.17 Write down the constrained coefficient matrix and right hand side vector for the problem in Exercise 2.5.12.

Example 2.5.7 Let us consider the problem on $(0, \ell)$:

$$-\kappa_0 u'' = f(x), \qquad \kappa_0 u'(0) = F_0, \; \kappa_0 u'(\ell) = F_\ell$$

where κ_0 is a constant. Using one element of degree p and the linear mapping given by Equation (2.60), the system of equations is

$$\frac{\kappa_0}{\ell} \begin{bmatrix} 1 & -1 & 0 & \cdots & 0 \\ -1 & 1 & 0 & \cdots & 0 \\ 0 & 0 & 2 & \cdots & 0 \\ \vdots & & & \ddots & \\ 0 & \cdots & & & 2 \end{bmatrix} \begin{Bmatrix} a_1 \\ a_2 \\ a_3 \\ \vdots \\ a_{p+1} \end{Bmatrix} = \begin{Bmatrix} r_1 \\ r_2 \\ r_3 \\ \vdots \\ r_{p+1} \end{Bmatrix}$$

where the coefficient matrix is from Equation (2.73) and the right hand side vector is from Equations (2.80) and (2.82):

$$r_1 = \frac{\ell}{2} \int_{-1}^{+1} f(Q(\xi)) N_1(\xi) \, d\xi - F_0$$

$$r_2 = \frac{\ell}{2} \int_{-1}^{+1} f(Q(\xi)) N_2(\xi) \, d\xi + F_\ell$$

$$r_i = \frac{\ell}{2} \int_{-1}^{+1} f(Q(\xi)) N_i(\xi) \, d\xi, \qquad i = 3, 4, \ldots, p+1.$$

Observe that the second row is -1 times the first row, hence the coefficient matrix is singular. Therefore this problem does not have a unique solution and solutions exist only if the equations are consistent. On adding the first and second rows we get

$$r_1 + r_2 = \frac{\ell}{2} \int_{-1}^{+1} f(Q(\xi)) \underbrace{(N_1(\xi) + N_2(\xi))}_{=1} d\xi - F_0 + F_\ell = 0$$

which is equivalent to

$$\int_0^\ell f(x)\,dx - F_0 + F_\ell = 0. \qquad (2.85)$$

This is a restatement of Equation (2.50). In solving such problems we assign an arbitrary value to a_1 or a_2. The treatment of the problem is then similar to that discussed in Example 2.5.6.

2.5.8 Solution

Following the assembly of the coefficient matrix and enforcement of the essential boundary conditions (when applicable), the system of simultaneous equations is solved by one of several methods designed to exploit the symmetry and sparsity of the coefficient matrix. The solvers are classified into two main categories: direct and iterative solvers. The choice of solver in a particular application is influenced by the size of the problem and the available computational resources. Discussion of the various solvers used in FEA programs is beyond the scope of this introductory exposition. We refer to standard texts, such as [25] and [31].

When a direct solver is used then the numbering of the basis functions should be optimized so that the solver can exploit the sparsity of the coefficient matrix. Commercial finite element codes offer a variety of solvers and optimization schemes for numbering the basis functions or the elements, depending on the choice of the solver.

The solver produces the N unknown coefficients of the basis functions. The finite element solution is then

$$u_{FE} = \bar{u} + u^* = \sum_{i=1}^{N+N^*} a_i \varphi_i(x)$$

where N^* is the number of coefficients determined by the essential boundary conditions. If no essential boundary conditions were specified then $N^* = 0$. For instance, in Example 2.5.6, $N = 3$, $N^* = 2$. The exact and finite element solutions are shown in Figure 2.14.

In the assembly process the element-level numbering scheme, ranging from 1 to $p_k + 1$, was replaced by the global numbering scheme. This is now reversed and the element-level coefficients $a_i^{(k)}$ are stored with the topological data and other information, such as material properties, for each element. At the end of the solution process the finite element solution is available in the form

$$u_{FE} = \sum_{i=1}^{p_k+1} a_i^{(k)} N_i(\xi) \qquad k = 1, 2, \ldots, M(\Delta). \qquad (2.86)$$

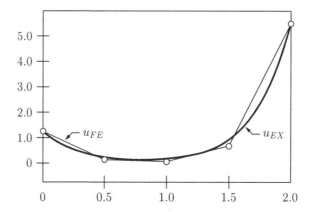

Figure 2.14 The exact and finite element solutions for Example 2.5.6.

The array of coefficients in Equation (2.86), the corresponding shape functions and the mapping functions contain all the information generated by a finite element solution.

2.5.9 Post-solution operations

The data of interest are typically functionals that depend on u and u'. These are computed from the finite element solution in the post-solution operations.

Computation of $u_{FE}(x_0)$

The computation of u_{FE} at some arbitrary point x_0 involves a search to identify the element I_k in which point x_0 lies and, using the inverse map given by Equation (2.61), identification of the point $\xi_0 \in I_{\text{st}}$ corresponding to x_0. Then $u_{FE}(x_0)$ is computed from

$$u_{FE}(x_0) = \sum_{i=1}^{p_k+1} a_i^{(k)} N_i(\xi_0).$$

Exercise 2.5.18 Assume that $(AE) = (AE)_k$ is constant on I_k. Show that when $c(x) = 0$ on I then $u_{FE}(x_k) = u_{EX}(x_k)$ for $k = 1, 2, \ldots, M(\Delta) + 1$. Hint: Make use of the Galerkin orthogonality, Theorem 2.4.1, and select $v = \varphi_k(x)$ where φ_k is defined in Figure 2.10.

Direct computation of $u'_{FE}(x_0)$

In many applications the first derivative of the solution is of interest. The first derivatives are computed from

$$\left(\frac{du_{FE}}{dx}\right)_{x=x_0} = \frac{2}{\ell_k} \sum_{i=1}^{p_k+1} a_i^{(k)} \left(\frac{dN_i}{d\xi}\right)_{\xi=\xi_0}. \tag{2.87}$$

Computation of the higher derivatives is analogous.

FEM IN ONE DIMENSION   63

Indirect computation of $u'_{FE}(x_0)$ at node points

The first derivative at node points is usually determined indirectly from the generalized formulation given by Equation (2.36). For example, to compute the first derivative at node x_k from the finite element solution computed for element k we select $v = N_1(\xi)$ and use Equation (2.36):

$$\int_{x_k}^{x_{k+1}} \left(\kappa u'_{FE}v' + cu_{FE}v\right) dx = \int_{x_k}^{x_{k+1}} fv \, dx + \left[\kappa u'_{FE}v\right]_{x=x_{k+1}} - \left[\kappa u'_{FE}v\right]_{x=x_k}. \quad (2.88)$$

Test functions used in post-solution operations for the computation of a particular functional are called extraction functions. Here $v = N_1(Q_k^{-1}(x))$ is an extraction function for the functional $-\left[\kappa u'_{FE}\right]_{x=x_k}$. This is because $v(x_k) = 1$ and $v(x_{k+1}) = 0$ and hence

$$-\left[\kappa u'_{FE}\right]_{x=x_k} = \int_{x_k}^{x_{k+1}} \left(\kappa u'_{FE}v' + cu_{FE}v\right) dx - \int_{x_k}^{x_{k+1}} fv \, dx$$

$$= \sum_{j=1}^{p_k+1} c_{1j}^{(k)} a_j^{(k)} - r_1^{(k)} \quad (2.89)$$

where by definition $c_{ij}^{(k)} = k_{ij}^{(k)} + m_{ij}^{(k)}$. This shows that the first derivative at node x_k can be computed from the element-level coefficient matrix and the right hand side vector.

Example 2.5.8 For the problem solved in Example 2.5.6 let us compute (a) the exact value of $u'(0)$; (b) $u'_{FE}(0)$ from Equation (2.87); and (c) $u'_{FE}(0)$ from Equation (2.89) using $v = N_1(\xi)$.

The exact solution of Equation (2.84) is

$$u_{EX} = 1.25 \cosh \sqrt{12}\, x + \frac{5.50 - 1.25 \cosh \sqrt{48}}{\sinh \sqrt{48}} \sinh \sqrt{12}\, x. \quad (2.90)$$

Therefore

$$u'_{EX}(0) = \sqrt{12}\, \frac{5.50 - 1.25 \cosh \sqrt{48}}{\sinh \sqrt{48}} = -4.292\,801.$$

Using the direct method of Equation (2.87) and the results from Example 2.5.6,

$$u'_{FE}(0) = 2.00(-a_1^{(1)} + a_2^{(1)}) = 2.00(-1.25 + 0.169\,859) = -2.160\,282.$$

The relative error is 49.7%. Using the indirect method of Equation (2.89) and the results from Example 2.5.6,

$$[-u'_{FE}]_{x=0} = 4a_1 - a_2 = 4 \times 1.25 - 0.169\,859 = 4.830\,141,$$

we have $u'_{FE}(0) = -4.830\,141$ (12.5% error).

64   AN OUTLINE OF THE FINITE ELEMENT METHOD

This example illustrates that it is much more efficient to compute the derivatives by the indirect method than by the direct method. Further details are given in Section 2.8.

Exercise 2.5.19 Compute $u'_{FE}(2)$ for Example 2.5.6 using the direct method given by Equation (2.87) and the indirect method. Compute the relative errors.

Exercise 2.5.20 Two elastic bars of length ℓ are constrained at the ends A and B and supported by a distributed spring characterized by the spring coefficient c (constant). Assume that the cross-sectional area A, the modulus of elasticity E and the coefficient of thermal expansion α are constants. The bars are separated by a gap Δ at a reference temperature T_0 (constant). The temperature of both bars is uniformly increased. The goal of computation is to determine the reaction forces developed at the supports A and B as a function of the temperature $T > T_0$.

(a) Explain how you would determine the temperature T_c at which the gap closes.

(b) Assuming that $T \leq T_c$, explain how you would compute the reaction force at A by the direct and indirect methods.

(c) Assuming that $T > T_c$, explain how you would compute the reaction force at A by the direct and indirect methods.

Exercise 2.5.21 Write an ad hoc computer program for solving the problem in Example 2.5.6 for an arbitrary number of elements. Plot the relative errors for $u'_{FE}(2)$ computed by (a) the direct method and (b) the indirect method versus the number of elements. Compare the number of elements needed for reducing the relative error to under 1%.

Nodal forces

The vector of nodal forces associated with element k, denoted by $\{f^{(k)}\}$, is defined as follows:

$$\{f^{(k)}\} = [K^{(k)}]\{a^{(k)}\} - \{\bar{r}^{(k)}\} \qquad k = 1, 2, \ldots, M(\Delta) \qquad (2.91)$$

where $[K^{(k)}]$ is the stiffness matrix, $\{a^{(k)}\}$ is the solution vector and $\{\bar{r}^{(k)}\}$ is the load vector corresponding to traction forces, concentrated forces and thermal loads acting on element k.

The sign convention for nodal forces is different from the sign convention for the bar force: whereas the bar force is positive when tensile, a nodal force is positive when acting in the direction of the positive coordinate axis.

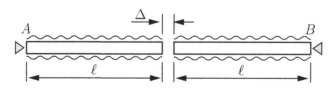

Figure 2.15 Exercise 2.5.20: notation.

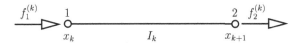

Figure 2.16 Exercise 2.5.22: notation.

Exercise 2.5.22 Assume that hierarchic basis functions based on Legendre polynomials are used. Show that when AE is constant and $c = 0$ on I_k then

$$f_1^{(k)} + f_2^{(k)} = r_1^{(k)} + r_2^{(k)}$$

independently of the polynomial degree p_k. Consider both thermal and traction loads. This exercise demonstrates that nodal forces are in equilibrium independently of the finite element solution. Therefore equilibrium of nodal forces is not an indicator of the quality of finite element solutions.

Exercise 2.5.23 Show that the sum of nodal forces computed for node k from element $k-1$ and from element k is zero unless a concentrated force F_k is acting on node k, in which case the sum is equal to F_k. Illustrate this by a simple example and demonstrate that the sum of forces acting on a node point is zero. Hint: Let $v = \varphi_k(x)$ shown in Figure 2.10.

Exercise 2.5.24 Refer to Exercise 2.5.12. Assume that the finite element solution is available. Write down an expression for the computation of nodal forces for element 2.

Remark 2.5.4 We have seen in Theorem 2.4.2 that the error in the energy norm depends on the choice of the finite element space S, which depends on the choice of discretization characterized by the mesh Δ, the polynomial degrees assigned to the elements **p** and the mapping functions **Q**. Of course, this is true for all data computed from the finite element solution. It was noted in Section 1.1.4 that it is necessary to verify that the data of interest are substantially independent of the discretization. It is of great practical importance to ensure that the relative errors in the data of interest are within acceptable bounds. Procedures for error estimation are discussed in Section 2.7.

Remark 2.5.5 Let us consider the problem of Equation (2.24) with $c = 0$ and $\kappa(x) = \kappa_k$ constant on I_k. Let us assume that the mapping is linear and $u(0) = \hat{u}_0$ is one of the boundary conditions. In this case the finite element solution at the node points is the exact solution. To show this we define $\varphi_k(x) \in S^0$ as follows:

$$\varphi_1(x) = 1 - \frac{x}{\ell_1} \quad \text{for} \quad 0 \leq x < x_2 = \ell_1$$

$$\varphi_{M(\Delta)+1}(x) = 1 - \frac{x - x_{M(\Delta)}}{\ell_{M(\Delta)}} \quad \text{for} \quad x_{M(\Delta)} \leq x \leq x_{M(\Delta)+1} = \ell$$

and for $k = 2, 3, \ldots, M(\Delta)$

$$\varphi_k(x) = \begin{cases} (x - x_{k-1})/\ell_{k-1} & \text{for } x_{k-1} \leq x \leq x_k \\ 1 - (x - x_k)/\ell_k & \text{for } x_k < x \leq x_{k+1}. \end{cases}$$

Using the Galerkin orthogonality we get

$$B(e, \varphi_1) = -\frac{\kappa_1}{\ell_1} \int_{x_1}^{x_2} e' \, dx = 0$$

$$B(e, \varphi_k) = \frac{\kappa_{k-1}}{\ell_{k-1}} \int_{x_{k-1}}^{x_k} e' \, dx - \frac{\kappa_k}{\ell_k} \int_{x_k}^{x_{k+1}} e' \, dx = 0, \quad k = 2, \ldots, M(\Delta)$$

$$B(e, \varphi_{M(\Delta)+1}) = \frac{\kappa_{M(\Delta)}}{\ell_{M(\Delta)}} \int_{x_{M(\Delta)}}^{x_{M(\Delta)+1}} e' \, dx = 0$$

where $e := u_{EX} - u_{FE}$. Therefore we have a system of $M(\Delta) + 1$ equations, the coefficient matrix of which is tridiagonal:

$$\frac{\kappa_1}{\ell_1} e(x_1) - \frac{\kappa_1}{\ell_1} e(x_2) = 0$$

$$\vdots$$

$$-\frac{\kappa_{k-1}}{\ell_{k-1}} e(x_{k-1}) + \left(\frac{\kappa_{k-1}}{\ell_{k-1}} + \frac{\kappa_k}{\ell_k}\right) e(x_k) - \frac{\kappa_k}{\ell_k} e(x_{k+1}) = 0$$

$$\vdots$$

$$-\frac{\kappa_{M(\Delta)}}{\ell_{M(\Delta)}} e(x_{M(\Delta)}) + \frac{\kappa_{M(\Delta)}}{\ell_{M(\Delta)}} e(x_{M(\Delta)+1}) = 0.$$

We have assumed that an essential boundary condition is prescribed at $x = 0$ (node 1). Then $e(x_1) = 0$ and hence $e(x_k) = 0$ for $k = 1, 2, \ldots, M(\Delta) + 1$. Alternatively, when an essential boundary condition is prescribed at $x = \ell$ (node $M(\Delta) + 1$) then $e(x_{M(\Delta)+1}) = 0$ and hence $e(x_k) = 0$ for $k = 1, 2, \ldots, M(\Delta) + 1$.

Exercise 2.5.25 Show that under the assumptions of Remark 2.5.5, $u_{FE}(x_k) = u_{EX}(x_k)$ when Robin boundary conditions are prescribed.

Exercise 2.5.26 Consider the problem $-u'' = \alpha(\alpha - 1)x^{\alpha-2}$, $\alpha > 1$, on $I = (0, 1)$ with the boundary conditions $u(0) = u(1) = 0$. The goal is to estimate $u'(0)$ using the FEM within an error tolerance of 1%. Let $\alpha = 1.05$. Using a uniform finite element mesh and $p = 1$, what is the number of elements $M(\Delta)$ needed when (a) $u'(0)$ is computed by the direct method and (b) $u'(0)$ is computed by the nodal force method? Hint: Make use of the fact that $u_{FE}(x_k) = u_{EX}(x_k)$ for $k = 1, 2, \ldots, M(\Delta) + 1$, see Remark 2.5.5.

2.6 Properties of the generalized formulation

Some of the key properties of the generalized formulation and the finite element solution are presented in the following. Although these properties are presented in the one-dimensional setting only, they are applicable to two and three dimensions as well, unless noted otherwise.

2.6.1 Uniqueness

The model problem given by Equation (2.24) has been replaced by the corresponding generalized formulation. The following theorem establishes that the solution of the generalized formulation is unique.

Theorem 2.6.1 The function $u \in \tilde{E}(I)$ that satisfies $B(u, v) = F(v)$ for all $v \in E^0(I)$ is unique in the space $E(I)$.

Proof: The theorem is proven by contradiction. Assume that there are two functions u_1, u_2 in $\tilde{E}(I)$, $u_1 \neq u_2$, that satisfy

$$B(u_1, v) = F(v)$$
$$B(u_2, v) = F(v)$$

for all $v \in E^0(I)$. Subtracting the second equation from the first we get

$$B(u_1 - u_2, v) = 0 \quad \text{for all } v \in E^0(I).$$

Since $(u_1 - u_2) \in E^0(I)$, we may select $v = u_1 - u_2$, in which case $B(u_1 - u_2, u_1 - u_2) = 0$. In view of Equation (2.53) this is equivalent to $\|u_1 - u_2\|_E = 0$, which contradicts the assumption that $u_1 \neq u_2$ in $\tilde{E}(I)$. $\square$

Remark 2.6.1 Note that uniqueness is understood in the space $E(I)$. Suppose that $c = 0$ in Equation (2.24) and Neumann boundary conditions are specified, subject to Equation (2.50). Then if u_1 is a solution then $u_2 = u_1 + C$ is also a solution, where C is an arbitrary constant. In this case the energy norm cannot distinguish between two functions that differ by an arbitrary constant. This is seen directly from the definition of the energy norm, given by Equation (2.53):

$$\|u_1 - u_2\|_E^2 = \frac{1}{2} \int_0^\ell (u_1' - u_2')^2 \, dx = 0.$$

Therefore the solution can be determined only up to an arbitrary constant.[16] In mechanics this has a simple physical interpretation: the constant C represents rigid body displacement.

---

[16] Such a norm is called a "seminorm."

2.6.2 Potential energy

An important property of the generalized formulation is that the solution minimizes a quadratic functional, called the potential energy. This is proven by the following theorem.

Theorem 2.6.2 The function $u \in \tilde{E}(I)$ that satisfies $B(u, v) = F(v)$ for all $v \in E^0(I)$ minimizes the quadratic functional $\pi(u)$, called the potential energy:

$$\pi(u) := \frac{1}{2} B(u, u) - F(u) \qquad (2.92)$$

on the space $\tilde{E}(I)$.

Proof: For any $v \in E^0(I)$, $\|v\|_E \neq 0$, we have

$$\pi(u+v) = \frac{1}{2} B(u+v, u+v) - F(u+v)$$

$$= \frac{1}{2} B(u, u) + B(u, v) + \frac{1}{2} B(v, v) - F(u) - F(v)$$

$$= \pi(u) + \underbrace{B(u, v) - F(v)}_{0} + \frac{1}{2} \underbrace{B(v, v)}_{>0}.$$

Therefore any admissible non-zero perturbation of u will increase $\pi(u)$. □

This theorem is known as the principle of minimum potential energy.

Remark 2.6.2 Whereas the strain energy is always positive, the potential energy may be positive, negative or zero.

2.6.3 Error in the energy norm

The relationship between the error in the energy norm and the error in potential energy is established by the following theorem. One of the methods used for estimating the error $e = u_{EX} - u_{FE}$ in the energy norm is based on this theorem.

Theorem 2.6.3

$$\|u_{EX} - u_{FE}\|_E^2 = \pi(u_{FE}) - \pi(u_{EX}). \qquad (2.93)$$

Proof: Writing $e = u_{EX} - u_{FE}$ and noting that $e \in E^0(I)$, we have

$$\pi(u_{FE}) = \pi(u_{EX} - e) = \frac{1}{2}B(u_{EX} - e, u_{EX} - e) - F(u_{EX} - e)$$

$$= \frac{1}{2}B(u_{EX}, u_{EX}) - F(u_{EX}) \underbrace{-B(u_{EX}, e) + F(e)}_{0} + \frac{1}{2}B(e, e)$$

$$= \pi(u_{EX}) + \|e\|_E^2$$

which is the same as Equation (2.93). $\square$

2.6.4 Continuity

By definition, $u(x)$ is continuous on $\bar{I} := \{0 \le x \le \ell\}$ if for any $\epsilon > 0$ we can find a $\delta(\epsilon)$ such that

$$|u(x_2) - u(x_1)| \le \epsilon \quad \text{if} \quad |x_2 - x_1| \le \delta(\epsilon), \quad x_1, x_2 \in \bar{I}. \tag{2.94}$$

First we show by an example that a discontinuous function cannot lie in the energy space. Specifically, let us consider the continuous function $u(x)$ shown in Figure 2.17 and compute the integral

$$\int_0^\ell (u')^2\, dx = \int_{x_0}^{x_0+\Delta x} \left(\frac{c_2 - c_1}{\Delta x}\right)^2 dx = \frac{(c_2 - c_1)^2}{\Delta x}.$$

Letting $\Delta x \to 0$, the function becomes discontinuous and the value of the integral is infinity.

With one qualification, any function $u(x) \in E(I)$ is continuous and bounded on $\bar{I}$ by the energy norm $\|u\|_E$. The qualification is that when $c(x) = 0$ and Neumann boundary conditions are specified, subject to Equation (2.50), then $u = C$ is a solution (where C is an arbitrary constant), hence $u(x)$ is not bounded by the energy norm. See Remark 2.6.1. The statement that $u(x) \in E(I)$ is bounded on $\bar{I}$ by $\|u\|_E$ is understood to mean that there exists a constant C, independent of $u(x)$, such that for any $x \in \bar{I}$ the following inequality holds: $|u| \le C\|u\|_E$.

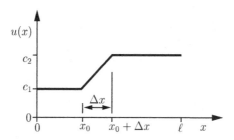

Figure 2.17 Notation.

70 AN OUTLINE OF THE FINITE ELEMENT METHOD

We now prove continuity for the case $\kappa = 1, c = 0$. Let us assume that $x_2 \geq x_1$. Then

$$u(x_2) - u(x_1) = \int_{x_1}^{x_2} u'(x)\,dx \tag{2.95}$$

where the prime (') represents differentiation with respect to x. Applying the Schwarz inequality (see Appendix A, Section A.8) we get

$$|u(x_2) - u(x_1)| \leq \left(\int_{x_1}^{x_2} dx\right)^{1/2} \left(\int_{x_1}^{x_2} (u'(x))^2 dx\right)^{1/2} = |x_2 - x_1|^{1/2}\sqrt{2}\,\|u\|_E. \tag{2.96}$$

Therefore if we select $\delta(\epsilon) < \epsilon^2/(2\,\|u\|_E^2)$ then the condition of continuity is satisfied. That $u(x)$ is bounded on $\bar{I}$ by $\|u\|_E$ follows directly from (2.96) for the case $u(x_1) = 0$. This theorem holds in one dimension only.

Remark 2.6.3 Whereas all functions $u \in E(I)$ are continuous and bounded, du/dx has to be neither continuous nor bounded. For this reason Neumann boundary conditions cannot be enforced by restriction.

Exercise 2.6.1 Consider functions of the form $u(x) = x^\alpha$ on the interval $0 < x < \ell$. Show that u is in the energy space only if $\alpha > 1/2$.

2.6.5 Convergence in the energy norm

In the example discussed in Section 2.2 the basis functions were selected to be polynomials that satisfied the homogeneous essential boundary conditions. The tacit assumption was that in some sense $u_n \to u$ as $n \to \infty$. Convergence in a normed linear space X is understood to mean that for any $\epsilon > 0$ there is an n, dependent on ϵ, such that $\|u - u_n\|_X < \epsilon$.

In the following we assume that $u''(x)$ is continuous and $\|u''(x)\|_{\max} \leq C < \infty$ on $\bar{I} := [0, \ell]$. Let us partition $\bar{I}$ into n segments and denote the kth node point by x_k. Let us interpolate $u(x)$ by a piecewise linear function $\bar{u}_n(x)$ such that $\bar{u}_n(x_k) = u(x_k)$. An example where $n = 4$ is shown in Figure 2.18(a). On the kth sub-interval $I_k := (x_k, x_{k+1})$ we have

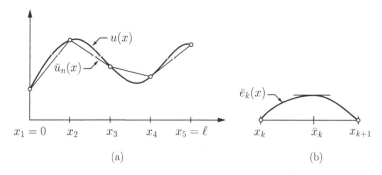

Figure 2.18 Linear interpolation: notation.

$$\bar{u}_n = \frac{x - x_k}{h_k} u(x_{k+1}) - \frac{x - x_{k+1}}{h_k} u(x_k) \qquad x \in I_k, \qquad k = 1, 2, \ldots, n$$

where $h_k := x_{k+1} - x_k$ is the length of the kth element. The maximum value of h_k will be denoted by h.

We first show that

$$\|u - \bar{u}_n\|_{\max} \leq \frac{h^2}{8} \|u''\|_{\max}. \qquad (2.97)$$

To prove this inequality, we consider the error on the kth sub-interval, shown in Figure 2.18(b). Since $\bar{e}_k(x) := u(x) - \bar{u}_n(x)$ for $x \in I_k$, $\bar{e}_k(x)$ vanishes at the endpoints of the sub-interval I_k. Furthermore, by assumption u'' is continuous on I_k, therefore there is a point $\bar{x}_k$ where $|\bar{e}_k|$ is maximum and $\bar{e}'_k(\bar{x}_k) = 0$. We expand $\bar{e}_k$ about this point into a Taylor series. Let us assume that $x_{k+1} - \bar{x}_k \leq h_k/2$ and write

$$\bar{e}_k(x_{k+1}) = 0 = \bar{e}_k(\bar{x}_k) + \underbrace{\bar{e}'_k(\bar{x}_k)(x_{k+1} - \bar{x}_k)}_{0} + \frac{1}{2}\bar{e}''_k(\xi_k)(x_{k+1} - \bar{x}_k)^2 \qquad \xi \in I_k$$

where the last term is the error term of the Taylor expansion. From this relationship we have

$$\max_{x \in I_k} |\bar{e}_k(x)| = |\bar{e}_k(\bar{x}_k)| = \frac{1}{2} |\bar{e}''_k(\xi_k)|(x_{k+1} - \bar{x}_k)^2 \qquad \xi_k \in I_k.$$

Since $\bar{u}''_n = 0$ we have $\bar{e}''_k = u''$. Also, since $x_{k+1} - \bar{x}_k \leq h_k/2$ we have

$$\max_{x \in I_k} |\bar{e}_k| \leq \frac{h_k^2}{8} \|u''\|_{\max} \qquad k = 1, 2, \ldots, n.$$

If $x_{k+1} - \bar{x}_k > h_k/2$ then we express $\bar{e}_k(x_k)$ and obtain the same result. On replacing h_k with h, Equation (2.97) follows directly from this result.

The estimate for $\|\bar{e}'_k\|_{L_2(I_k)}$ is obtained from the Taylor expansion of $\bar{e}'_k(x)$ at $x = \bar{x}_k$:

$$\bar{e}'_k(x) = \underbrace{\bar{e}'_k(\bar{x}_k)}_{0} + \bar{e}''_k(\xi_k)(x - \bar{x}_k) = u''(\xi_k)(x - \bar{x}_k).$$

Therefore

$$|\bar{e}'_k| \leq \|u''\|_{\max} |x - \bar{x}_k|$$

and

$$\int_{x_k}^{x_{k+1}} (\bar{e}'_k)^2 \, dx \leq \|u''\|_{\max}^2 \int_{x_k}^{x_{k+1}} (x - \bar{x}_k)^2 \, dx = \|u''\|_{\max}^2 \frac{h_k^3}{3}.$$

72 AN OUTLINE OF THE FINITE ELEMENT METHOD

On the entire domain,

$$\int_0^\ell (\bar{e}'_k)^2 \, dx \leq \|u''\|_{\max}^2 \sum_{k=1}^{M(\Delta)} \frac{h_k^3}{3} \leq \frac{1}{3} h^2 \|u''\|_{\max}^2 \sum_{k=1}^{M(\Delta)} h_k.$$

Since $\sum_{k=1}^{M(\Delta)} h_k = \ell$, we have

$$\int_0^\ell (\bar{e}'_k)^2 \, dx \equiv \|u' - \bar{u}'_n\|_{L_2(I)}^2 \leq h^2 \frac{\ell}{3} \|u''\|_{\max}^2. \tag{2.98}$$

We are now in a position to obtain an estimate for the error of the interpolant in the energy norm. By definition,

$$\|u - \bar{u}_n\|_{E(I)}^2 = \frac{1}{2} \int_0^\ell \left(\kappa (u' - \bar{u}'_n)^2 + c(u - \bar{u}_n)^2 \right) dx.$$

Using Equations (2.97) and (2.98), for sufficiently small h we have

$$\|u - \bar{u}_n\|_{E(I)} \leq Ch \|u''\|_{\max}$$

where C is a constant that depends on κ, c, ℓ but is independent of h and u.

Theorem 2.4.2 (see page 43) states that the finite element solution minimizes the error in the energy norm on the space $\tilde{S}(I)$. Therefore $\|u - u_n\|_{E(I)} \leq \|u - \bar{u}_n\|_{E(I)}$ and

$$\|u - u_n\|_{E(I)} \leq Ch \|u''\|_{\max}. \tag{2.99}$$

Error estimates of this type are called a priori estimates.[17] This estimate shows that the finite element solution converges to the exact solution in the energy norm given the assumption that u'' is continuous and bounded on $\bar{I}$. It also shows how fast the error in the energy norm decreases as the mesh is refined so that the size of the largest element h approaches zero. This estimate holds for all h, and for sufficiently small h the inequality (2.99) becomes an approximate equality. Convergence can be proven for any $u \in E(I)$, see, for example, [70]. A brief overview of a priori estimates for two- and three-dimensional problems is presented in Section 6.4.

Remark 2.6.4 It is possible to prove convergence of the finite element solution in other norms, such as $\|u' - u'_n\|_{\max}$ and $\|u' - u'_n\|_{L_2(I)}$, but this is beyond the scope of this book.

Exercise 2.6.2 The estimate (2.99) was derived for linear shape functions. Obtain an analogous estimate for quadratic shape functions under the assumption that $|u'''(x)| \leq C < \infty$.

Exercise 2.6.3 Repeat Exercise 2.6.2 assuming that $|u''(x)| \leq C < \infty$ but $|u'''(x)|$ can be arbitrarily large.

---

[17] A priori estimates are obtained through deductive reasoning, based on certain characteristics of a problem class. In this instance the problem class is characterized by $u''(x)$ being continuous and bounded on $\bar{I}$.

2.7 Error estimation based on extrapolation

Computed values of the potential energy corresponding to a hierarchic sequence of finite element spaces can be used for estimating the error in the energy norm by extrapolation. By Theorem 2.6.3 we have

$$\|u_{EX} - u_{FE}\|_E^2 = \pi(u_{FE}) - \pi(u_{EX}). \tag{2.100}$$

For a large and important class of problems the error in the energy norm is proportional to $N^{-\beta}$ when N is sufficiently large:

$$\|u_{EX} - u_{FE}\|_E \approx \frac{k}{N^\beta} \tag{2.101}$$

where k is some positive constant. An error estimate can be based on this relationship. Using Equation (2.100) we obtain

$$\pi(u_{FE}) - \pi(u_{EX}) \approx \frac{k^2}{N^{2\beta}}. \tag{2.102}$$

There are three unknowns: $\pi(u_{EX})$, k and β. Assume that we have a sequence of solutions corresponding to $S_{p-2} \subset S_{p-1} \subset S_p$. Let us denote the corresponding potential energy values by π_{p-2}, π_{p-1}, π_p and the degrees of freedom by N_{p-2}, N_{p-1}, N_p. We will denote $\pi := \pi(u_{EX})$. With this notation we have

$$\pi_p - \pi \approx \frac{k^2}{N_p^{2\beta}} \tag{2.103}$$

$$\pi_{p-1} - \pi \approx \frac{k^2}{N_{p-1}^{2\beta}}. \tag{2.104}$$

On dividing Equation (2.103) by (2.104), we get

$$\frac{\pi_p - \pi}{\pi_{p-1} - \pi} \approx \left(\frac{N_{p-1}}{N_p}\right)^{2\beta} \quad \text{or} \quad \log \frac{\pi_p - \pi}{\pi_{p-1} - \pi} \approx 2\beta \log \frac{N_{p-1}}{N_p}.$$

Repeating for $p-1$ and $p-2$, it is possible to eliminate 2β to obtain

$$\frac{\pi_p - \pi}{\pi_{p-1} - \pi} \approx \left(\frac{\pi_{p-1} - \pi}{\pi_{p-2} - \pi}\right)^Q \quad \text{where} \quad Q := \log \frac{N_{p-1}}{N_p} \left(\log \frac{N_{p-2}}{N_{p-1}}\right)^{-1}. \tag{2.105}$$

This equation can be solved for π to obtain an estimate for the exact value of the potential energy.

With the estimated value of π it is possible to estimate the relative error in the energy norm using Equation (2.100). By definition, the relative error in the energy norm is

$$(e_r)_E = \frac{\|u_{EX} - u_{FE}\|_E}{\|u_{EX}\|_E} \tag{2.106}$$

where $\|u_{EX}\|_E$ is estimated from u_{FE} corresponding to the finite element space of the largest number of degrees of freedom:

$$\|u_{EX}\|_E \approx \sqrt{\frac{1}{2} B(u_{FE}, u_{FE})}.$$

The relative error is usually reported as percent error. This estimator has been tested against a number of problems for which the exact solutions are known. It was found that the estimator works well for a wide range of problems, including most problems of practical interest, but it cannot be guaranteed to work for all problems.

The quality of an estimator is measured by the effectivity index θ, defined as the estimated error divided by the true error:

$$\theta := \frac{(\|u_{EX} - u_{FE}\|_E)_{\text{est.}}}{(\|u_{EX} - u_{EX}\|_E)_{\text{true}}}. \tag{2.107}$$

Of course, the effectivity index can be computed only for those problems for which the exact solution is known. Evaluation of an estimator involves the solution of a variety of such problems. An estimator is generally considered to be reliable if $0.8 < \theta < 1.2$ for most problems.

Remark 2.7.1 Let us divide Equation (2.101) by $\|\mathbf{u}_{EX}\|_E$ to obtain

$$\frac{\|u_{EX} - u_{FE}\|_E}{\|\mathbf{u}_{EX}\|_E} = (e_r)_E \approx \frac{\bar{k}}{N^\beta}$$

where $\bar{k} := k/\|\mathbf{u}_{EX}\|_E$. Therefore

$$\log(e_r)_E \approx \log \bar{k} - \beta \log N.$$

If we plot $(e_r)_E$ vs. N on a log–log scale then, if the assumption (2.101) is correct, we will see a straight line of slope $-\beta$. The convergence (2.101) is called algebraic convergence and β is called the rate of convergence.

2.7.1 The root-mean-square measure of stress

When $c = 0$ and the boundary conditions do not include spring boundary conditions then the error in the energy norm is closely related to the root-mean-square (RMS) error in stress. We define the RMS measure of stress $S = S(\sigma)$ in one dimension as follows:

$$S(\sigma) := \sqrt{\frac{1}{V} \int_0^\ell \sigma^2 A\, dx} \quad \text{where} \quad V := \int_0^\ell A\, dx. \tag{2.108}$$

From this definition and the definition of strain energy for the bar we have

$$S^2(\sigma) = \frac{1}{V} \int_0^\ell \sigma^2 A\, dx = \frac{1}{V} \int_0^\ell A(Eu')^2\, dx \leq \frac{2}{V} \max_{x \in I} E(x) \|u\|_{E(I)}^2$$

$$\geq \frac{2}{V} \min_{x \in I} E(x) \|u\|_{E(I)}^2. \tag{2.109}$$

Therefore

$$\sqrt{\frac{2}{V} \min_{x \in I} E(x)} \, \|u\|_{E(I)} \leq S(\sigma) \leq \sqrt{\frac{2}{V} \max_{x \in I} E(x)} \, \|u\|_{E(I)}. \tag{2.110}$$

In the general case when $c \neq 0$ and/or $k_0 \neq 0$, $k_\ell \neq 0$ then there is a positive constant C that depends on $E(x)$, $c(x)$, k_0, k_ℓ and V such that

$$S(\sigma) \leq C \|u\|_{E(I)}. \tag{2.111}$$

Exercise 2.7.1 Denote the relative error in the RMS stress by $(e_r)_S$, the maximum (resp. minimum) of $E(x)$, $x \in I$, by $E_{\max}$ (resp. $E_{\min}$) and assume that $c = 0$ and the boundary conditions do not include spring boundary conditions. Show that

$$\sqrt{E_{\min}/E_{\max}}\, (e_r)_E \leq (e_r)_S \leq \sqrt{E_{\max}/E_{\min}}\, (e_r)_E. \tag{2.112}$$

This exercise demonstrates that when E is constant, $c = 0$ and either force or displacement boundary conditions are prescribed, then the relative error in the RMS stress equals the relative error in the energy norm.

2.8 Extraction methods

We have seen in Example 2.5.8 that the indirect method used for computing $u'_{FE}(0)$ yielded a much smaller error than the direct method. The reasons for this are discussed in the following.

Let us say we are interested in computing some functional Q and let us assume that we can find a function w such that the exact value of Q, denoted by Q_{EX}, is

$$Q_{EX} = B(u_{EX}, w) - F(w). \tag{2.113}$$

The function w is called the extraction function. It does not have to lie in the energy space; however, the indicated operations must be defined. The finite element approximation of Q_{EX} is denoted by Q_{FE}:

$$Q_{FE} = B(u_{FE}, w) - F(w). \tag{2.114}$$

Therefore the error in Q is

$$Q_{EX} - Q_{FE} = B(e_u, w) \qquad (2.115)$$

where $e_u := u_{EX} - u_{FE}$. Note that e_u lies in $E^0(I)$. Define $g_{EX} \in E^0(I)$ such that

$$B(g_{EX}, v) = B(w, v) \qquad \text{for all } v \in E^0(I). \qquad (2.116)$$

The function g_{EX} is the projection of w onto the space $E^0(I)$. By selecting $v = e_u$ and using the symmetry of the bilinear form, Equation (2.115) can be written as

$$Q_{EX} - Q_{FE} = B(e_u, g_{EX}). \qquad (2.117)$$

Let g_{FE} be the projection of g_{EX} onto the test space $S^0(I)$:

$$B(g_{FE}, v) = B(g_{EX}, v) \qquad \text{for all } v \in S^0(I). \qquad (2.118)$$

Since $v = g_{FE} \in S^0(I)$, by the Galerkin orthogonality (see Equation (2.57)) we have $B(e_u, g_{FE}) = 0$. Therefore we can rewrite Equation (2.117) as

$$Q_{EX} - Q_{FE} = B(e_u, g_{EX}) - \underbrace{B(e_u, g_{FE})}_{0}. \qquad (2.119)$$

Denoting $e_g := g_{EX} - g_{FE}$ we obtain

$$Q_{EX} - Q_{FE} = B(e_u, e_g). \qquad (2.120)$$

Using the Schwarz inequality (see Section A.8) we get

$$|Q_{EX} - Q_{FE}| = |B(e_u, e_g)| \le 2\|e_u\|_{E(I)} \|e_g\|_{E(I)}. \qquad (2.121)$$

This equation shows that the error in Q depends on the error in the finite element solution e_u and the error e_g. Therefore the finite element space has to be designed such that both errors are small. Note that g_{EX} does not have to be known, it is of theoretical importance only.

If $\|e_g\|_{E(I)}$ converges to zero at approximately the same rate as $\|e_u\|_{E(I)}$ then $|Q_{EX} - Q_{FE}|$ converges to zero at about the same rate as the error in strain energy, which is twice the rate of convergence of the error in the energy norm. A method of computation for some functional is said to be superconvergent when the data of interest converge to their limit value at approximately the same rate as the strain energy.

Remark 2.8.1 Superconvergent methods utilize extraction functions constructed so as to approximate the appropriate Green's function.[18] For details we refer to [5].

---

[18] George Green (1793–1841).

2.9 Laboratory exercises

The following chapters deal with the FEM in two and three dimensions. To perform basic exercises, the reader will need to use the student edition of FEA software product StressCheck,[19] which is provided with this book.

At this point the reader should become acquainted with the key features of StressCheck. The best way to start is to read the Getting Started Guide, which can be found under the main menu heading "Help." Chapter 2 of this guide provides information about the most important features of the user interface. Chapter 4 is a tutorial that provides information about the preparation of input data for problems in two- and three-dimensional elasticity, execution of the solution, and post-processing procedures.

Chapter 3 of the Getting Started Guide provides an overview of StressCheck's Handbook Library. The Handbook Library contains a number of problems defined in terms of parameters in much the same way as in conventional engineering handbooks. The principal difference is that here the solutions are obtained by the FEM. This allows problems of far greater complexity and variety to be formulated in terms of parameters. The finite element meshes change automatically with the geometric parameters, therefore handbook users do not need to be concerned with mesh generation. The reader is encouraged to explore the handbook library and use it for guidance when formulating and solving exercise problems in the following chapters.

Having gained some familiarity with StressCheck, the reader will find detailed information in the Master Guide.[20] The Master Guide is composed of four parts. Part 1, the Users' Guide, provides detailed information about the user interface, post-processing and the handbook framework. Part 2, the Modeling Guide, explains procedures for the creation of geometric entities in two and three dimensions and for automatic generation of finite element meshes. Part 3, the Analysis Guide, provides instructions on the preparation of data for the various types of analysis supported by StressCheck. Part 4, the Advanced Guide, provides information about fracture mechanics applications, nonlinear analysis procedures, the solvers, and other topics that are of interest to advanced users. Specific topics can be located by means of the Index which can be accessed through the Bookmarks section of the Master Guide.

2.10 Chapter summary

Fundamental concepts, procedures and definitions, essential for understanding the finite element method, were presented in a simple setting:

1. The generalized formulation, its dependence on the boundary conditions, treatment of natural and essential conditions, definitions of the energy space, various norms and the potential energy are fundamental to the finite element method.

2. The approximate solution and hence the error of approximation is determined by the finite element space characterized by the finite element mesh, the polynomial degrees of the elements and the mapping functions.

---

[19] StressCheck is a trademark of Engineering Software Research and Development, Inc., St. Louis, Missouri.
[20] The Master Guide can be found under the main menu heading "Help."

3. The finite element solution is unique and minimizes the error in energy norm, see Theorem 2.4.2 on page 43.

4. All information generated by the finite element method resides in the standard basis functions, called shape functions, their coefficients and the mapping functions.

5. The errors in the data of interest depend on how the data are computed from the finite element solution. In computing the first derivative the indirect method was substantially more accurate than the direct method.

3

Formulation of mathematical models

The formulation of mathematical models is introduced on the basis of linear heat conduction, linear elastostatic and viscous flow problems. Linear models are used very frequently in engineering practice. Nevertheless they should be viewed as special cases of nonlinear models that account for material nonlinearities, geometric nonlinearities, mechanical contact, radiation, etc. Nonlinear models are discussed in Chapter 10.

For the purposes of the following discussion a mathematical model is understood to be a statement of a mathematical problem in the form of one or more ordinary or partial differential equations and specification of the solution domain and boundary conditions. This is called the strong form of the model. Beginning with Chapter 4, generalized formulations, also known as weak forms, will be discussed.

3.1 Notation

The Euclidean space in n dimensions will be denoted by $\mathbb{R}^n$. The Cartesian[1] coordinate axes will be labeled x, y, z (in cylindrical systems r, θ, z) and a vector u in $\mathbb{R}^n$ will be denoted either by $\vec{u}$ or $\mathbf{u}$. For example, $\vec{u} \equiv \mathbf{u} \equiv \{u_x \, u_y \, u_z\}^T$ represents a vector in $\mathbb{R}^3$.

We will employ the index notation also where the Cartesian coordinate axes are labeled $x = x_1$, $y = x_2$, $z = x_3$. The index notation will be introduced gradually in parallel with the conventional vector notation so that readers can become familiar with it. The basic rules of index notation are summarized in the following:

1. A free index is understood to range from one to three (in two spatial dimensions from one to two).

---

[1] René Descartes (in Latin: Renatus Cartesius; 1596–1650).

Introduction to Finite Element Analysis: Formulation, Verification and Validation, First Edition.
Barna Szabó and Ivo Babuška.   © 2011 John Wiley & Sons, Ltd. Published 2011 by John Wiley & Sons, Ltd.

80   FORMULATION OF MATHEMATICAL MODELS

2. The position vector of a point is x_i whereas in conventional notation it is $\{x\ y\ z\}^T$. A general vector $\vec{a} \equiv \{a_x\ a_y\ a_z\}^T$ is written simply as a_i.

3. Two free indices represent a matrix. The size of the matrix depends on the range of indices. Thus, in three dimensions,

$$a_{ij} \equiv \begin{bmatrix} a_{11} & a_{12} & a_{13} \\ a_{21} & a_{22} & a_{23} \\ a_{31} & a_{32} & a_{33} \end{bmatrix} \equiv \begin{bmatrix} a_{xx} & a_{xy} & a_{xz} \\ a_{yx} & a_{yy} & a_{yz} \\ a_{zx} & a_{zy} & a_{zz} \end{bmatrix}.$$

The identity matrix is represented by the Kronecker[2] delta δ_{ij}, defined as follows:

$$\delta_{ij} = \begin{cases} 1 & \text{if } i = j \\ 0 & \text{if } i \neq j. \end{cases}$$

4. Indices following a comma represent differentiation with respect to the variables identified by the indices. For example, if $u(x_i)$ is a scalar function then

$$u_{,2} \equiv \frac{\partial u}{\partial x_2}, \qquad u_{,23} \equiv \frac{\partial^2 u}{\partial x_2 \partial x_3}.$$

The gradient of u is simply $u_{,i}$.

5. Repeated indices represent summation. For example, the scalar product of two vectors a_i and b_j is $a_i b_i = a_1 b_1 + a_2 b_2 + a_3 b_3$. If $u_i = u_i(x_k)$ is a vector function in three-dimensional space then

$$u_{i,i} \equiv \frac{\partial u_1}{\partial x_1} + \frac{\partial u_2}{\partial x_2} + \frac{\partial u_3}{\partial x_3}$$

is the divergence of u_i. Repeated indices are also called "dummy indices," since summation is performed and therefore the index designation is immaterial. For example, $a_i b_i = a_k b_k$. The product of two matrices a_{ij} and b_{ij} is $c_{ij} = a_{ik} b_{kj}$.

6. The transformation rules for Cartesian vectors and tensors are presented in Appendix C.

Example 3.1.1 The divergence theorem in index notation is

$$\int_\Omega u_{i,i}\, dV = \int_{\partial\Omega} u_i n_i\, dS \qquad (3.1)$$

where u_i and $u_{i,i}$ are continuous on the domain Ω and its boundary $\partial\Omega$, n_i is the outward unit normal vector to the boundary, dV is the differential volume and dS is the differential surface. We will use the divergence theorem in the derivation of generalized formulations.

Exercise 3.1.1 Outline a derivation of the divergence theorem in two dimensions. Hint: Review the derivation of Green's theorem and cast it in the form of Equation (3.1).

---

[2] Leopold Kronecker (1823–1891).

3.2 Heat conduction

The problem of heat conduction is representative of an important class of problems in engineering and physics, called potential flow problems, that include, for example, the flow of viscous fluids in porous media and electrostatic phenomena. Furthermore, the mathematical formulation of potential flow problems has analogies with some physical phenomena that are not related to potential flow problems, such as the small deflection of membranes and torsion of elastic bars.

Mathematical models of heat conduction are based on two fundamental relationships: the conservation law and Fourier's law. These are described in the following:

1. **The conservation law** states that the quantity of heat entering any volume element of the conducting medium equals the quantity of heat exiting the volume element plus the quantity of heat retained in the volume element. The heat retained causes a change in temperature in the volume element which is proportional to the specific heat of the conducting medium c (in J/(kg K) units) multiplied by the density ρ (in kg/m^3 units). The temperature will be denoted by $u(x, y, z, t)$ where t is time.

   The heat flow rate across a unit area is represented by a vector quantity called heat flux. The heat flux is measured in W/m^2 units and will be denoted by $\vec{q} = \vec{q}(x, y, z, t) := \{q_x(x, y, z, t) \, q_y(x, y, z, t) \, q_z(x, y, z, t)\}^T$. In addition to heat flux entering and leaving the volume element, heat may be generated within the volume element, for example, from chemical reactions. The heat generated per unit volume and unit time will be denoted by Q (in W/m^3 units).

   Applying the conservation law to the volume element shown in Figure 3.1, we obtain

$$\Delta t[q_x \Delta y \Delta z - (q_x + \Delta q_x)\Delta y \Delta z + q_y \Delta x \Delta z - (q_y + \Delta q_y)\Delta x \Delta z$$
$$+ q_z \Delta x \Delta y - (q_z + \Delta q_z)\Delta x \Delta y + Q \Delta x \Delta y \Delta z] = c\rho \Delta u \Delta x \Delta y \Delta z. \quad (3.2)$$

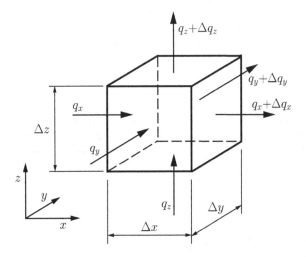

Figure 3.1 Control volume and notation for heat conduction.

82   FORMULATION OF MATHEMATICAL MODELS

Assuming that u and $\vec{q}$ are continuous and differentiable, and neglecting terms that go to zero faster than $\Delta x, \Delta y, \Delta z, \Delta t$, we get

$$\Delta q_x = \frac{\partial q_x}{\partial x}\Delta x, \quad \Delta q_y = \frac{\partial q_y}{\partial y}\Delta y, \quad \Delta q_z = \frac{\partial q_z}{\partial z}\Delta z, \quad \Delta u = \frac{\partial u}{\partial t}\Delta t.$$

On factoring $\Delta x \Delta y \Delta z \Delta t$ the conservation law is obtained:

$$-\frac{\partial q_x}{\partial x} - \frac{\partial q_y}{\partial y} - \frac{\partial q_z}{\partial z} + Q = c\rho \frac{\partial u}{\partial t}. \tag{3.3}$$

In index notation

$$-q_{i,i} + Q = c\rho \frac{\partial u}{\partial t}. \tag{3.4}$$

2. **Fourier's law** states that the heat flux vector is related to the temperature gradient in the following way:

$$q_x = -\left(k_{xx}\frac{\partial u}{\partial x} + k_{xy}\frac{\partial u}{\partial y} + k_{xz}\frac{\partial u}{\partial z}\right) \tag{3.5}$$

$$q_y = -\left(k_{yx}\frac{\partial u}{\partial x} + k_{yy}\frac{\partial u}{\partial y} + k_{yz}\frac{\partial u}{\partial z}\right) \tag{3.6}$$

$$q_z = -\left(k_{zx}\frac{\partial u}{\partial x} + k_{zy}\frac{\partial u}{\partial y} + k_{zz}\frac{\partial u}{\partial z}\right) \tag{3.7}$$

where the coefficients $k_{xx}, k_{xy}, \ldots, k_{zz}$ are called coefficients of thermal conductivity (measured in W/(m K) units). It is customary to write $k_x := k_{xx}$, $k_y := k_{yy}$, $k_z := k_{zz}$. The coefficients of thermal conductivity will be assumed to be independent of the temperature u, unless otherwise stated. Denoting the matrix of coefficients by $[K]$, Fourier's law of heat conduction can be written as

$$\vec{q} = -[K]\operatorname{grad} u. \tag{3.8}$$

The matrix of coefficients $[K]$ is symmetric and positive definite. The negative sign indicates that the direction of heat flow is opposite to the direction of the temperature gradient; that is, the direction of heat flow is from high to low temperature. In index notation Equation (3.8) is written as

$$q_i = -k_{ij}u_{,j}. \tag{3.9}$$

For isotropic materials $k_{ij} = k\delta_{ij}$.

3.2.1 The differential equation

Combining Equations (3.3) through (3.7), we get

$$\frac{\partial}{\partial x}\left(k_x \frac{\partial u}{\partial x} + k_{xy}\frac{\partial u}{\partial y} + k_{xz}\frac{\partial u}{\partial z}\right) + \frac{\partial}{\partial y}\left(k_{yx}\frac{\partial u}{\partial x} + k_y\frac{\partial u}{\partial y} + k_{yz}\frac{\partial u}{\partial z}\right)$$
$$+ \frac{\partial}{\partial z}\left(k_{zx}\frac{\partial u}{\partial x} + k_{zy}\frac{\partial u}{\partial y} + k_z\frac{\partial u}{\partial z}\right) + Q = c\rho\frac{\partial u}{\partial t} \qquad (3.10)$$

which can be written in the following compact form:

$$\operatorname{div}([K]\operatorname{grad} u) + Q = c\rho\frac{\partial u}{\partial t}. \qquad (3.11)$$

In index notation

$$(k_{ij}u_{,j})_{,i} + Q = c\rho\frac{\partial u}{\partial t}. \qquad (3.12)$$

In many practical problems u is independent of time, that is, the right hand side of Equation (3.11) and hence Equation (3.12) is zero. Such problems are called stationary or steady state problems. The solution of a stationary problem can be viewed as the solution of some time-dependent problem with time-independent boundary conditions at $t = \infty$.

In formulating Equation (3.11) we assumed that k_{ij} are differentiable functions. In many practical problems the solution domain is composed of subdomains Ω_i that have different material properties. In such cases Equation (3.11) is valid on each subdomain and on the boundaries of adjoining subdomains continuous temperature and flux are prescribed.

For a complete definition of a mathematical model initial and boundary conditions have to be specified. This is discussed in the following section.

3.2.2 Boundary and initial conditions

The solution domain will be denoted by Ω and its boundary by $\partial\Omega$. We will consider three kinds of boundary conditions:

1. **Prescribed temperature**. The temperature $u = \hat{u}$ is prescribed on boundary region $\partial\Omega_u$.

2. **Prescribed flux**. The flux vector component normal to the boundary, denoted by q_n, is prescribed on the boundary region $\partial\Omega_q$. By definition,

$$q_n := \vec{q}\cdot\vec{n} \equiv -([K]\operatorname{grad} u)\cdot\vec{n} \equiv -k_{ij}u_{,j}n_i \qquad (3.13)$$

where $\vec{n} \equiv n_i$ is the (outward) unit normal to the boundary. The prescribed flux on $\partial\Omega_q$ will be denoted by $\hat{q}_n$.

3. **Convective heat transfer**. On boundary region $\partial\Omega_c$ the flux vector component q_n is proportional to the difference between the temperature of the boundary and the

84 FORMULATION OF MATHEMATICAL MODELS

temperature of a convective medium:

$$q_n = h_c(u - u_c) \qquad (x, y, z) \in \partial\Omega_c \tag{3.14}$$

where h_c is the coefficient of convective heat transfer in W/(m^2 K) units and u_c is the (known) temperature of the convective medium.

The sets $\partial\Omega_u$, $\partial\Omega_q$ and $\partial\Omega_c$ are non-overlapping and collectively cover the entire boundary. Any of the sets may be empty.

The boundary conditions are generally time dependent. For time dependent problems an initial condition has to be prescribed on Ω: $u(x, y, z, 0) = U(x, y, z)$.

It is possible to show that Equation (3.10), subject to the enumerated boundary conditions, has a unique solution. Stationary problems also have unique solutions with the exception that when flux is prescribed over the entire boundary of Ω then the following condition must be satisfied:

$$\int_\Omega Q \, dV = \int_{\partial\Omega} q_n \, dS. \tag{3.15}$$

This is easily seen by integrating

$$(k_{ij} u_{,j})_{,i} + Q = 0 \tag{3.16}$$

on Ω and using the divergence theorem, Equation (3.1) and the definition (3.13).

Note that if u_i is a solution of Equation (3.16) then $u_i + C$ is also a solution, where C is an arbitrary constant. Therefore the solution is unique up to an arbitrary constant.

In addition to the three types of boundary conditions discussed in this section, radiation may have to be considered. When two bodies exchange heat by radiation then the flux is proportional to the difference of the fourth power of their absolute temperatures, therefore radiation is a nonlinear boundary condition. The boundary region subject to radiation, denoted by $\partial\Omega_r$, may overlap $\partial\Omega_c$. Radiation is discussed in Section 10.1.1.

In the following it will be assumed that the coefficients of thermal conductivity, the flux prescribed on Ω_q, and h_c and u_c prescribed on Ω_c are independent of the temperature u. In general, this assumption is justified in a narrow range of temperature only.

Exercise 3.2.1 Discuss the physical meaning of Equation (3.15).

Exercise 3.2.2 Show that in cylindrical coordinates r, θ, z the conservation law is of the form

$$-\frac{1}{r}\frac{\partial(r q_r)}{\partial r} - \frac{1}{r}\frac{\partial q_\theta}{\partial \theta} - \frac{\partial q_z}{\partial z} + Q = c\rho \frac{\partial u}{\partial t}. \tag{3.17}$$

Use two methods: (a) apply the conservation law to a differential volume element in cylindrical coordinates and (b) transform Equation (3.3) to cylindrical coordinates.

Exercise 3.2.3 Show that there are three mutually perpendicular directions (called principal directions) such that the heat flux is proportional to the (negative) gradient vector.

Hint: Consider steady state heat conduction and let

$$[K] \operatorname{grad} u = \lambda \operatorname{grad} u$$

and then show that the principal directions are defined by the normalized characteristic vectors.

Remark 3.2.1 The result of Exercise 3.2.3 implies that the general form of matrix $[K]$ can be obtained by rotation from orthotropic material axes.

Exercise 3.2.4 List all of the physical assumptions incorporated into the mathematical model represented by Equation (3.10) and the boundary conditions described in this section.

3.2.3 Symmetry, antisymmetry and periodicity

A scalar function is said to be symmetric with respect to a plane if in symmetrically located points with respect to the plane the function has equal values. On a plane of symmetry $q_n = 0$. A function is said to be antisymmetric with respect to a plane if in symmetrically located points with respect to the plane the function has equal absolute values but of opposite sign. On a plane of antisymmetry $u = 0$.

In many instances the domain and the coefficients have one or more planes of symmetry and the source function and boundary conditions are either symmetric or antisymmetric with respect to the planes of symmetry. In such cases it is often advantageous to exploit symmetry and antisymmetry in the formulation of the problem.

When Ω, $[K]$, Q and the boundary conditions are periodic then a periodic sector of Ω has at least one periodic boundary segment pair denoted by $\partial \Omega_p^+$ and $\partial \Omega_p^-$. On corresponding points of a periodic boundary segment pair, $P^+ \in \partial \Omega_p^+$ and $P^- \in \partial \Omega_p^-$, the boundary conditions are $u(P^+) = u(P^-)$ and $q_n^+ = -q_n^-$. Periodic, symmetric and antisymmetric boundary conditions are illustrated in the following example.

Example 3.2.1 Figure 3.2 represents a plate-like body of constant thickness. It can be divided into five sectors as illustrated. Let us assume that on the cylindrical boundary represented by the inner circle ($\partial \Omega^{(i)}$) constant flux is prescribed and on the boundary represented by the outer circle ($\partial \Omega^{(o)}$) constant temperature is prescribed. The planar surfaces parallel to the xy plane are perfectly insulated. On the surfaces represented by the elliptical boundaries let

$$q_n = \begin{cases} C \sin \varphi & \text{for } -\pi/2 \leq \varphi \leq \pi/2 \\ 0 & \text{for } \pi/2 < \varphi < 3\pi/2 \end{cases}$$

where C is constant and the formula is understood to be given in the local coordinate system of each ellipse. In this case the solution is periodic and the solution obtained for one sector can be extended to the other sectors. Periodic boundary conditions are prescribed on $\partial \Omega_p^+$ and $\partial \Omega_p^-$.

If $q_n = C \sin \varphi$ is prescribed on the elliptical boundaries in the local coordinate system of each ellipse then the solution is symmetric with respect to the y-axis, see Figure 3.2(b). If $q_n = C \cos \varphi$ is prescribed on the elliptical boundaries in the local coordinate system of each ellipse then the solution is antisymmetric with respect to the y-axis.

86   FORMULATION OF MATHEMATICAL MODELS

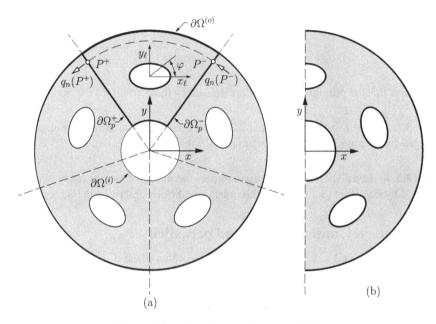

Figure 3.2 Notation for Example 3.2.1.

Exercise 3.2.5 Consider the domain shown in Figure 3.2(a). Assume that $u = 0$ on the boundary represented by the inner circle ($\partial\Omega^{(i)}$) and $q_n = 0$ on the boundaries represented by ellipses. Prescribe sinusoidal fluxes on the outer boundary ($\partial\Omega^{(o)}$) that will result in (a) periodic, (b) symmetric and (c) antisymmetric solutions.

3.2.4 Dimensional reduction

In many important practical applications reduction of the number of dimensions is possible without significantly affecting the data of interest. In other words, a mathematical model in one or two dimensions may be an acceptable replacement for the fully three-dimensional model.

Planar problems

Consider the plate-like body shown in Figure 3.3. The thickness t_z will be assumed constant unless otherwise stated. The mid-surface is the solution domain which will be denoted by Ω.

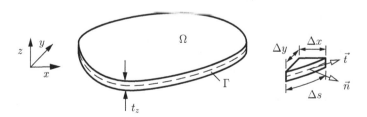

Figure 3.3 Notation for two-dimensional models.

The solution domain lies in the xy plane. The boundary points of Ω (shown as a dashed line) will be denoted by Γ. The unit outward normal to the boundary is denoted by $\vec{n}$.

The heat conduction problem in two dimensions is

$$\frac{\partial}{\partial x}\left(k_x\frac{\partial u}{\partial x}+k_{xy}\frac{\partial u}{\partial y}\right)+\frac{\partial}{\partial y}\left(k_{yx}\frac{\partial u}{\partial x}+k_y\frac{\partial u}{\partial y}\right)+\bar{Q}=c\rho\frac{\partial u}{\partial t} \qquad (3.18)$$

where the meaning of $\bar{Q}$ depends on the boundary conditions prescribed on the top and bottom surfaces ($z = \pm t_z/2$) as described in the following.

Equation (3.18) represents one of two cases:

1. The thickness is large in comparison with the other dimensions and both the material properties and boundary conditions are independent of z, that is, $u(x, y, z) = u(x, y)$. This is equivalent to the case of finite thickness with the top and bottom surfaces ($z = \pm t_z/2$) perfectly insulated and $\bar{Q} = Q$.

2. The thickness is small in relation to the other dimensions. In this case the two-dimensional solution is an approximation of the three-dimensional solution that can be interpreted as the first term in the expansion of the three-dimensional solution with respect to the z coordinate. The definition of $\bar{Q}$ depends on the boundary conditions on the top and bottom surfaces as explained below:

    (a) Prescribed flux. Let us denote the heat flux prescribed on the top (resp. bottom) surface by $\hat{q}_n^+$ (resp. $\hat{q}_n^-$). Note that positive $\hat{q}_n$ is heat flux exiting the body. The amount of heat exiting the body over a small area ΔA per unit time is $(\hat{q}_n^+ + \hat{q}_n^-)\Delta A$. Dividing by $\Delta A t_z$, the heat generated per unit volume becomes

    $$\bar{Q} = Q - (\hat{q}_n^+ + \hat{q}_n^-)\frac{1}{t_z}. \qquad (3.19)$$

    (b) Convective heat transfer. Let us denote the coefficient of convective heat transfer at $z = t_z/2$ (resp. $z = -t_z/2$) by h_c^+ (resp. h_c^-) and the corresponding temperature of the convective medium by u_c^+ (resp. u_c^-); then the amount of heat exiting the body over a small area ΔA per unit time is $[h_c^+(u - u_c^+) + h_c^-(u - u_c^-)]\Delta A$. Therefore the heat generated per unit volume is changed to

    $$\bar{Q} = Q - [h_c^+(u - u_c^+) + h_c^-(u - u_c^-)]\frac{1}{t_z}. \qquad (3.20)$$

Of course, combinations of these boundary conditions are possible. For example, flux may be prescribed on the top surface and convective boundary conditions may be prescribed on the bottom surface.

In the two-dimensional formulation u is assumed to be constant through the thickness. Therefore, prescribing a temperature has meaning only if the temperature is the same on the top and bottom surfaces, as well as on the side surface, in which case the solution is the prescribed temperature.

Exercise 3.2.6 Using the control volume shown in Figure 3.4, derive Equation (3.18) from first principles.

88   FORMULATION OF MATHEMATICAL MODELS

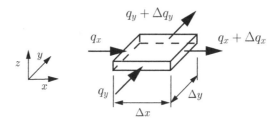

Figure 3.4 Control volume and notation for heat conduction in two dimensions.

Exercise 3.2.7 Assume that $t_z = t_z(x, y) > 0$ is a continuous and differentiable function and the maximum value of t_z is small in comparison with the other dimensions. Assume further that convective heat transfer occurs on the surfaces $z = \pm t_z/2$ with $h_c^+ = h_c^- = h_c$ and $u_c^+ = u_c^- = u_c$. Using a control volume similar to that shown in Figure 3.4, but accounting for variable thickness, show that in this case the conservation law for heat conduction in two dimensions is

$$-\frac{\partial}{\partial x}(t_z q_x) - \frac{\partial}{\partial y}(t_z q_y) - 2h_c(u - u_c) + Qt_z = c\rho t_z \frac{\partial u}{\partial t}. \tag{3.21}$$

Remark 3.2.2 In Exercise 3.2.7 the thickness t_z was assumed to be continuous and differentiable on Ω. If t_z is continuous and differentiable over two or more subdomains of Ω, but discontinuous on the boundaries of the subdomains, then Equation (3.21) is applicable on each subdomain subject to the requirement that $q_n t_z$ is continuous on the boundaries of the subdomains.

Example 3.2.2 The plan view of a conducting medium is shown in Figure 3.5. The thickness t_z is constant. The top and bottom surfaces ($z = \pm t_z/2$) are perfectly insulated. On the side surfaces ($x = \pm b$, $y = \pm c$) the temperature is constant ($u = \hat{u}_0$). Let $k_x = k_y = k$, $k_{xy} = 0$ and $Q = Q_0$ where k and Q_0 are constants. The goal is to determine the stationary temperature distribution. The mathematical problem is to solve

$$k\left(\frac{\partial^2 u}{\partial x^2} + \frac{\partial^2 u}{\partial y^2}\right) + Q_0 = 0$$

on the rectangular domain shown in Figure 3.5 with $u = \hat{u}_0$ on the boundary.

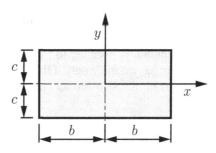

Figure 3.5 Solution domain for Example 3.2.2.

The solution of this problem can be determined by classical methods:[3]

$$u = \hat{u}_0 + 16\frac{Q_0}{k}\frac{b^2}{\pi^3}\sum_{n=1,3,5,\ldots}^{\infty}\frac{(-1)^{(n-1)/2}}{n^3}\left(1 - \frac{\cosh n\pi y/2b}{\cosh n\pi c/2b}\right)\cos n\pi x/2b. \quad (3.22)$$

This infinite series converges absolutely. It is seen that the classical solution of this seemingly simple problem is rather complicated and in fact the exact solution can be computed only approximately, although the truncation error can be made arbitrarily small by computing a sufficiently large number of terms of the infinite series.

Exercise 3.2.8 Assume that the coefficients of thermal conductivity k_x, k_{xy}, k_y are given. Show that in a Cartesian coordinate system $x'y'$, rotated counterclockwise by the angle α relative to the xy system, the coefficients will be

$$\begin{bmatrix} k_{x'} & k_{x'y'} \\ k_{y'x'} & k_{y'} \end{bmatrix} = \begin{bmatrix} \cos\alpha & \sin\alpha \\ -\sin\alpha & \cos\alpha \end{bmatrix}\begin{bmatrix} k_x & k_{xy} \\ k_{yx} & k_y \end{bmatrix}\begin{bmatrix} \cos\alpha & -\sin\alpha \\ \sin\alpha & \cos\alpha \end{bmatrix}.$$

Hint: A scalar $\{a\}^T[K]\{a\}$, where $\{a\}$ is an arbitrary vector, is invariant under coordinate transformation by rotation.

Axisymmetric problems

In many important practical problems the topological description, the material properties and the boundary conditions are axially symmetric. A simple example is a straight pipe of constant wall thickness. In such cases the problem is usually formulated in cylindrical coordinates and, since the solution is independent of the circumferential variable, the number of dimensions is reduced to two. In the following the z-axis will be the axis of symmetry and the radial (resp. circumferential) coordinates will be denoted by r (resp. θ). Referring to the result of Exercise 3.2.2, and letting $q_\theta = 0$, the conservation law is

$$-\frac{1}{r}\frac{\partial(rq_r)}{\partial r} - \frac{\partial q_z}{\partial z} + Q = c\rho\frac{\partial u}{\partial t}.$$

Substituting the axisymmetric form of Fourier's law,

$$q_r = -k_r\frac{\partial u}{\partial r}, \qquad q_z = -k_z\frac{\partial u}{\partial z}$$

we get the formulation of the axisymmetric heat conduction problem in cylindrical coordinates:

$$\frac{1}{r}\frac{\partial}{\partial r}\left(rk_r\frac{\partial u}{\partial r}\right) + \frac{\partial}{\partial z}\left(k_z\frac{\partial u}{\partial z}\right) + Q = c\rho\frac{\partial u}{\partial t}. \quad (3.23)$$

---

[3] See, for example, Timoshenko, S. and Goodier, J. N., *Theory of Elasticity*, 2nd edition, McGraw-Hill, New York, 1951, pp. 275–276.

90 FORMULATION OF MATHEMATICAL MODELS

One or more segments of the boundary may lie on the z-axis. It is implied in the formulation that the boundary condition is the zero-flux condition on those segments. Therefore it would not be meaningful to prescribe essential boundary conditions on those boundary segments. To show this, consider an axisymmetric problem of heat conduction, the solution of which is independent of z. For simplicity we assume that $k_r = 1$. In this case the problem is essentially one-dimensional:

$$\frac{1}{r}\frac{d}{dr}\left(r\frac{du}{dr}\right) = 0 \qquad r_i < r < r_o.$$

Assuming that the boundary conditions $u(r_i) = \hat{u}_i$, $u(r_o) = \hat{u}_o$ are prescribed, the exact solution of this problem is

$$u(r) = \frac{\hat{u}_o - \hat{u}_i}{\ln r_o - \ln r_i}\ln r + \frac{\hat{u}_i \ln r_o - \hat{u}_o \ln r_i}{\ln r_o - \ln r_i}. \tag{3.24}$$

Consider now the solution at an arbitrary fixed point $r = \varrho$ where $r_i < \varrho < r_o$ and let $r_i \to 0$:

$$\lim_{r_i \to 0} u(\varrho) = \lim_{r_i \to 0} \left(\frac{\hat{u}_o - \hat{u}_i}{\ln r_o/\ln r_i - 1}\frac{\ln \varrho}{\ln r_i} + \frac{\hat{u}_i \ln r_o/\ln r_i - \hat{u}_o}{\ln r_o/\ln r_i - 1}\right) = \hat{u}_o.$$

Therefore the solution is independent of $\hat{u}_i$ when $r_i = 0$. It is left to the reader in Exercise 3.2.9 to show that $du/dr \to 0$ as $r_i \to 0$, hence the boundary condition at $r = 0$ is the zero-flux boundary condition.

Exercise 3.2.9 Refer to the solution given by Equation (3.24). Show that for any $\varrho > 0$

$$\left(\frac{du}{dr}\right)_{r=\varrho} \to 0 \quad \text{as } r_i \to 0$$

independently of $\hat{u}_i$ and $\hat{u}_o$.

Exercise 3.2.10 Derive Equation (3.23) by considering a control volume in cylindrical coordinates and using the assumption that the temperature is independent of the circumferential variable.

Exercise 3.2.11 Consider water flowing in a stainless steel pipe. The temperature of the water is 80 °C. The outer surface of the pipe is cooled by air flow. The temperature of the air is 20 °C. The outer diameter of the pipe is 0.20 m and its wall thickness is 0.01 m. (a) Assuming that convective heat transfer occurs on both the inner and outer surfaces of the pipe and u is a function of r only, formulate the mathematical model for stationary heat transfer. (b) Assume that the coefficient of thermal conductivity of stainless steel is 20 W/m K. Using $h_c^{(w)} = 750$ W/m² K for the water and $h_c^{(a)} = 10$ W/m² K for the air, determine the temperature of the external surface of the pipe and the rate of heat loss per unit length.

Heat conduction in a bar

One-dimensional models of heat conduction are discussed in this section. The solution domain is a bar, one end of which is located at $x = 0$, the other end at $x = \ell$, and the cross-sectional area $A = A(x) > 0$ is a continuous and differentiable function.

One-dimensional formulations are justified when the solution of the three-dimensional problem represented by the one-dimensional model is a function of x only and/or the cross-section A is small. If the solution is known to be a function of x only then the bar is perfectly insulated along its length. If the cross-section is small then the boundary conditions on the surface of the bar are transferred to the differential equation.

We will assume in the following that the cross-sectional area is small and convective heat transfer may occur along the bar, as described in Section 3.2.4. In this case the conservation law is

$$-\frac{\partial (Aq)}{\partial x} - c_b(u - u_a) + QA = c\rho A \frac{\partial u}{\partial t}$$

where $c_b = c_b(x)$ is the coefficient of convective heat transfer of the bar (in W/(m K) units) obtained from h_c by integration:

$$c_b = \oint h_c \, ds$$

with the contour integral taken along the perimeter of the cross-section. Therefore the differential equation of heat conduction in a bar is

$$\frac{\partial}{\partial x}\left(Ak\frac{\partial u}{\partial x}\right) - c_b(u - u_a) + QA = c\rho A \frac{\partial u}{\partial t}. \qquad (3.25)$$

One of the boundary conditions described in Section 3.2 is prescribed at $x = 0$ and $x = \ell$.

Example 3.2.3 Consider stationary heat flow in a partially insulated bar of length ℓ and constant cross-section A. The coefficients k and c_b are constant and $Q = 0$. Therefore Equation (3.25) can be cast in the following form:

$$u'' - \lambda^2(u - u_a) = 0, \qquad \lambda^2 := \frac{c_b}{Ak}.$$

If the temperature u_a is a linear function of x, that is, $u_a(x) = a + bx$, and the boundary conditions are $u(0) = \hat{u}_0$, $q(\ell) = \hat{q}_\ell$ then the solution of this problem is

$$u = C_1 \cosh \lambda x + C_2 \sinh \lambda x + a + bx$$

where

$$C_1 = \hat{u}_0 - a, \quad C_2 = -\frac{1}{\lambda \cosh \lambda \ell}\left(\frac{\hat{q}_\ell}{k} + (\hat{u}_0 - a)\lambda \sinh \lambda \ell + b\right).$$

Exercise 3.2.12 Solve the problem described in Example 3.2.3 using the following boundary conditions: $q(0) = \hat{q}_0$, $q(\ell) = h_\ell(u(\ell) - U_a)$ where $\hat{q}_0$, h_ℓ, U_a are given data.

Exercise 3.2.13 A perfectly insulated bar of constant cross-section, length ℓ, thermal conductivity k, density ρ and specific heat c is subject to the initial condition $u(x, 0) = U_0$ (constant) and the boundary conditions $u(0, t) = u(\ell, t) = 0$. Assuming that $Q = 0$, verify that the solution of this problem is

$$u = 2U_0 \sum_{n=1}^{\infty} \frac{1 - \cos(n\pi)}{n\pi} \exp\left(-\frac{n^2\pi^2 k}{\ell^2 c\rho} t\right) \sin\left(n\pi \frac{x}{\ell}\right).$$

It is sufficient to show that the differential equation, the boundary conditions and the initial condition are satisfied.

Remark 3.2.3 As noted in Section 2.1.4, more than one physical phenomenon can be modeled by the same differential equation. For example, flow of viscous fluids in porous media is based on the conservation law where the flux represents the flow rate of an incompressible viscous fluid per unit area (m/s units) and Q represents a distributed source or sink injecting or extracting fluid (1/s units). The potential function u represents the piezometric head and the relationship between the piezometric gradient and the flux is formally identical to Fourier's law given by Equation (3.8), except that it is called Darcy's law,[4] and the elements of matrix $[K]$ are called coefficients of permeability (m/s units).

The boundary conditions are analogous to the linear boundary conditions described in Section 3.2; the piezometric head, the flux, or a linear combination of the flux and piezometric head may be prescribed. In viscous flow problems there is no physical analogy to radiation. On the other hand, one of the boundaries may be a free surface. On a free surface the piezometric head equals the elevation head, that is, u equals the elevation with respect to a datum plane. Furthermore, under stationary conditions the flux vector is tangential to the free surface. The position of the free surface is unknown a priori and must be determined by an iterative process.

3.3 The scalar elliptic boundary value problem

In view of Remark 3.2.3, a generic treatment of diverse physical phenomena is possible when their common mathematical basis is exploited. We will be concerned with the following model problem:

$$-\text{div}\,([\kappa]\,\text{grad}\,u) + cu = f(x, y, z) \qquad (x, y, z) \in \Omega \qquad (3.26)$$

where

$$[\kappa] := \begin{bmatrix} \kappa_x & \kappa_{xy} & \kappa_{xz} \\ \kappa_{yx} & \kappa_y & \kappa_{yz} \\ \kappa_{zx} & \kappa_{zy} & \kappa_z \end{bmatrix} \qquad (3.27)$$

---

[4] Henry Philibert Gaspard Darcy (1803–1858).

is a positive-definite matrix and $c = c(x, y, z) \geq 0$. In index notation Equation (3.26) reads

$$-(\kappa_{ij}u_{,j})_{,i} + cu = f. \tag{3.28}$$

We will be concerned with the following boundary conditions:

1. **Essential or Dirichlet boundary condition**: $u = \hat{u}$ is prescribed on boundary region $\partial\Omega_u$. When $\hat{u} = 0$ on $\partial\Omega_u$ then the Dirichlet boundary condition is said to be homogeneous.

2. **Neumann boundary condition**: $-[\kappa]\,\mathrm{grad}\,u \cdot \vec{n} = \hat{q}_n$ is prescribed on boundary region $\partial\Omega_q$. In this expression $\vec{n}$ is the unit outward normal to the boundary, shown in Figure 3.3. When $\hat{q}_n = 0$ on $\partial\Omega_q$ then the Neumann boundary condition is said to be homogeneous.

3. **Robin boundary condition**: $-[\kappa]\,\mathrm{grad}\,u \cdot \vec{n} = h_R(u - u_R)$ is given on boundary segment $\partial\Omega_R$. In this expression $h_R > 0$ and u_R are given functions. When $u_R = 0$ on $\partial\Omega_R$ then the Robin boundary condition is said to be homogeneous.

The boundary segments $\partial\Omega_u$, $\partial\Omega_q$, $\partial\Omega_R$ and $\partial\Omega_p$ are non-overlapping and collectively cover the entire boundary $\partial\Omega$. Any of the boundary segments may be empty.

We will consider restrictions of Equation (3.26) to one and two dimensions as well. In one dimension we will use

$$-\frac{d}{dx}\left(\kappa\frac{du}{dx}\right) + cu \equiv -(\kappa u')' + cu = f(x) \tag{3.29}$$

on the domain $I = (0, \ell)$ with boundary conditions prescribed on the endpoints.

3.4 Linear elasticity

Mathematical models of linear elastostatic and elastodynamic problems are based on three fundamental relationships: the strain–displacement equations, the stress–strain relationships and the equilibrium equations. The unknowns are the components of the displacement vector

$$\begin{aligned}\vec{u} &= u_x(x, y, z)\,\vec{e}_x + u_y(x, y, z)\,\vec{e}_y + u_z(x, y, z)\,\vec{e}_z \\ &\equiv \{u_x(x, y, z)\ u_y(x, y, z)\ u_z(x, y, z)\}^T \\ &\equiv u_i(x_j).\end{aligned} \tag{3.30}$$

1. **Strain–displacement relationships**. We will introduce the infinitesimal strain–displacement relationships here. A detailed derivation of these relationships is presented in Section 10.2.1. By definition, the infinitesimal normal strain components are

$$\epsilon_x \equiv \epsilon_{xx} := \frac{\partial u_x}{\partial x}, \quad \epsilon_y \equiv \epsilon_{yy} := \frac{\partial u_y}{\partial y}, \quad \epsilon_z \equiv \epsilon_{zz} := \frac{\partial u_z}{\partial z} \tag{3.31}$$

and the shear strain components are

$$\epsilon_{xy} = \epsilon_{yx} \equiv \frac{\gamma_{xy}}{2} := \frac{1}{2}\left(\frac{\partial u_x}{\partial y} + \frac{\partial u_y}{\partial x}\right)$$

$$\epsilon_{yz} = \epsilon_{zy} \equiv \frac{\gamma_{yz}}{2} := \frac{1}{2}\left(\frac{\partial u_y}{\partial z} + \frac{\partial u_z}{\partial y}\right) \quad (3.32)$$

$$\epsilon_{zx} = \epsilon_{xz} \equiv \frac{\gamma_{zx}}{2} := \frac{1}{2}\left(\frac{\partial u_z}{\partial x} + \frac{\partial u_x}{\partial z}\right)$$

where $\gamma_{xy}, \gamma_{yz}, \gamma_{zx}$ are called the engineering shear strain components. In index notation, the state of (infinitesimal) strain at a point is characterized by the strain tensor

$$\epsilon_{ij} := \frac{1}{2}\left(u_{i,j} + u_{j,i}\right). \quad (3.33)$$

2. **Stress–strain relationships**. Mechanical stress is defined as force per unit area ($N/m^2 \equiv Pa$). Since 1 pascal (Pa) is a very small stress, the usual unit of mechanical stress is the megapascal (MPa) which can be interpreted either as $10^6 \, N/m^2$ or as $1 \, N/mm^2$.

The usual engineering notation for stress components is illustrated on an infinitesimal volume element shown in Figure 3.6. The indexing rules are as follows: faces to which the positive x-, y-, z-axes are normal are called positive faces, the opposite faces are called negative faces. The normal stress components are denoted by σ, the shear stresses components by τ. The normal stress components are assigned one subscript only, since the orientation of the face and the direction of the stress component are the same. For example, σ_x is the stress component acting on the faces to which the x-axis is normal and the stress component is acting in the positive (resp. negative) coordinate direction on the positive (resp. negative) face. For the shear stresses, the first index refers to the coordinate direction of the normal to the face on which the shear stress

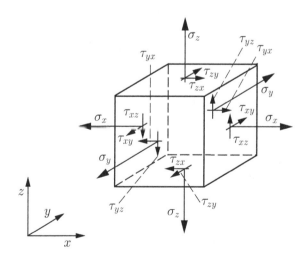

Figure 3.6 Notation for stress components.

is acting. The second index refers to the coordinate direction in which the shear stress component is acting.

On a positive (resp. negative) face the positive stress components are oriented in the positive (resp. negative) coordinate directions. The reason for this is that if we subdivide a solid body into infinitesimal hexahedral volume elements similar to the element shown in Figure 3.6, then a negative face will be coincident with each positive face. By the action–reaction principle, the forces acting on coincident faces must be equal in magnitude and opposite in sign.

An alternative notation is $\sigma_{xx} \equiv \sigma_x$, $\sigma_{xy} \equiv \tau_{xy}$, etc., which corresponds directly to the index notation σ_{ij}.

The mechanical properties of isotropic elastic materials are characterized by the modulus of elasticity $E > 0$, Poisson's ratio[5] ν, and the coefficient of thermal expansion $\alpha > 0$. The stress–strain relationships, known as Hooke's law,[6] are

$$\epsilon_x = \frac{1}{E}\left(\sigma_x - \nu\sigma_y - \nu\sigma_z\right) + \alpha T_\Delta \tag{3.34}$$

$$\epsilon_y = \frac{1}{E}\left(-\nu\sigma_x + \sigma_y - \nu\sigma_z\right) + \alpha T_\Delta \tag{3.35}$$

$$\epsilon_z = \frac{1}{E}\left(-\nu\sigma_x - \nu\sigma_y + \sigma_z\right) + \alpha T_\Delta \tag{3.36}$$

$$\gamma_{xy} \equiv 2\epsilon_{xy} = \frac{2(1+\nu)}{E}\tau_{xy} \tag{3.37}$$

$$\gamma_{yz} \equiv 2\epsilon_{yz} = \frac{2(1+\nu)}{E}\tau_{yz} \tag{3.38}$$

$$\gamma_{zx} \equiv 2\epsilon_{zx} = \frac{2(1+\nu)}{E}\tau_{zx} \tag{3.39}$$

where $T_\Delta = T_\Delta(x, y, z)$ is the temperature change with respect to a reference temperature at which the strain is zero. The strain components represent the total strain; αT_Δ is the thermal strain. Mechanical strain is defined as the total strain minus the thermal strain.

In index notation Hooke's law can be written as

$$\epsilon_{ij} = \frac{1+\nu}{E}\sigma_{ij} - \frac{\nu}{E}\sigma_{kk}\delta_{ij} + \alpha T_\Delta \delta_{ij}. \tag{3.40}$$

The inverse is

$$\sigma_{ij} = \lambda \epsilon_{kk}\delta_{ij} + 2G\epsilon_{ij} - (3\lambda + 2G)\alpha T_\Delta \delta_{ij} \tag{3.41}$$

where λ and G, called the Lamé constants,[7] are defined by

$$\lambda := \frac{E\nu}{(1+\nu)(1-2\nu)}, \qquad G := \frac{E}{2(1+\nu)}. \tag{3.42}$$

---

[5] Simeon Denis Poisson (1781–1840).
[6] Robert Hooke (1635–1703).
[7] Gabriel Lamé (1795–1870).

96   FORMULATION OF MATHEMATICAL MODELS

G is also called the shear modulus and modulus of rigidity. Since λ and G are positive, the range of Poisson's ratio is $-1 < \nu < 1/2$. Typically $0 \leq \nu < 1/2$.

The generalized Hooke's law states that the components of the stress tensor are linearly related to the mechanical strain tensor:

$$\sigma_{ij} = C_{ijkl}(\epsilon_{kl} - \alpha_{kl}T_\Delta) \tag{3.43}$$

where C_{ijkl} and α_{ij} are Cartesian tensors. By symmetry considerations the maximum number of independent elastic constants that characterize C_{ijkl} is 21. The symmetric tensor α_{ij} is characterized by six independent coefficients of thermal expansion. This is the most general form of anisotropy in linear elasticity.

According to Equation (3.43) the elastic body is stress-free when the mechanical strain tensor is zero. There may be, however, an initial stress field $\sigma_{ij}^{(0)}$ present in the reference configuration. The initial stress field, called residual stress, must satisfy the equilibrium equations and stress-free boundary conditions [47]. Residual stresses may be introduced by manufacturing processes,[8] forging, machining, and various types of cold-working operations [48]. In composite materials there is a large difference in the coefficients of thermal expansion of the fiber and matrix. Residual stresses are introduced when a part cools following curing operations at elevated temperatures. When residual stresses are present we have

$$\sigma_{ij} = \sigma_{ij}^{(0)} + C_{ijkl}(\epsilon_{kl} - \alpha_{kl}T_\Delta). \tag{3.44}$$

Accurate determination of residual stresses is generally difficult.

3. **Equilibrium**. Considering the dynamic equilibrium of a volume element, similar to that shown in Figure 3.6, except that the edges are of length $\Delta x, \Delta y, \Delta z$, six equations of equilibrium can be written: the resultants of the forces and moments must vanish. Assuming that the material is not loaded by distributed moments (body moments), consideration of moment equilibrium leads to the conclusion that $\tau_{xy} = \tau_{yx}, \tau_{yz} = \tau_{zy}, \tau_{zx} = \tau_{xz}$; that is, the stress tensor is symmetric. Assuming further that the components of the stress tensor are continuous and differentiable, application of d'Alembert's principle[9] and consideration of force equilibrium leads to three partial differential equations:

$$\frac{\partial \sigma_x}{\partial x} + \frac{\partial \tau_{xy}}{\partial y} + \frac{\partial \tau_{xz}}{\partial z} + F_x - \varrho \frac{\partial^2 u_x}{\partial t^2} = 0 \tag{3.45}$$

$$\frac{\partial \tau_{xy}}{\partial x} + \frac{\partial \sigma_y}{\partial y} + \frac{\partial \tau_{yz}}{\partial z} + F_y - \varrho \frac{\partial^2 u_y}{\partial t^2} = 0 \tag{3.46}$$

$$\frac{\partial \tau_{xz}}{\partial x} + \frac{\partial \tau_{yz}}{\partial y} + \frac{\partial \sigma_z}{\partial z} + F_z - \varrho \frac{\partial^2 u_z}{\partial t^2} = 0 \tag{3.47}$$

where F_x, F_y, F_z are the components of the body force vector (in N/m^3 units), ϱ is the specific density (in kg/m$^3 \equiv$ N s^2/m^4 units). These equations are called the equations

---

[8] For example, 7050-T7451 aluminum plates are hot-rolled, quenched, over aged and stretched by the imposition of 1.5 to 3.0% strain in the rolling direction.

[9] Jean Le Rond d'Alembert (1717–1783).

of motion. In index notation

$$\sigma_{ij,j} + F_i = \varrho \frac{\partial^2 u_i}{\partial t^2}. \qquad (3.48)$$

For elastostatic problems the time derivative is zero and the boundary conditions are independent of time. This yields the equations of static equilibrium:

$$\sigma_{ij,j} + F_i = 0. \qquad (3.49)$$

3.4.1 The Navier equations

The equations of motion, called the Navier equations,[10] are obtained by substituting Equation (3.41) into Equation (3.48). In elastodynamics the effects of temperature are usually negligible, hence we will assume $T_\Delta = 0$:

$$G u_{i,jj} + (\lambda + G) u_{j,ji} + F_i = \varrho \frac{\partial^2 u_i}{\partial t^2}. \qquad (3.50)$$

In elastostatic problems we have

$$G u_{i,jj} + (\lambda + G) u_{j,ji} + F_i = (3\lambda + 2G)\alpha (T_\Delta)_{,i}. \qquad (3.51)$$

Exercise 3.4.1 Derive Equation (3.51) by substituting Equation (3.41) into Equation (3.49). Indicate the rules under which the indices are changed to obtain Equation (3.51). Hint: $(\epsilon_{kk}\delta_{ij})_{,j} = (u_{k,k}\delta_{ij})_{,j} = u_{k,ki} = u_{j,ji}$.

Exercise 3.4.2 Derive the equilibrium equations from first principles.

Exercise 3.4.3 In deriving Equations (3.50) and (3.51) it was assumed that λ and G are constants. Formulate the analogous equations assuming that λ and G are smooth functions of $x_i \in \Omega$.

Exercise 3.4.4 Assume that Ω is the union of two or more subdomains and the material properties are constants on each subdomain but vary from subdomain to subdomain. Formulate the elastostatic problem for this case.

3.4.2 Boundary and initial conditions

As in the case of heat conduction, we will consider three kinds of boundary conditions: prescribed displacements, prescribed tractions and spring boundary conditions. Tractions are forces per unit area acting on the boundary. Prescribed displacements and tractions are often specified in a normal–tangent reference frame:

---

[10] Claude Louis Marie Henri Navier (1785–1836).

98   FORMULATION OF MATHEMATICAL MODELS

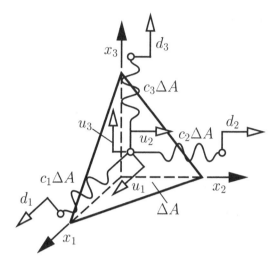

Figure 3.7 Spring boundary condition: schematic representation.

1. **Prescribed displacement.** One or more components of the displacement vector is prescribed on all or part of the boundary. This is called a kinematic boundary condition.

2. **Prescribed traction.** One or more components of the traction vector is prescribed on all or part of the boundary. The definition of traction vector is given in Appendix C.1.

3. **Linear spring.** A linear relationship is prescribed between the traction and displacement vector components. The general form of this relationship is

$$T_i = c_{ij}(d_j - u_j) \tag{3.52}$$

where T_i is the traction vector, c_{ij} is a positive-definite matrix that represents the distributed spring coefficients, d_j is a prescribed function that represents displacement imposed on the spring and u_j is the (unknown) displacement vector function on the boundary. The spring coefficients c_{ij} (in N/m^3 units) may be functions of the position x_k but are independent of the displacement u_i. This is called a "Winkler[11] spring."

A schematic representation of this boundary condition on an infinitesimal boundary surface element is shown in Figure 3.7 under the assumption that c_{ij} is a diagonal matrix and therefore three spring coefficients $c_1 := c_{11}$, $c_2 := c_{22}$ and $c_3 := c_{33}$ characterize the elastic properties of the boundary condition.

Figure 3.7 should be interpreted to mean that the imposed displacement d_i will cause a differential force ΔF_i to act on the centroid of the surface element. Suspending the summation rule, the magnitude of ΔF_i is

$$\Delta F_i = c_i \Delta A (d_i - u_i), \quad i = 1, 2, 3$$

---

[11] Emil Winkler (1835–1888).

where u_i is the displacement of the surface element. The corresponding traction vector is

$$T_i = \lim_{\Delta A \to 0} \frac{\Delta F_i}{\Delta A} = c_i(d_i - u_i), \qquad i = 1, 2, 3.$$

The enumerated boundary conditions may occur in any combination. For example, the displacement vector component u_1, the traction vector component T_2 and a linear combination of T_3 and u_3 may be prescribed on a boundary segment.

In engineering practice boundary conditions are most conveniently prescribed in the normal–tangent reference frame. The normal is uniquely defined on smooth surfaces but the tangential coordinate directions are not. It is necessary to specify the tangential coordinate directions with respect to the reference frame used. The required coordinate transformations are discussed in Appendix C.

The boundary conditions are generally time dependent. For time-dependent problems the initial conditions, that is, the initial displacement and velocity fields, denoted respectively by $U(x, y, z)$ and $V(x, y, z)$, have to be prescribed:

$$\vec{u}(x, y, z, 0) = \vec{U}(x, y, z) \quad \text{and} \quad \left(\frac{\partial \vec{u}}{\partial t}\right)_{(x,y,z,0)} = \vec{V}(x, y, z).$$

Exercise 3.4.5 Assume that the following boundary conditions are given in the normal–tangent reference frame x_i', x_1' being coincident with the normal: $T_1' = c_1'(d_1' - u_1')$; $T_2' = T_3' = 0$. Using the transformation $x_i' = g_{ij} x_j$, determine the boundary conditions in the x_i coordinate system. (See Appendix C, Section C.3.)

3.4.3 Symmetry, antisymmetry and periodicity

Symmetry and antisymmetry of a vector function with respect to a line are illustrated in Figure 2.3. The definitions of symmetry and antisymmetry of a vector function in three dimensions are analogous. The corresponding vector components parallel to a plane of symmetry (resp. antisymmetry) have the same absolute value and the same (resp. opposite) sign. The corresponding vector components normal to a plane of symmetry (resp. antisymmetry) have the same absolute value and opposite (resp. same) sign.

In a plane of symmetry the normal displacement and the shearing traction components are zero. In a plane of antisymmetry the normal traction is zero and the in-plane components of the displacement vector are zero.

When the solution is periodic on Ω then a periodic sector of Ω has at least one periodic boundary segment pair denoted by $\partial \Omega_p^+$ and $\partial \Omega_p^-$. On corresponding points of a periodic boundary segment pair, $P^+ \in \partial \Omega_p^+$ and $P^- \in \partial \Omega_p^-$, the normal component of the displacement vector and the periodic in-plane components of the displacement vector have the same value. The normal component of the traction vector and the periodic in-plane components of the traction vector have the same absolute value but opposite sign.

Exercise 3.4.6 A homogeneous isotropic elastic body with Poisson's ratio zero occupies the domain $\Omega = \{x, y, z \mid |x| < a, |y| < b, |z| < c\}$. Define tractions on the boundaries of Ω such that the tractions satisfy equilibrium and the plane $z = 0$ is a plane of (a) symmetry, (b) antisymmetry.

Exercise 3.4.7 Consider the domain shown in Figure 3.2. Assume that $T_n = 0$ and $T_t = \tau_o$ where τ_o is constant, is prescribed on $\partial\Omega^{(o)}$, $T_n = T_t = 0$ on the elliptical boundaries and $u_n = u_t = 0$ on $\partial\Omega^{(i)}$. Specify periodic boundary conditions on $\partial\Omega_p^+$ and $\partial\Omega_p^-$.

3.4.4 Dimensional reduction

Due to the complexity of three-dimensional problems in elasticity, dimensional reduction is widely used. Various kinds of dimensional reduction are possible in elasticity, such as planar, axisymmetric, shell, plate, beam and bar models. Each of these model types is sufficiently important to have generated a substantial technical literature. In the following models for planar and axisymmetric problems, axially loaded bars are discussed. Models for beams, plates and shells will be discussed separately.

Planar elastostatic models

Let us consider a prismatic body of length ℓ. The material points occupy the domain Ω_ℓ, defined as follows:

$$\Omega_\ell = \{(x, y, z) \mid (x, y) \in \omega, -\ell/2 < z < \ell/2, \ell > 0\} \qquad (3.53)$$

where $\omega \in \mathbb{R}^2$ is a bounded domain. The lateral boundary of the body is denoted by

$$\Gamma_\ell = \{(x, y, z) \mid (x, y) \in \partial\omega, -\ell/2 < z < \ell/2, \ell > 0\} \qquad (3.54)$$

and the faces are denoted by

$$\gamma_\pm = \{(x, y, z) \mid (x, y) \in \omega, z = \pm\ell/2\}. \qquad (3.55)$$

The notation is shown in Figure 3.8. The diameter of ω will be denoted by d_ω.

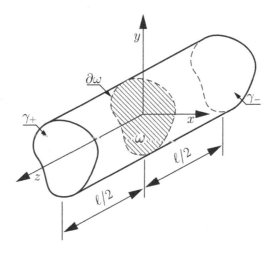

Figure 3.8 Notation.

The material properties, volume forces and temperature change acting on Ω_ℓ and tractions acting on Γ_ℓ will be assumed to be independent of z. Therefore the xy plane is a plane of symmetry. It is assumed that tractions are specified on the entire boundary Γ_ℓ. This is usually called the first fundamental boundary value problem of elasticity.

When the tractions acting on $\gamma_\pm$ are zero and $\ell/d_\omega \ll 1$ then such problems are usually formulated on ω as plane stress problems. When the normal displacements and shearing tractions acting on $\gamma_\pm$ are zero then such problems are formulated on ω as plane strain problems, independently of the ℓ/d_ω ratio. When $1 \ll \ell/d_\omega$ and the tractions acting on $\gamma_\pm$ are zero then the three-dimensional thermoelasticity problem on Ω_ℓ is called the generalized plane strain problem. In a generalized plane strain problem the stress resultants corresponding to σ_z must be zero:

$$\int_{\partial \omega} \sigma_z \, dz dy = 0 \quad \int_{\partial \omega} \sigma_z x \, dz dy = 0 \quad \int_{\partial \omega} \sigma_z y \, dz dy = 0.$$

For details we refer to [14].

Planar elastostatic models are specializations of the Navier equations. The mid-plane is understood to be a plane of symmetry, that is, $u_z(x, y, 0) = 0$. The formulation is written in unabridged notation in the following:

1. **The linear strain–displacement relationships**. These are

$$\epsilon_x := \frac{\partial u_x}{\partial x} \tag{3.56}$$

$$\epsilon_y := \frac{\partial u_y}{\partial y} \tag{3.57}$$

$$\gamma_{xy} := \frac{\partial u_x}{\partial y} + \frac{\partial u_y}{\partial x}. \tag{3.58}$$

In two-dimensional problems the shear strain components γ_{yz} and γ_{zx} and the corresponding shear stress components τ_{yz}, τ_{zx} are zero.

2. **Stress–strain relationships**. We will be concerned with two special cases, the case of plane stress ($\sigma_z = \tau_{yz} = \tau_{zx} = 0$) and the case of plane strain ($\epsilon_z = \gamma_{yz} = \gamma_{zx} = 0$). Applying the appropriate restrictions to the three-dimensional Hooke's law, from Equation (3.34) and Equation (3.35), we obtain for plane stress

$$\begin{Bmatrix} \sigma_x \\ \sigma_y \\ \tau_{xy} \end{Bmatrix} = \frac{E}{1-\nu^2} \begin{bmatrix} 1 & \nu & 0 \\ \nu & 1 & 0 \\ 0 & 0 & \frac{1-\nu}{2} \end{bmatrix} \begin{Bmatrix} \epsilon_x \\ \epsilon_y \\ \gamma_{xy} \end{Bmatrix} - \frac{E\alpha T_\Delta}{1-\nu} \begin{Bmatrix} 1 \\ 1 \\ 0 \end{Bmatrix} \tag{3.59}$$

and, letting $\epsilon_z = 0$ in Equation (3.36), we obtain for in-plane strain

$$\begin{Bmatrix} \sigma_x \\ \sigma_y \\ \tau_{xy} \end{Bmatrix} = \begin{bmatrix} \lambda + 2G & \lambda & 0 \\ \lambda & \lambda + 2G & 0 \\ 0 & 0 & G \end{bmatrix} \begin{Bmatrix} \epsilon_x \\ \epsilon_y \\ \gamma_{xy} \end{Bmatrix} - (3\lambda + 2G)\alpha T_\Delta \begin{Bmatrix} 1 \\ 1 \\ 0 \end{Bmatrix}. \tag{3.60}$$

102   FORMULATION OF MATHEMATICAL MODELS

3. **The static equations of equilibrium.** These are

$$\frac{\partial \sigma_x}{\partial x} + \frac{\partial \tau_{xy}}{\partial y} + F_x = 0 \tag{3.61}$$

$$\frac{\partial \tau_{yx}}{\partial x} + \frac{\partial \sigma_y}{\partial y} + F_y = 0 \tag{3.62}$$

where F_x, F_y are the components of body force vector $\vec{F}(x, y)$ (in N/m^3 units).

The Navier equations are obtained by substituting the stress–strain and strain–displacement relationships into the equilibrium equations. For plane strain

$$(\lambda + G)\frac{\partial}{\partial x}\left(\frac{\partial u_x}{\partial x} + \frac{\partial u_y}{\partial y}\right) + G\left(\frac{\partial^2 u_x}{\partial x^2} + \frac{\partial^2 u_x}{\partial y^2}\right) = \frac{E\alpha}{1 - 2\nu}\frac{\partial T_\Delta}{\partial x} - F_x \tag{3.63}$$

$$(\lambda + G)\frac{\partial}{\partial y}\left(\frac{\partial u_x}{\partial x} + \frac{\partial u_y}{\partial y}\right) + G\left(\frac{\partial^2 u_y}{\partial x^2} + \frac{\partial^2 u_y}{\partial y^2}\right) = \frac{E\alpha}{1 - 2\nu}\frac{\partial T_\Delta}{\partial y} - F_y. \tag{3.64}$$

The boundary conditions are most conveniently written in the normal–tangent (nt) reference frame illustrated in Figure 3.3. The relationship between the xy and nt components of displacement and traction is established by the rules of vector transformation, described in Appendix C, Section C.3. The linear boundary described in Section 3.4.2 is directly applicable in two dimensions.

Exercise 3.4.8   Derive Equation (3.59) and Equation (3.60) from Hooke's law.

Exercise 3.4.9   Show that for plane stress the Navier equations are

$$\frac{E}{2(1 - \nu)}\frac{\partial}{\partial x}\left(\frac{\partial u_x}{\partial x} + \frac{\partial u_y}{\partial y}\right) + G\left(\frac{\partial^2 u_x}{\partial x^2} + \frac{\partial^2 u_x}{\partial y^2}\right) = \frac{E\alpha}{1 - \nu}\frac{\partial T_\Delta}{\partial x} - F_x \tag{3.65}$$

$$\frac{E}{2(1 - \nu)}\frac{\partial}{\partial y}\left(\frac{\partial u_x}{\partial x} + \frac{\partial u_y}{\partial y}\right) + G\left(\frac{\partial^2 u_y}{\partial x^2} + \frac{\partial^2 u_y}{\partial y^2}\right) = \frac{E\alpha}{1 - \nu}\frac{\partial T_\Delta}{\partial y} - F_y. \tag{3.66}$$

Exercise 3.4.10   Denote the components of the unit normal vector to the boundary by n_x and n_y. Show that

$$T_x = T_n n_x - T_t n_y$$
$$T_y = T_n n_y + T_t n_x$$

where T_n (resp. T_t) is the normal (resp. tangential) component of the traction vector.

Axisymmetric elastostatic models

Axial symmetry exists when the solution domain can be generated by sweeping a plane figure around an axis, known as the axis of symmetry, and the material properties, loading and constraints are axially symmetric. For example, pipes, cylindrical and spherical pressure vessels are often idealized in this way.

The radial, circumferential and axial coordinates are denoted by r, θ and z respectively and the displacement, stress, strain and traction components are labeled with corresponding subscripts. The problem is formulated in terms of the displacement vector components $u_r(r, z)$ and $u_z(r, z)$.

1. **The linear strain–displacement relationships.** These are [87]

$$\epsilon_r := \frac{\partial u_r}{\partial r} \tag{3.67}$$

$$\epsilon_\theta := \frac{u_r}{r} \tag{3.68}$$

$$\epsilon_z := \frac{\partial u_z}{\partial z} \tag{3.69}$$

$$\epsilon_{rz} = \frac{\gamma_{rz}}{2} := \frac{1}{2}\left(\frac{\partial u_r}{\partial z} + \frac{\partial u_z}{\partial r}\right). \tag{3.70}$$

2. **Stress–strain relationships.** For isotropic materials the stress–strain relationship is obtained from Equation (3.41):

$$\begin{Bmatrix} \sigma_r \\ \sigma_\theta \\ \sigma_z \\ \tau_{rz} \end{Bmatrix} = \begin{bmatrix} \lambda+2G & \lambda & \lambda & 0 \\ \lambda & \lambda+2G & \lambda & 0 \\ \lambda & \lambda & \lambda+2G & 0 \\ 0 & 0 & 0 & G \end{bmatrix} \begin{Bmatrix} \epsilon_r \\ \epsilon_\theta \\ \epsilon_z \\ \gamma_{rz} \end{Bmatrix} - \frac{E\alpha T_\Delta}{1-2\nu} \begin{Bmatrix} 1 \\ 1 \\ 1 \\ 0 \end{Bmatrix}. \tag{3.71}$$

3. **Equilibrium.** The elastostatic equations of equilibrium are [87]

$$\frac{1}{r}\frac{\partial(r\sigma_r)}{\partial r} + \frac{\partial \tau_{rz}}{\partial z} - \frac{\sigma_\theta}{r} + F_r = 0 \tag{3.72}$$

$$\frac{1}{r}\frac{\partial(r\tau_{rz})}{\partial r} + \frac{\partial \sigma_z}{\partial z} + F_z = 0. \tag{3.73}$$

Exercise 3.4.11 Write down the Navier equations for the axisymmetric model.

3.5 Incompressible elastic materials

When $\nu \to 1/2$ then $\lambda \to \infty$, therefore the relationship represented by Equation (3.41) breaks down. Referring to Equation (3.40), the sum of normal strain components is related to the sum of normal stress components by

$$\epsilon_{kk} = \frac{1-2\nu}{E}\sigma_{kk} + 3\alpha T_\Delta.$$

104 FORMULATION OF MATHEMATICAL MODELS

The sum of normal strain components is called the volumetric strain and will be denoted by $\epsilon_{vol} := \epsilon_{kk}$. Defining

$$\sigma_0 := \frac{1}{3}\sigma_{kk},$$

we get

$$\epsilon_{vol} \equiv \epsilon_{kk} = \frac{3(1-2\nu)}{E}\sigma_0 + 3\alpha T_\Delta \qquad (3.74)$$

where the first term on the right is the mechanical strain and the second term is the thermal strain. For incompressible materials, that is, when $\nu = 1/2$, $\epsilon_{vol} = 3\alpha T_\Delta$ is independent of σ_0. Therefore σ_0 cannot be computed using Hooke's law. Substituting Equation (3.74) into Equation (3.41) and letting $\nu = 1/2$ we get

$$\sigma_{ij} = \sigma_0 \delta_{ij} + \frac{2E}{3}(\epsilon_{ij} - \alpha T_\Delta \delta_{ij}). \qquad (3.75)$$

Substituting into Equation (3.49), and assuming that E and α are constant,

$$(\sigma_0)_{,i} + \frac{2E}{3}(\epsilon_{ij,j} - \alpha(T_\Delta)_{,i}) + F_i = 0.$$

Writing

$$\epsilon_{ij,j} = \frac{1}{2}(u_{i,j} + u_{j,i})_{,j} = \frac{1}{2}(u_{i,jj} + u_{j,ij})$$

and interchanging the order of differentiation in the second term we get

$$u_{j,ij} = u_{j,ji} = (u_{j,j})_{,i} = (\epsilon_{jj})_{,i} = 3\alpha(T_\Delta)_{,i}.$$

Therefore, for incompressible materials,

$$\epsilon_{ij,j} = \frac{1}{2}u_{i,jj} + \frac{3}{2}\alpha(T_\Delta)_{,i}.$$

The problem is to determine u_i such that

$$(\sigma_0)_{,i} + \frac{E}{3}\left(u_{i,jj} + \alpha(T_\Delta)_{,i}\right) + F_i = 0 \qquad (3.76)$$

subject to the condition of incompressibility, that is, the condition that volumetric strain can be caused by temperature change but not by mechanical stress,

$$u_{i,i} = 3\alpha T_\Delta \qquad (3.77)$$

and the appropriate boundary conditions. If displacement boundary conditions are prescribed over the entire boundary ($\partial\Omega_u = \partial\Omega$) then the problem is solvable only if the prescribed displacements are consistent with the incompressibility condition:

$$\int_{\partial\Omega} u_i n_i \, dS = 3 \int_{\Omega} \alpha T_\Delta \, dV.$$

This follows directly from integrating Equation (3.77) and applying the divergence theorem.

Exercise 3.5.1 An incompressible bar of constant cross-section and length ℓ is subjected to uniform temperature change (i.e., T_Δ is constant). The centroidal axis of the bar is coincident with the x_1-axis. The boundary conditions are $u_1(0) = u_1(\ell) = 0$. The body force vector is zero. Explain how Equations (3.75) and (3.77) are applied in this case to find that $\sigma_{11} = -E\alpha T_\Delta$.

Exercise 3.5.2 Write the equations of incompressible elasticity in unabridged notation.

3.6 Stokes' flow

The flow of viscous fluids at very low Reynolds numbers[12] ($Re < 1$) is modeled by the Stokes[13] equations. There is a very close analogy between the Stokes equations and the equations of incompressible elasticity discussed in Section 3.5. In fluid mechanics the average normal stress is the pressure, $p := -\sigma_0$. The vector u_i represents the components of the velocity vector and the shear modulus of the incompressible elastic solid $E/3$ is replaced by the coefficient of dynamic viscosity μ (measured in N s/m^2 units):

$$\mu u_{i,jj} = p_{,i} - F_i \quad (3.78)$$

$$u_{i,i} = 0. \quad (3.79)$$

Exercise 3.6.1 Write the Stokes equations in unabridged notation.

Exercise 3.6.2 Assume that velocities are prescribed over the entire boundary for a Stokes problem (i.e., $\partial\Omega_u = \partial\Omega$). What condition must be satisfied by the prescribed velocities?

Remark 3.6.1 In this chapter we treated the physical properties such as coefficients of heat conduction, the surface coefficient, the modulus of elasticity and Poisson's ratio as given constants. Readers should be mindful of the fact that physical properties are empirical data inferred from experimental observations. Due to variations in experimental conditions and other factors, these data are not known precisely and are always subject to restrictions. For example, stress is proportional to strain up to the proportional limit only; the flux is proportional to the temperature gradient only within a narrow range of temperatures. In fact, the coefficient of thermal conductivity (k) is typically temperature dependent. For example, in the case of AISI 304 stainless steel k changes from about 15 to about 20 W/m °C in the

---

[12] Osborne Reynolds (1842–1912).
[13] George Gabriel Stokes (1819–1903).

temperature range of 0 to 400 °C [11]. For a narrow range of temperature (say 100 to 200 °C) the size of the uncertainty in k is about the same as the change in the mean value. Therefore, using a constant value for k this range may be "good enough." Ignoring temperature dependence for a much wider range of temperatures can lead to substantial errors.

Taking into account the temperature dependence of the coefficient of thermal conductivity leads to the formulation of nonlinear problems, the solutions of which are found by iteratively solving sequences of linear problems. This is discussed in Section 10.1.2.

Analysts usually rely on data published in various handbooks. The published data can vary widely, however. For example, Babuška and Silva, having consulted seven references, found that in these references the coefficient of thermal conductivity of pure iron varies between 71.8 and 80.4 W/(m K) [11]. Published data are typically accompanied by legal and technical disclaimers.

3.7 The hierarchic view of mathematical models

Up to this point we have considered the formulation of mathematical models in the form of linear partial differential equations and linear boundary conditions. Linear models of heat conduction and elasticity should be viewed as special cases of models with less restrictive assumptions, such as models that account for material nonlinearities, which in turn should be viewed as special cases of models that also account for radiation, mechanical contact, large strains and displacements, etc. In this sense, any mathematical model should be viewed as a special case of a more comprehensive model. This is the hierarchic view of mathematical models.

Many possible mathematical models can be constructed for the representation of material laws. For example, many models have been proposed to represent elastic–plastic, viscoelastic and viscoplastic behavior. Each can be viewed as special case of a more elaborate model and therefore a member of a hierarchic sequence.

A model based on the assumptions of linear elasticity is a special case of any model that accounts for plastic deformation; however, there are many models of plastic deformation that are not members of the same hierarchic sequence. Therefore alternative sequences of hierarchic models can be constructed for the simulation of a physical reality.

The effects of nonlinearities and uncertainties in material properties and boundary conditions on the data of interest are evaluated in the process of conceptualization discussed in Section 1.1.1. This requires that appropriate software tools be available for the consideration of nonlinear effects. An introductory treatment of nonlinear formulations is presented in Chapter 10.

Ideally the choice of a mathematical model for a particular purpose involves informed judgment based on systematic consideration of alternative choices and the quality and reliability of available information. In practice the choice of mathematical models is often influenced by the subjective preferences of the analyst and the limitations of the available software tools and other computational resources.

3.8 Chapter summary

The formulation of mathematical models for linear problems in heat conduction and elasticity was described. A mathematical model was understood to be one or more partial differential equations and the prescribed initial and boundary conditions. Alternative formulations will be presented in Chapter 4.

The model of heat conduction is based on the conservation law, a fundamental law of physics, and some empirical relationships between the derivatives of the temperature u and the flux vector which are subject to certain limitations. For example, the coefficients of thermal conductivity generally depend on the value of u as well as the gradient of u. Only within a certain range of u and its gradient can these coefficients be treated as constants.

Similarly, the model for elastic bodies is based on the conservation of momentum (in statics, the equations of equilibrium) and an empirical linear relationship between the stress and strain tensors. The linear relationship holds for small strains only. It is important to bear the limitations of a particular mathematical model in mind when considering it in the context of an engineering decision-making process. A mathematical model must never be confused with the physical reality that it was conceived to simulate.

Dimensional reduction should be understood as a special case of the fully three-dimensional formulation.

It was noted that more than one physical phenomenon can be represented by a partial differential equation. This allows generic mathematical treatment of various physical phenomena.

4

Generalized formulations

The idea of a generalized formulation was introduced in the one-dimensional setting, and its usefulness for obtaining approximate solutions by the finite element method was demonstrated in Chapter 2. In this chapter generalized formulations in two and three dimensions are described.

The examples and exercises in this chapter are introductions to the use of the finite element method for solving linear problems in heat transfer and elasticity. It is assumed that the reader is familiar with at least one finite element software product. The solutions presented here were obtained with StressCheck.

4.1 The scalar elliptic problem

Multiplying Equation (3.28) by a test function v and integrating, we get

$$-\int_\Omega (\kappa_{ij} u_{,j})_{,i} v\, dV + \int_\Omega cuv\, dV = \int_\Omega fv\, dV. \qquad (4.1)$$

Clearly, this equation must hold for arbitrary v, provided that the indicated operations are defined. The first integral can be written as

$$\int_\Omega (\kappa_{ij} u_{,j})_{,i} v\, dV = \int_\Omega (\kappa_{ij} u_{,j} v)_{,i}\, dV - \int_\Omega \kappa_{ij} u_{,j} v_{,i}\, dV.$$

Applying the divergence theorem (Equation (3.1)) we get

$$\int_\Omega (\kappa_{ij} u_{,j} v)_{,i}\, dV = \int_{\partial\Omega} \kappa_{ij} u_{,j} n_i v\, dS$$

where n_i is the unit normal vector to the boundary surface. Therefore Equation (4.1) can be written in the following form:

$$-\int_{\partial\Omega} \kappa_{ij} u_{,j} n_i v\, dS + \int_{\Omega} \kappa_{ij} u_{,j} v_{,i}\, dV + \int_{\Omega} cuv\, dV = \int_{\Omega} fv\, dV. \qquad (4.2)$$

It is customary to write

$$q_i := -\kappa_{ij} u_{,j} \quad \text{and} \quad q_n := q_i n_i.$$

With this notation we have

$$\int_{\Omega} \kappa_{ij} u_{,j} v_{,i}\, dV + \int_{\Omega} cuv\, dV = \int_{\Omega} fv\, dV - \int_{\partial\Omega} q_n v\, dS. \qquad (4.3)$$

This is the generalization of Equation (2.36) to two and three dimensions. As we have seen in Section 2.3, the specific statement of a generalized formulation depends on the boundary conditions. In the general case $u = \hat{u}$ is prescribed on $\partial\Omega_u$ (Dirichlet boundary condition), $q_n = \hat{q}_n$ is prescribed on $\partial\Omega_q$ (Neumann boundary condition) and $q_n = h_R(u - u_R)$ is prescribed on Ω_R (Robin boundary condition), see Section 3.3. We now define the bilinear form as

$$B(u, v) := \int_{\Omega} \kappa_{ij} u_{,j} v_{,i}\, dV + \int_{\Omega} cuv\, dV + \int_{\partial\Omega_R} h_R uv\, dS \qquad (4.4)$$

and the linear functional as

$$F(v) := \int_{\Omega} fv\, dV - \int_{\partial\Omega_q} q_n v\, dS + \int_{\partial\Omega_R} h_R u_R v\, dS. \qquad (4.5)$$

Of course, if $\partial\Omega_R$ is empty then the last terms in Equations (4.4) and (4.5) are not present. $F(v)$ (resp. $B(u, v)$) satisfies all of the properties listed in Section A.3 (resp. Section A.4). The space $E(\Omega)$ is defined by

$$E(\Omega) := \{u \mid B(u, u) \leq C < \infty\}$$

and the norm

$$\|u\|_E := \sqrt{\frac{1}{2} B(u, u)}$$

is associated with $E(\Omega)$. The space of admissible functions is defined by

$$\tilde{E}(\Omega) := \{u \mid u \in E(\Omega),\ u = \hat{u} \text{ on } \partial\Omega_u\}.$$

Here we assume that, corresponding to any $u = \hat{u}$ specified on $\partial\Omega_u$, there is a $u^\star \in E(\Omega)$ such that $u^\star = \hat{u}$ on $\partial\Omega_u$. This imposes certain restrictions on $\hat{u}$ and ensures that $\tilde{E}(\Omega)$ is not empty. The space of test functions is defined by

$$E^0(\Omega) := \{u \mid u \in E(\Omega),\ u = 0 \text{ on } \partial\Omega_u\}.$$

The generalized formulation is now stated as follows: "Find $u \in \tilde{E}(\Omega)$ such that $B(u, v) = F(v)$ for all $v \in E^0(\Omega)$." A function u that satisfies this condition is called a generalized solution.

The generalized formulation is often based on Theorem 2.6.2. The exact solution of the generalized formulation minimizes the potential energy defined by Equation (2.92) on the space $\tilde{E}(I)$. Alternatively, the potential energy is defined by

$$\Pi(u) := \frac{1}{2}\int_\Omega \kappa_{ij} u_{,i} u_{,j}\, dV + \frac{1}{2}\int_\Omega cu^2\, dV + \frac{1}{2}\int_{\partial\Omega_R} h_R(u - u_R)^2\, dS$$
$$- \int_\Omega fu\, dV + \int_{\partial\Omega_q} q_n u\, dS. \tag{4.6}$$

Note that $\pi(u)$ differs from $\Pi(u)$ only by a constant. Therefore the minimum $\pi(u)$ is the same as the minimum of $\Pi(u)$.

Exercise 4.1.1 Consider the generalized formulation of steady state heat conduction in cylindrical coordinates in the special case when the solution depends only on the radial variable r:

$$\int_{r_i}^{r_o} k(r) \frac{du}{dr}\frac{dv}{dr} r\, dr = \left(rk\frac{du}{dr}v\right)_{r=r_o} - \left(rk\frac{du}{dr}v\right)_{r=r_i}.$$

(a) Derive this formulation from Equation (3.23) and (b) apply this formulation to a long pipe of internal radius r_i, external radius r_o using the boundary conditions $u(r_i) = \hat{u}_i$ and

$$q_n := -k\frac{du}{dr} = h_c(u - u_c) \quad \text{at } r = r_o.$$

Exercise 4.1.2 Following the proof of Theorem 2.6.2, show that the generalized formulation minimizes $\Pi(u)$ given by Equation (4.6).

4.1.1 Continuity

In two and three dimensions $u \in E(\Omega)$ is not necessarily continuous or bounded. For example, the function $u := \log |\log r|$ where $r := (x^2 + y^2)^{1/2}$ is discontinuous and unbounded at the point $r = 0$ yet it lies in $E(\Omega)$. This has important implications: recall that in discussing concentrated forces in Section 2.5.5 continuity of the test function v was invoked. The same argument cannot be extended to two and three dimensions, hence concentrated forces and concentrated fluxes are inadmissible data. Similarly, point constraints are inadmissible except for the enforcement of the uniqueness of the solution when Neumann conditions are specified on the entire boundary.

Exercise 4.1.3 Let $r := (x^2 + y^2)^{1/2}$ and $\Omega := \{r \mid r \leq \rho_0 < 1\}$. Show that $u_1 := \log r$ does not lie in $E(\Omega)$ but $u_2 := \log |\log r|$ does. Hint: $u_2 = \log(-\log r)$ on Ω.

4.1.2 Existence

To guarantee the existence of a generalized solution, it is sufficient to show that:

(a) There exist positive real numbers α, β such that for any $x_i \in \Omega$

$$\alpha \|x\|^2 \leq \kappa_{ij} x_i x_j \leq \beta \|x\|^2$$

where κ_{ij} is the matrix in Equation (3.27), x_i is a vector of real numbers and $\|x\|^2 := x_i x_i$.

(b) Further, $0 \leq c(x_i) \leq \gamma, 0 \leq h_R \leq \Theta$ where $\gamma, \Theta \in \mathbb{R}^1$ and

$$\int_\Omega f^2 \, dV < \infty, \quad \int_{\Omega_q} q_n^2 \, dS < \infty, \quad \int_{\Omega_R} u_R^2 \, dS < \infty.$$

(c) When $\partial \Omega_q = \partial \Omega$ (i.e., $\partial \Omega_u$ and $\partial \Omega_R$ are empty) and $c = 0$, then the specified functions f and q_n have to satisfy the following condition:

$$\int_\Omega f \, dV - \int_{\partial \Omega} q_n \, dS = 0. \tag{4.7}$$

This is because $E^0(\Omega) = E(\Omega)$ and hence $v = C$ (where C is an arbitrary constant) lies in the test space. Therefore in Equation (4.4), $B(u, C) = 0$ and consequently $F(C)$ has to be zero also:

$$F(C) = C \left(\int_\Omega f \, dV - \int_{\partial \Omega} q_n \, dS \right) = 0.$$

If this condition is satisfied then, as shown in Theorem 2.6.1 in Section 2.6, the solution is unique up to an arbitrary constant.

(d) When $\partial \Omega_u \neq \emptyset$ then $\hat{u}$ must be such that $\tilde{E}$ is not empty. In one dimension this condition was satisfied because any $u \in E(I)$ is continuous on $\bar{I}$. In two and three dimensions not every $u \in E(\Omega)$ is continuous on $\bar{\Omega}$ (see, for example, Exercise 4.1.3). It is necessary to prescribe $\hat{u}$ only on boundary segments that have non-zero length in two dimensions and non-zero area in three dimensions. In other words, $\hat{u}$ cannot be prescribed at a point in two and three dimensions, or along a line in three dimensions.

Exercise 4.1.4 Explain the physical meaning of Equation (4.7).

4.1.3 Approximation by the finite element method

When seeking an approximate solution by the finite element method, denoted by u_{FE}, we construct a finite-dimensional subspace of $E(\Omega)$, called the finite element space

THE SCALAR ELLIPTIC PROBLEM   113

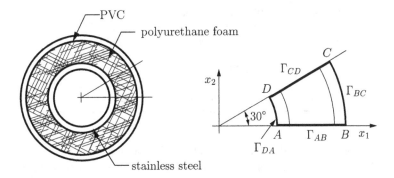

Figure 4.1 Solution domain and finite element mesh for Example 4.1.1.

$S = S(\Omega, \Delta, \mathbf{p}, \mathbf{Q})$, which is analogous to the finite element space in one dimension, described in Section 2.5.3. Finite element spaces in two and three dimensions are described in Chapter 5. The subspace $\tilde{S}$ (resp. S^0) is the corresponding finite-dimensional space of admissible functions (resp. test functions). The approximation problem is formulated as follows: "Find $u_{FE} \in \tilde{S}$ such that $B(u_{FE}, v) = F(v)$ for all $v \in S^0$."

Example 4.1.1 A stainless steel pipe is insulated by polyurethane (PU) foam which is encased in a polyvinyl chloride (PVC) pipe. The outside diameter of the stainless steel (ss) pipe is 250.0 mm, the wall thickness is 22 mm. The outside diameter of the PVC casing is 450 mm, the wall thickness is 20 mm. The stainless steel pipe is carrying a hot liquid at 400 K. The PVC casing is cooled by convection. The physical properties are: $k_{PVC} = 0.14$ W/m K, $h_{PVC} = 6.5$ W/(m² K), $k_{ss} = 17$ W/m K, $k_{PU} = 0.025$ W/m K. The temperature of the convective medium is 300 K. Determine the temperature distribution and the rate of heat loss per unit length of pipe.

Due to circular symmetry, this is essentially a one-dimensional problem; however, we will formulate it in two dimensions. Utilizing symmetry, we define the solution domain as a 30° sector $ABCD$, shown in Figure 4.1.

The materials are assumed to be isotropic, therefore $\kappa_{ij} = k\delta_{ij}$ where $k = k(x, y)$ is the thermal conductivity, a piecewise constant function. In this case $c = 0$ and no heat is generated in the materials, hence Equation (4.3) is simplified to

$$\int_\Omega k u_{,i} v_{,i}\, dx_1 dx_2 = -\int_\Gamma q_n v\, ds. \quad (4.8)$$

The boundary conditions are listed in Table 4.1. Note that segments AB and CD lie on lines of symmetry. Therefore $q_n = 0$ and the line integral on the right hand side is zero on these boundary segments. On segment DA the boundary condition $u = 400$ K is enforced by restriction on the space of admissible functions and the test functions $v \in E^0(\Omega)$ are zero, therefore the line integral on the right hand side is zero on this boundary segment. On segment BC $q_n = 6.5(u - 300)$. Therefore Equation (4.8) can be written as

$$\underbrace{\int_\Omega k u_{,i} v_{,i}\, dx_1 dx_2 + 6.5 \int_{\Gamma_{BC}} uv\, ds}_{B(u,v)} = \underbrace{6.5 \times 300 \int_{\Gamma_{BC}} v\, ds}_{F(v)}. \quad (4.9)$$

Table 4.1 Example 4.1.1: boundary conditions.

Segments	Description	Condition
AB, CD	Symmetry	$q_n = 0$
BC	Convection	$q_n = 6.5(u - 300)$
DA	Temperature	$u = 400$

The definitions for $B(u, v)$ and $F(v)$ are indicated in Equation (4.9).

To complete the generalized formulation of the problem, we first define the energy space $E(\Omega)$:

$$E(\Omega) := \{u \mid B(u, u) \leq C < \infty\}$$

where C is some positive constant. The norm associated with $E(\Omega)$, called the energy norm, is defined by

$$\|u\|_E := \sqrt{\frac{1}{2} B(u, u)}.$$

The space of admissible functions $\tilde{E}(\Omega)$ is defined by

$$\tilde{E}(\Omega) := \{u \mid u \in E(\Omega), \ u = 400 \text{ on } \Gamma_{DA}\}$$

and the space of test functions $E^0(\Omega)$ is defined by

$$E^0(\Omega) := \{u \mid u \in E(\Omega), \ u = 0 \text{ on } \Gamma_{DA}\}.$$

The generalized formulation is stated as follows: "Find $u \in \tilde{E}(\Omega)$ such that $B(u, v) = F(v)$ for all $v \in E^0(\Omega)$."

In the finite element approximation the solution domain $ABCD$ was partitioned into three elements with curved boundaries corresponding to the three materials (see Figure 4.1). A sequence of seven solutions was obtained using a sequence of finite element spaces $S_1 \subset S_2 \subset \cdots \subset S_7$ characterized by uniform p-distributions ranging from 1 to 7 (trunk space)[1] on the three-element mesh.

The relative error in the energy norm was estimated by extrapolation. The estimated relative error in the energy norm vs. the number of degrees of freedom N is shown in Figure 4.2(a). The estimated error in the energy norm is seen to decrease very rapidly; at $p = 3$ it is already well under 1%.

The heat loss rate per 30° sector per unit length is computed from

$$HL_{30°} = \int_{\Gamma_{BC}} q_n \, ds$$

---

[1] The term "trunk space" is defined in Section 5.2.1.

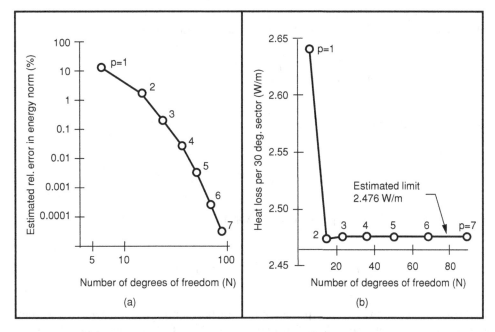

Figure 4.2 (a) Estimated relative error in energy norm (%) vs. N and (b) computed heat loss rate per 30° sector (W/m) in Example 4.1.1.

by numerical quadrature. The convergence of the heat loss $HL_{30°}$ (W/m) is shown in Figure 4.2(b). The estimated limit value with respect to $N \to \infty$ is 2.476 W/m. Of course, this number must be multiplied by 12 to obtain the heat loss rate per unit length of pipe (29.7 W/m).

Exercise 4.1.5 Assume that the value of the coefficient of thermal conductivity of the polyurethane foam (k_{PU}) in Example 4.1.1 is not known precisely. It is known only to be in the range 0.02 to 0.03 W/m K. Compute the corresponding range of the heat loss rate per unit length of pipe.

Exercise 4.1.6 Formulate and solve the problem of Example 4.1.1 as an axisymmetric problem.

4.2 The principle of virtual work

Consider the equations of equilibrium (3.48) without inertia forces:

$$\sigma_{ij,j} + F_i = 0. \tag{4.10}$$

Multiply Equation (4.10) by a test function v_i and integrate on Ω:

$$\int_\Omega \sigma_{ij,j} v_i \, dV + \int_\Omega F_i v_i \, dV = 0. \tag{4.11}$$

Observe that if the equilibrium equation (4.10) is satisfied then Equation (4.11) holds for arbitrary v_i, subject to the condition that the indicated operations are defined. We write

$$\int_\Omega \sigma_{ij,j} v_i \, dV = \int_\Omega (\sigma_{ij} v_i)_{,j} \, dV - \int_\Omega \sigma_{ij} v_{i,j} \, dV$$

and use the divergence theorem to obtain

$$\int_\Omega \sigma_{ij,j} v_i \, dV = \int_{\partial\Omega} \sigma_{ij} n_j v_i \, dS - \int_\Omega \sigma_{ij} v_{i,j} \, dV.$$

Noting that $\sigma_{ij} n_j = T_i$ (see Equation (C.3) in Appendix C), Equation (4.11) can be written as

$$\int_\Omega \sigma_{ij} v_{i,j} \, dV = \int_\Omega F_i v_i \, dV + \int_{\partial\Omega} T_i v_i \, dS. \tag{4.12}$$

This equation has a physical interpretation: the test function v_i can be interpreted as some arbitrary displacement field, independent of the applied body force F_i and traction T_i. For this reason v_i is called a "virtual displacement." The terms on the right hand side represent the work done by the body force and the traction forces acting on the body, collectively called "external forces." The left hand side represents the work done by the internal stresses. To see this, refer to Figure 4.3 and assume that vertex A of the infinitesimal hexahedral element, the coordinates of which are x_i, is subjected to a virtual displacement v_i. Then, since v_i is continuous and differentiable, the face located at $x_1 + dx_1$ will be displaced, relative to point A, by $v_{i,1} dx_1$ and the virtual work done by σ_{11} is

$$dW_{\sigma_{11}} = \underbrace{(\sigma_{11} dx_2 dx_3)}_{\text{force}} \underbrace{(v_{1,1} dx_1)}_{\text{displacement}} = \sigma_{11} v_{1,1} dV.$$

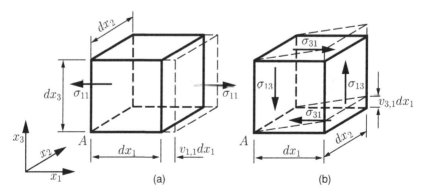

Figure 4.3 Virtual displacements corresponding to (a) σ_{11} and (b) σ_{13}.

Similarly, the virtual work done by σ_{13} is

$$dW_{\sigma_{13}} = \underbrace{(\sigma_{13}dx_2dx_3)}_{\text{force}}\underbrace{(v_{3,1}dx_1)}_{\text{displacement}} = \sigma_{13}v_{3,1}dV$$

etc. The principle of virtual work states that "the virtual work of external forces is equal to the virtual work of internal stresses." Note that since this result is based on the equilibrium Equation (4.10), it is independent of the material properties and therefore holds for any continuum.

Equation (4.12) is the generic form of the principle of virtual work. Particular statements of the principle of virtual work depend on the material properties and the boundary conditions. Often an alternative form of (4.12) is used: observe that the sum $\sigma_{ij}v_{i,j}$ equals the sum $\sigma_{11}v_{1,1} + \sigma_{22}v_{2,2} + \sigma_{33}v_{3,3}$ plus the sum of pairs like $\sigma_{12}v_{1,2} + \sigma_{21}v_{2,1}$. Since $\sigma_{ij} = \sigma_{ji}$, this can be written as

$$\sigma_{12}v_{1,2} + \sigma_{21}v_{2,1} = \sigma_{12}\frac{1}{2}(v_{1,2} + v_{2,1}) + \sigma_{21}\frac{1}{2}(v_{2,1} + v_{1,2}) = \sigma_{12}\epsilon_{12}^{(v)} + \sigma_{21}\epsilon_{21}^{(v)}$$

where the superscript (v) indicates that these are the infinitesimal strain terms corresponding to the test function v_i, that is,

$$\epsilon_{ij}^{(v)} := \frac{1}{2}(v_{i,j} + v_{j,i}).$$

Therefore

$$\sigma_{ij}v_{i,j} = \sigma_{ij}\epsilon_{ij}^{(v)}$$

and Equation (4.12) can be written as

$$\int_\Omega \sigma_{ij}\epsilon_{ij}^{(v)}\,dV = \int_\Omega F_i v_i\,dV + \int_{\partial\Omega} T_i v_i\,dS. \tag{4.13}$$

This equation has great importance in the formulation of mathematical models in continuum mechanics. In this chapter we will consider its application to problems in linear elasticity.

Exercise 4.2.1 Starting from Equation (3.3), derive the counterpart of Equation (4.13) for the stationary heat conduction problem.

4.3 Elastostatic problems

On substituting Equation (3.43) into Equation (4.13) we obtain

$$\int_\Omega C_{ijkl}\epsilon_{ij}^{(v)}\epsilon_{kl}\,dV = \int_\Omega F_i v_i\,dV + \int_{\partial\Omega} T_i v_i\,dS + \int_\Omega C_{ijkl}\epsilon_{ij}^{(v)}\alpha_{kl}T_\Delta\,dV. \tag{4.14}$$

This is an application of the principle of virtual work to elastostatic problems. Specific statements of the principle of virtual work depend on the boundary conditions. Let $\partial\Omega_u$

denote the boundary region where $u_i = \hat{u}_i$ is prescribed; let $\partial\Omega_T$ denote the boundary region where $T_i = \hat{T}_i$ is prescribed; and let $\partial\Omega_s$ denote the boundary region where $T_i = k_{ij}(d_j - u_j)$ (i.e., spring boundary condition) is prescribed. Let us define

$$B(\mathbf{u}, \mathbf{v}) := \int_\Omega C_{ijkl} \epsilon_{ij}^{(v)} \epsilon_{kl} \, dV + \int_{\partial\Omega_s} k_{ij} u_j v_i \, dS \qquad (4.15)$$

and

$$F(\mathbf{v}) := \int_\Omega F_i v_i \, dV + \int_{\partial\Omega_T} \hat{T}_i v_i \, dS + \int_{\partial\Omega_s} k_{ij} d_j v_i \, dS$$
$$+ \int_\Omega C_{ijkl} \epsilon_{ij}^{(v)} \alpha_{kl} T_\Delta \, dV \qquad (4.16)$$

where $\mathbf{u} \equiv u_i$ and $\mathbf{v} \equiv v_i$. In the interest of simplicity we will assume that the material constants E, ν and α are piecewise constant. The space $E(\Omega)$, called the energy space, is defined by

$$E(\Omega) := \{\mathbf{u} \mid B(\mathbf{u}, \mathbf{u}) \leq C < \infty\} \qquad (4.17)$$

and the norm

$$\|\mathbf{u}\|_E := \sqrt{\frac{1}{2} B(\mathbf{u}, \mathbf{u})} \qquad (4.18)$$

is associated with $E(\Omega)$. The space of admissible functions is defined by

$$\tilde{E}(\Omega) := \{u_i \mid u_i \in E(\Omega), \; u_i = \hat{u}_i \text{ on } \partial\Omega_u\}.$$

Note that this definition imposes a restriction on the prescribed displacement conditions: for the reasons discussed in Section 4.1.2, there has to be a $u_i \in E(\Omega)$ so that $u_i = \hat{u}_i$ on $\partial\Omega_u$.

The space of test functions is defined by

$$E^0(\Omega) := \{u_i \mid u_i \in E(\Omega), \; u_i = 0 \text{ on } \partial\Omega_u\}.$$

The generalized formulation based on the principle of virtual work is stated as follows: "Find $\mathbf{u} \in \tilde{E}(\Omega)$ such that $B(\mathbf{u}, \mathbf{v}) = F(\mathbf{v})$ for all $\mathbf{v} \in E^0(\Omega)$."

Exercise 4.3.1 When the material is isotropic we can substitute Equation (3.41) into Equation (4.12) to obtain

$$\int_\Omega \left(\lambda \epsilon_{kk} \epsilon_{ii}^{(v)} + 2G \epsilon_{ij} \epsilon_{ij}^{(v)} \right) dV = \int_\Omega F_i v_i \, dV + \int_{\partial\Omega} T_i v_i \, dS$$
$$+ \int_\Omega \frac{E}{1-2\nu} \alpha T_\Delta \epsilon_{ii}^{(v)} \, dV. \qquad (4.19)$$

We define the differential operator matrix $[D]$ and the material stiffness matrix $[E]$ as follows:

$$[D] := \begin{bmatrix} \frac{\partial}{\partial x} & 0 & 0 \\ 0 & \frac{\partial}{\partial y} & 0 \\ 0 & 0 & \frac{\partial}{\partial z} \\ \frac{\partial}{\partial y} & \frac{\partial}{\partial x} & 0 \\ 0 & \frac{\partial}{\partial z} & \frac{\partial}{\partial y} \\ \frac{\partial}{\partial z} & 0 & \frac{\partial}{\partial x} \end{bmatrix} \quad [E] := \begin{bmatrix} \lambda + 2G & \lambda & \lambda & 0 & 0 & 0 \\ \lambda & \lambda + 2G & \lambda & 0 & 0 & 0 \\ \lambda & \lambda & \lambda + 2G & 0 & 0 & 0 \\ 0 & 0 & 0 & G & 0 & 0 \\ 0 & 0 & 0 & 0 & G & 0 \\ 0 & 0 & 0 & 0 & 0 & G \end{bmatrix}.$$

Furthermore, we denote $\mathbf{u} \equiv \{u\} := \{u_x \; u_y \; u_z\}^T$ and $\mathbf{v} \equiv \{v\} := \{v_x \; v_y \; v_z\}^T$. Show that Equation (4.19) can be written in the following form:

$$\int_\Omega ([D]\{v\})^T [E][D]\{u\} \, dV = \int_\Omega \{v\}^T \{F\} \, dV + \int_{\partial\Omega} \{v\}^T \{T\} \, dS$$

$$+ \int_\Omega \left\{ \frac{\partial v_x}{\partial x} \; \frac{\partial v_y}{\partial y} \; \frac{\partial v_z}{\partial z} \right\} \begin{Bmatrix} 1 \\ 1 \\ 1 \end{Bmatrix} \frac{E \alpha T_\Delta}{1 - 2\nu} \, dV \quad (4.20)$$

where $\{F\} := \{F_x \; F_y \; F_z\}^T$ is the body force vector and $\{T\} := \{T_x \; T_y \; T_z\}^T$ is the traction vector.

Remark 4.3.1 In the general anisotropic case, represented by Equation (4.14), we have

$$\int_\Omega ([D]\{v\})^T [E][D]\{u\} \, dV = \int_\Omega \{v\}^T \{F\} \, dV + \int_{\partial\Omega} \{v\}^T \{T\} \, dS$$

$$+ \int_\Omega ([D]\{v\})^T [E]\{\alpha\} T_\Delta \, dV \quad (4.21)$$

where the material stiffness matrix $[E]$ is a symmetric positive-definite matrix with at most 21 independent coefficients and $\{\alpha\} := \{\alpha_{11} \; \alpha_{22} \; \alpha_{33} \; 2\alpha_{12} \; 2\alpha_{23} \; 2\alpha_{31}\}^T$.

4.3.1 Uniqueness

The generalized formulation based on the principle of virtual work is unique in the energy space $E(\Omega)$. The proof of uniqueness given by Theorem 2.6.1 in Section 2.6 is applicable to the elasticity problem in three dimensions.

Uniqueness in the energy space does not necessarily mean uniqueness of the displacement field $\mathbf{u}$. When $\partial\Omega_u$ and $\partial\Omega_s$ are both empty then there are six linearly independent test functions in $E^0(\Omega) = E(\Omega)$ for which $\epsilon_{ij}^{(v)} = 0$ and hence $B(\mathbf{u}, \mathbf{v}) = 0$. Three of these functions correspond to rigid body displacements:

$$\epsilon_{11}^{(v)} = 0: \quad v_i^{(1)} = c_1\{1\ 0\ 0\}^T$$

$$\epsilon_{22}^{(v)} = 0: \quad v_i^{(2)} = c_2\{0\ 1\ 0\}^T$$

$$\epsilon_{33}^{(v)} = 0: \quad v_i^{(3)} = c_3\{0\ 0\ 1\}^T$$

and three correspond to infinitesimal rigid body rotations:

$$\epsilon_{12}^{(v)} = 0: \quad v_i^{(4)} = c_4\{-x_2\ x_1\ 0\}^T \quad \text{rotation about } x_3$$

$$\epsilon_{23}^{(v)} = 0: \quad v_i^{(5)} = c_5\{0\ -x_3\ x_2\}^T \quad \text{rotation about } x_1$$

$$\epsilon_{31}^{(v)} = 0: \quad v_i^{(6)} = c_6\{x_3\ 0\ -x_1\}^T \quad \text{rotation about } x_2$$

where $c_1, c_2, \ldots, c_6$ are arbitrary constants. Consequently the body force vector and the surface tractions must satisfy the following six conditions:

$$F(\mathbf{v}) = 0: \quad \int_\Omega F_i v_i^{(k)} dV + \int_{\partial\Omega} T_i v_i^{(k)} dS = 0 \quad k = 1, 2, \ldots, 6. \tag{4.22}$$

The physical interpretation of these conditions is that the body must be in equilibrium, that is, the sum of forces and the sum of moments must be zero.

The solution is unique up to rigid body displacements. In order to ensure uniqueness of the solution, "rigid body constraints" are imposed; that is, the rigid body displacement functions are eliminated from the space of test functions:

$$E^0(\Omega) = \{v_i \mid v_i \in E(\Omega),\ v_i^{(k)} = 0,\ k = 1, 2, \ldots, 6\}. \tag{4.23}$$

The values of rigid body displacements are arbitrary, therefore the space of admissible functions is

$$\tilde{E}(\Omega) = \{u_i \mid u_i \in E(\Omega),\ u_i^{(k)} = \hat{u}_i^{(k)},\ k = 1, 2, \ldots, 6\} \tag{4.24}$$

where $\hat{u}_i^{(k)}$ are arbitrary rigid body displacements, usually chosen to be zero.

Rigid body constraints are enforced by setting six displacement components in at least three points to arbitrary values. The usual procedure is as follows. Three non-collinear points, labeled A, B, C in Figure 4.4, are selected arbitrarily, subject only to the restriction that the displacement function must be continuous at those points. A Cartesian coordinate system is associated with these points such that points A and B lie on axis X_1, axis X_3 is perpendicular to the plane defined by the points ABC and axis X_2 is perpendicular to axes X_1, X_3. The displacement components $U_1^{(A)}$, $U_2^{(A)}$, $U_3^{(A)}$, $U_2^{(B)}$, $U_3^{(B)}$ and $U_3^{(C)}$, shown in Figure 4.4, are assigned arbitrary values, usually zero. This will ensure that the coefficients of the rigid body displacement functions are uniquely defined.

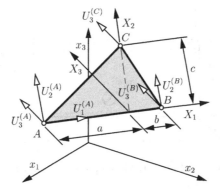

Figure 4.4 Rigid body constraints: notation.

For example, letting

$$U_1^{(A)} = C_1 - C_4 X_2^{(A)} + C_6 X_3^{(A)} = 0$$
$$U_2^{(A)} = C_2 + C_4 X_1^{(A)} - C_5 X_3^{(A)} = 0$$
$$U_3^{(A)} = C_3 + C_5 X_2^{(A)} - C_6 X_1^{(A)} = 0$$

prevents rigid body translation. Letting

$$U_2^{(B)} = C_2 + C_4 X_1^{(B)} - C_5 X_3^{(B)} = 0$$
$$U_3^{(B)} = C_3 + C_5 X_2^{(B)} - C_6 X_1^{(B)} = 0$$

prevents rigid body rotation about axes X_3 and X_2 respectively, and letting

$$U_3^{(C)} = C_3 + C_5 X_2^{(C)} - C_6 X_1^{(C)} = 0$$

prevents rigid body rotation about axis X_1. This is equivalent to writing

$$\begin{bmatrix} 1 & 0 & 0 & 0 & 0 & 0 \\ 0 & 1 & 0 & -a & 0 & 0 \\ 0 & 0 & 1 & 0 & 0 & a \\ 0 & 1 & 0 & b & 0 & 0 \\ 0 & 0 & 1 & 0 & 0 & -b \\ 0 & 0 & 1 & 0 & c & 0 \end{bmatrix} \begin{Bmatrix} C_1 \\ C_2 \\ C_3 \\ C_4 \\ C_5 \\ C_6 \end{Bmatrix} = 0 \qquad (4.25)$$

where the determinant of the coefficient matrix is $c(a+b)^2$ which is nonzero for any choice of the non-collinear points A, B, C and therefore all rigid body displacement modes vanish. Of course, there are other ways of preventing or specifying rigid body displacements. It is necessary only that the solution at the points where rigid body constraints are prescribed must be continuous and the determinant of the coefficient matrix, analogous to that in Equation (4.25), must be non-zero.

122 GENERALIZED FORMULATIONS

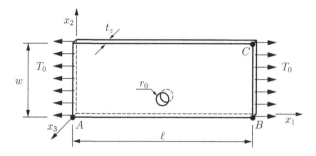

Figure 4.5 Illustration for Example 4.3.1 and Exercise 4.3.2.

In many cases $\partial\Omega_u$ and/or $\partial\Omega_s$ are not empty but the prescribed conditions do not provide a sufficient number of constraints to prevent all rigid body displacements and/or rotations. In those cases it is necessary to provide a sufficient number of rigid body constraints to prevent rigid body motion. Such a case is illustrated in Example 4.3.2.

Example 4.3.1 A thin elastic plate-like body with a circular hole, shown in Figure 4.5, is loaded by constant normal tractions T_0. The equilibrium conditions (4.22) are obviously satisfied. In this case there are six rigid body modes and the space of admissible functions is given by Equation (4.24) and the space of test functions by Equation (4.23). Letting

$$u_1^{(A)} = u_2^{(A)} = u_3^{(A)} = u_2^{(B)} = u_3^{(B)} = u_3^{(C)} = 0$$

the rigid body constraints are enforced.

Example 4.3.2 If the center of the circular hole in the thin elastic plate-like body of Example 4.3.1 is located at $x_1 = \ell/2$ then the solution is symmetric with respect to the plane $x_1 = \ell/2$ and the problem may be formulated on the half domain shown in Figure 4.6(a). On a

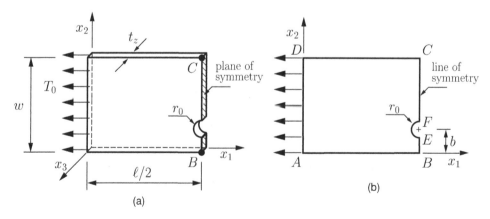

Figure 4.6 (a) Domain, Example 4.3.2; (b) planar model, Example 4.3.3.

plane of symmetry the normal displacement component u_n and the shearing stress components are zero.

Setting $u_n = u_1 = 0$ on the plane of symmetry prevents displacement in the x_1 direction and rotations about the x_2- and x_3-axes, but does not prevent displacements in the x_2 and x_3 coordinate directions, nor rotation about the x_1-axis. One may set

$$u_2^{(B)} = u_3^{(B)} = u_3^{(C)} = 0$$

to prevent rigid body displacements not prevented by the plane of symmetry.

Remark 4.3.2 In planar problems the mid-plane is treated as a plane of symmetry, hence there are at most three rigid body displacement modes: the in-plane displacement components and rotation about an arbitrary normal to the mid-plane. The problem in Example 4.3.2 can be formulated as a planar problem. In that case the implied zero normal displacement of the mid-plane prevents displacement in the x_3 direction as well as rotation about the x_1- and x_2-axes. Therefore, only rigid body constraint has to be imposed, for example, by letting $u_2^{(A)} = 0$. Note that in Figure 4.6(b) the coordinate axes x_1, x_2 are understood to lie in the mid-plane.

Exercise 4.3.2 Refer to Figure 4.5. Relocate point A to $\{0\ 0\ 0\}$ and point C to $\{\ell\ w\ 0\}$. Assume that the following constraints are specified:

$$u_1^{(A)} = u_2^{(A)} = u_3^{(A)} = u_2^{(B)} = u_3^{(B)} = u_3^{(C)} = 0.$$

Write down the system of equations analogous to Equation (4.25) and determine whether the coefficient matrix is of full rank.

Exercise 4.3.3 Refer to the notation in Figure 4.7. The stress resultants acting on the circular boundary of a cylindrical body at $z = 0$ are the force vector components denoted by F_x, F_y, F_z and the moment vector components denoted M_x, M_y, M_z.

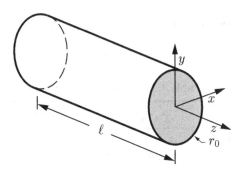

Figure 4.7 Notation for Exercise 4.3.3.

Using formulas available in introductory texts on the strength of materials, the traction vector components corresponding to F_y, F_z, M_x and M_z acting on the circular boundary at $z = 0$ are

$$T_x(x, y) = -\frac{2M_z y}{\pi r_0^4} \qquad (4.26)$$

$$T_y(x, y) = \frac{4F_y}{3\pi r_0^2}(1 - (y/r_0)^2) + \frac{2M_z x}{\pi r_0^4} \qquad (4.27)$$

$$T_z(x, y) = \frac{F_z}{\pi r_0^2} + \frac{4M_x y}{\pi r_0^4}. \qquad (4.28)$$

1. Assume that only the bending moment M_x and the shearing force F_y are non-zero. Determine the traction vector components acting on the cross-section at $z = -\ell$ and, using symmetry and antisymmetry as appropriate, sketch the smallest domain on which this problem can be solved. Specify the necessary rigid body constraints.

2. Assume that only the bending moment M_x, the shearing force F_y and the twisting moment M_z are non-zero. Determine the traction vector components acting on the cross-section at $z = -\ell$ and, using symmetry and antisymmetry as appropriate, sketch the smallest domain on which this problem can be solved. Specify the necessary rigid body constraints.

3. Assume that the axial force F_z, the bending moment M_x, the twisting moment M_z and the shearing force F_y are non-zero. Determine the traction vector components acting on the cross-section at $z = -\ell$ and, using symmetry and antisymmetry as appropriate, sketch the smallest domain on which this problem can be solved. Specify the necessary rigid body constraints.

Example 4.3.3 Let us formulate the problem shown in Figure 4.6 as a plane stress problem. The solution domain is shown in Figure 4.6(b) and the boundary conditions are given in Table 4.2. The stress–strain law is given by Equation (3.59).

In this case the virtual work of internal stresses is

$$B(\mathbf{u}, \mathbf{v}) := \int_\Omega ([D]\{v\})^T [E][D]\{u\} t_z \, dx_1 dx_2$$

Table 4.2 Example 4.3.3: boundary conditions.

Segments	Description	Condition
AB, CD, EF	Free	$T_n = T_t = 0$
DA	Traction loading	$T_n = T_0$, $T_t = 0$
BE, FC	Symmetry	$u_n = T_t = 0$

where t_z is the thickness, matrix $[D]$ is a differential operator matrix and the material stiffness matrix $[E]$ is defined by Equation (3.59)

$$[D] := \begin{bmatrix} \dfrac{\partial}{\partial x_1} & 0 \\ 0 & \dfrac{\partial}{\partial x_2} \\ \dfrac{\partial}{\partial x_2} & \dfrac{\partial}{\partial x_1} \end{bmatrix} \quad [E] := \dfrac{E}{1-\nu^2} \begin{bmatrix} 1 & \nu & 0 \\ \nu & 1 & 0 \\ 0 & 0 & \dfrac{1-\nu}{2} \end{bmatrix}$$

and $\{v\} := \{v_1\ v_2\}^T$, $\{u\} := \{u_1\ u_2\}^T$. The virtual work of external forces is

$$F(\mathbf{v}) = -\int_{\Gamma_{DA}} T_0 v_1 t_z\, ds.$$

The space $E(\Omega)$ is defined by Equation (4.17). Since all of the prescribed displacements are zero, the space of admissible functions is the same as the space of test functions:

$$\tilde{E}(\Omega) = E^0(\Omega) := \{u_i \mid u_i \in E(\Omega),\ u_n = 0 \text{ on } \Gamma_{BE} \text{ and } \Gamma_{FC}\}.$$

The generalized formulation of this problem is: "Find $u_i \in \tilde{E}(\Omega)$ such that $B(\mathbf{u}, \mathbf{v}) = F(\mathbf{v})$ for all $v_i \in E^0(\Omega)$." The solution of this problem exists (because the equilibrium condition is satisfied) and is unique up to one rigid body displacement $u^{(2)}$. It can be shown that the solution is continuous on $\bar{\Omega}$, therefore it is possible to set $u^{(2)} = 0$ by letting $u_2 = 0$ at a point (e.g., at the point C in Figure 4.6).

The functions that lie in a finite element space are always continuous on $\bar{\Omega}$. Rigid body constraints are enforced by setting one or more components of the displacement vector at node point(s) to an arbitrary value, usually to zero. This is permissible provided that the equilibrium conditions are satisfied.

Let us now solve this problem with the following data: $\ell = 100$ mm, $w = 40$ mm, $t_z = 2$ mm, $r_0 = 3.5$ mm, $b = 10$ mm, $T_0 = 75$ MPa. The material properties are: $E = 70 \times 10^3$ MPa, $\nu = 0.35$. These material properties are typical for the aluminum alloy 6061-T6, the yield strength of which is 265 MPa. The goal of analysis is to determine the location and magnitude of maximum principal stress with a reasonable certainty that the relative error is not greater than 2%.

To solve this problem by the finite element method, subspaces for $E(\Omega)$ are constructed. In this example, a 10-element mesh, shown in the inset in Figure 4.8, was constructed. The polynomial degrees were uniformly distributed and ranged from $p = 1$ to $p = 8$ (trunk space)[2]. The solutions were performed with StressCheck. The convergence of the first principal stress with respect to the number of degrees of freedom (N) and its location are shown in Figure 4.8. It is seen that the first principal stress is virtually independent of N for $p \geq 6$. The estimate of the limit value is based on the assumption that the absolute value of the error in σ_1 is proportional to N^{-1}.

---

[2] The definition for the term "trunk space" is given in Section 5.2.1 in Chapter 5.

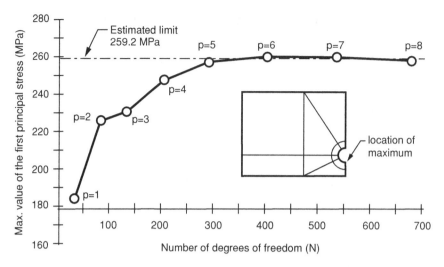

Figure 4.8 Example 4.3.3: convergence of the first principal stress.

Remark 4.3.3 The solution of the problem in Example 4.3.3 is available in standard engineering handbooks, see for example [60], [95]. The solution is based on the assumption that the length ℓ is large in comparison with the other dimensions so that it does not affect the solution. The stress distribution corresponding to the solution is independent of the material properties.

Example 4.3.4 In Example 4.3.2 it was possible to reduce the size of the problem by utilizing symmetry. In some cases it is possible to take advantage of antisymmetry. On a plane of antisymmetry the tangential components of the displacement vector and the normal stress are zero.

Consider, for example, an annular plate with internal radius r_i, external radius r_o, loaded by constant shearing tractions on the inner and outer surfaces, as shown in Figure 4.9. The thickness is constant. Any straight line passing through the origin is a line of antisymmetry.

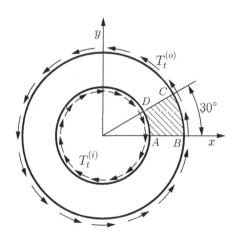

Figure 4.9 Annular plate for Example 4.3.4.

Table 4.3 Example 4.3.4: boundary conditions.

Segments	Description	Condition
AB, CD	Antisymmetry	$T_n = u_t = 0$
BC	Traction loading	$T_n = 0, \ T_t = T_t^{(o)}$
DA	Traction loading	$T_n = 0, \ T_t = T_t^{(i)}$
Point B	Rigid body constraint	$u_y^{(B)} = 0$

Therefore one can define the solution domain as the shaded sector $ABCD$ indicated in Figure 4.9. The solution domain is bounded by two lines of antisymmetry. Note that setting the tangential displacements to zero on boundary segments AB and CD does not prevent rigid body rotation about the origin. Therefore it is necessary to impose a rigid body constraint at one point (e.g., point B) in the circumferential direction. The boundary conditions are listed in Table 4.3.

Exercise 4.3.4 Consider the annular plate loaded by constant shearing tractions on the inner and outer surfaces, shown in Figure 4.9. Let $r_i = 50$ mm, $r_o = 100$ mm, The thickness is constant: $t_z = 5$ mm. The material properties are: $E = 200 \times 10^3$ MPa, $\nu = 0.3$. Let $T_t^{(i)} = 120$ MPa.

Determine the (constant) traction $T_t^{(o)}$ acting on the outer surface so that the tractions satisfy the static equilibrium equations and obtain finite element solutions using (a) a sequence of uniformly refined triangular elements with $p = 1$; (b) a sequence of uniformly refined quadrilateral elements with $p = 1$; and (c) one element using $p = 1, 2, \ldots, 8$. Plot the circumferential displacement of point A vs. the number of degrees of freedom.

Exercise 4.3.5 A rectangular plate of constant thickness is subjected to linearly varying normal tractions as shown in Figure 4.10. This is a classical problem of elasticity, representative of beams in pure bending. Taking advantage of symmetry and antisymmetry, define the boundary conditions for the domain $ABCD$. Check whether rigid body displacement is possible and specify the appropriate constraint(s) if necessary.

4.3.2 The principle of minimum potential energy

By definition the potential energy is the functional

$$\pi(\mathbf{u}) := \frac{1}{2} B(\mathbf{u}, \mathbf{u}) - F(\mathbf{u}). \tag{4.29}$$

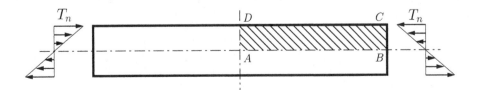

Figure 4.10 Plate for Exercise 4.3.5.

128   GENERALIZED FORMULATIONS

The principle of minimum potential energy states that the exact solution of the generalized formulation based on the principle of virtual work is the minimizer of the potential energy on the space of admissible functions:

$$\pi(\mathbf{u}_{EX}) = \min_{\mathbf{u} \in \tilde{E}(\Omega)} \pi(\mathbf{u}). \qquad (4.30)$$

The proof given by Theorem 2.6.2 in Section 2.6 is directly applicable to the problem of elasticity. This theorem is valid even if the definition of the potential energy is modified by an arbitrary constant. Specifically, referring to Equations (4.15) and (4.16), we define the potential energy for the problem of elasticity as follows:

$$\Pi(\mathbf{u}) := \frac{1}{2} \int_{\Omega} C_{ijkl}(\epsilon_{ij} - \alpha_{ij}T_\Delta)(\epsilon_{kl} - \alpha_{kl}T_\Delta) dV$$
$$+ \frac{1}{2} \int_{\partial\Omega_s} k_{ij}(u_i - d_i)(u_j - d_j) dS$$
$$- \int_{\Omega} F_i u_i \, dV - \int_{\partial\Omega_T} T_i u_i \, dS. \qquad (4.31)$$

The advantage of this definition over the definition given by Equation (4.29) is that in the special cases when a free body is subjected to a temperature change ($\epsilon_{ij} = \alpha_{ij}T_\Delta$), or a body with a spring boundary condition is given a rigid body displacement ($u_i = d_i$), then $\Pi(\mathbf{u}) = 0$, whereas $\pi(\mathbf{u}) \neq 0$.

In the finite element method $\tilde{E}(\Omega)$ is replaced by a finite-dimensional subspace $\tilde{S}$:

$$\pi(\mathbf{u}_{FE}) = \min_{\mathbf{u} \in \tilde{S}(\Omega)} \pi(\mathbf{u}). \qquad (4.32)$$

In Example 4.3.3 a sequence of hierarchic finite element spaces was used with the polynomial degree p ranging from 1 to 8. When a sequence of spaces is hierarchic (i.e., $S_1 \subset S_2 \subset \ldots$) then the potential energy converges monotonically.

Exercise 4.3.6 Compare the definition of $\pi(\mathbf{u})$ given by Equation (4.29) with the definition of $\Pi(\mathbf{u})$ given by Equation (4.31) and show that the two definitions differ by a constant, defined as follows:

$$\Pi(\mathbf{u}) - \pi(\mathbf{u}) = \frac{1}{2} \int_{\Omega} C_{ijkl} \alpha_{ij} \alpha_{kl} T_\Delta^2 \, dV + \frac{1}{2} \int_{\partial\Omega_s} k_{ij} d_i d_j \, dS.$$

Exercise 4.3.7 Show that for isotropic elastic materials with $\nu \neq 0$, $\Pi(\mathbf{u})$ can be written in the following form:

$$\Pi(\mathbf{u}) := \frac{1}{2} \int_{\Omega} \left[\lambda \left(\epsilon_{kk} - \frac{1+\nu}{\nu} \alpha T_\Delta \right)^2 + 2G \epsilon_{ij} \epsilon_{ij} - \frac{E}{\nu} (\alpha T_\Delta)^2 \right] dV$$
$$+ \frac{1}{2} \int_{\partial\Omega_s} k_{ij}(u_i - d_i)(u_j - d_j) \, dS - \int_{\Omega} F_i u_i \, dV - \int_{\partial\Omega_T} T_i u_i \, dS \qquad (4.33)$$

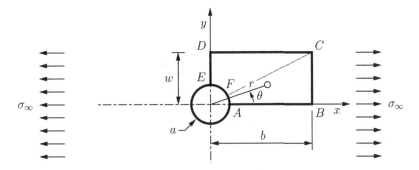

Figure 4.11 Circular hole in an infinite plate subjected to unidirectional tension (σ_∞).

and verify that if an unconstrained elastic body is subjected to a temperature change T_Δ (i.e., $\epsilon_{ij} = \alpha T_\Delta \delta_{ij}$) then $\Pi(\mathbf{u}) = 0$.

Example 4.3.5 In this example we illustrate the performance of the error estimator based on extrapolation on the basis of a two-dimensional problem, the exact solution of which is an analytic function. We will be concerned with the solution in the neighborhood of a circular hole in an infinite plate subjected to unidirectional tension, shown schematically in Figure 4.11.

We "remove" the domain $ABCDE$ shown in Figure 4.11 from the infinite plate and impose on edges BC and CD the stress distribution of the classical solution for the infinite plate:[3]

$$\sigma_x = \sigma_\infty \left[1 - \frac{a^2}{r^2} \left(\frac{3}{2} \cos 2\theta + \cos 4\theta \right) + \frac{3}{2} \frac{a^4}{r^4} \cos 4\theta \right] \quad (4.34)$$

$$\sigma_y = \sigma_\infty \left[-\frac{a^2}{r^2} \left(\frac{1}{2} \cos 2\theta - \cos 4\theta \right) - \frac{3}{2} \frac{a^4}{r^4} \cos 4\theta \right] \quad (4.35)$$

$$\tau_{xy} = \sigma_\infty \left[-\frac{a^2}{r^2} \left(\frac{1}{2} \sin 2\theta + \sin 4\theta \right) + \frac{3}{2} \frac{a^4}{r^4} \sin 4\theta \right] \quad (4.36)$$

where σ_∞ is the uniaxial stress in the plate; a is the radius of the hole; and r and θ are polar coordinates, defined as shown in Figure 4.11. On boundary segments AB and DE symmetry conditions are imposed. The circular arc EA is stress free. We select $b = w = 4a$ and compute the solution for plane strain conditions with $\nu = 0.3$.

The components of the displacement vector, with the rigid body displacement and the rigid body rotation terms set to zero, are

$$u_x = \frac{\sigma_\infty a}{8G} \left[\frac{r}{a}(\kappa + 1)\cos\theta + 2\frac{a}{r}((1+\kappa)\cos\theta + \cos 3\theta) - 2\frac{a^3}{r^3}\cos 3\theta \right] \quad (4.37)$$

$$u_y = \frac{\sigma_\infty a}{8G} \left[\frac{r}{a}(\kappa - 3)\sin\theta + 2\frac{a}{r}((1-\kappa)\sin\theta + \sin 3\theta) - 2\frac{a^3}{r^3}\sin 3\theta \right] \quad (4.38)$$

---

[3] See, for example, [45].

where G is the modulus of rigidity and κ is a constant which depends only on Poisson's ratio ν:

$$\kappa := \begin{cases} \dfrac{3-\nu}{1+\nu} & \text{for plane stress} \\ 3-4\nu & \text{for plane strain.} \end{cases} \quad (4.39)$$

Because we know the exact solution, the exact value of the potential energy can be computed. Noting that $\mathbf{u}_{EX} \in E^0(\Omega)$,

$$B(\mathbf{u}_{EX}, \mathbf{u}_{EX}) = F(\mathbf{u}_{EX})$$

and the exact value of the potential energy is

$$\Pi(\mathbf{u}_{EX}) = \frac{1}{2}B(\mathbf{u}_{EX}, \mathbf{u}_{EX}) - F(\mathbf{u}_{EX}) = -\frac{1}{2}F(\mathbf{u}_{EX}) = -\|\mathbf{u}_{EX}\|_E^2. \quad (4.40)$$

Therefore $\Pi(\mathbf{u}_{EX})$ can be determined from computing $F(\mathbf{u}_{EX})$:

$$F(\mathbf{u}_{EX}) = \int_0^w (\sigma_x u_x + \tau_{xy} u_y) t_z \, dy + \int_0^b (\tau_{xy} u_x + \sigma_y u_y) t_z \, dx$$

$$= 15.3873 \frac{\sigma_\infty^2 a^2 t_z}{E} \quad (4.41)$$

where $w = b = 4a$ was used. This computation was performed with Mathematica.

The finite element computations were performed with StressCheck, using only two finite elements. The vertices of the finite elements are labeled $ABCF$ and $CDEF$ in Figure 4.11. Line CF is normal to the circle. The load vectors were computed from the tractions corresponding to the stress components given by Equations (4.34) through (4.36) using 14-point Gaussian quadrature on the element sides corresponding to boundary segments BC and CD.

The number of degrees of freedom N, the normalized potential energy computed from the finite element solution, the estimated and true rates of convergence 2β, the estimated and true relative errors in energy norm $(e_r)_E$, and the effectivity index θ are given for $\nu = 0.3$ in Table 4.4.

The estimated relative error in the energy norm was computed using the best available estimate of the exact value of the potential energy, that is, the sequence of potential energies computed for $p = 6$ to $p = 8$. The procedure is described in Section 2.7. For p ranging from 1 to 6 the estimated and true relative errors are very close. For $p = 7$ and $p = 8$ the estimates are not as accurate. This is because the estimate is based on the assumption that the error in the energy norm is proportional to $N^{-\beta}$. In this case, however, the error goes to zero faster than $N^{-\beta}$, therefore the error in the energy norm is overestimated, whereas the rate of convergence is underestimated. This point is addressed in Chapter 6.

Exercise 4.3.8 Consider the problem

$$-\Delta u = f(x, y) \quad \text{on } \bar{\Omega} = \{x, y \mid 0 \le x \le a, \ 0 \le y \le b\}$$

Table 4.4 Example 4.3.5: circular hole in an infinite plate. Plane strain, $v = 0.3$. Estimated and true relative errors in energy norm. Effectivity index (θ).

			β		$(e_r)_E$ (%)		
p	N	$\dfrac{\Pi(\mathbf{u}_{FE})E}{\sigma_\infty^2 a^2 t_z}$	Est.	True	Est.	True	θ
1	8	−7.367 67	—	—	20.59	20.59	1.00
2	20	−7.547 40	0.44	0.44	13.80	13.79	1.00
3	32	−7.620 09	0.73	0.73	9.80	9.78	1.00
4	48	−7.659 04	0.92	0.93	6.73	6.71	1.00
5	68	−7.678 76	1.19	1.21	4.44	4.40	1.01
6	92	−7.688 05	1.56	1.62	2.77	2.70	1.03
7	120	−7.691 65	1.77	1.94	1.73	1.61	1.07
8	152	−7.692 95	1.77	2.22	1.14	0.96	1.19
∞	∞	−7.693 65			0	0	

where $f(x, y) = \sin(m\pi x/a)\sin(n\pi y/b)$. The boundary condition is $u = 0$ on $\partial\Omega$. The exact solution of this problems is

$$u_{EX} = \frac{1}{\pi^2} \frac{a^2 b^2}{m^2 b^2 + n^2 a^2} \sin(m\pi x/a)\sin(n\pi y/b) \quad (4.42)$$

and the potential energy of the exact solution is

$$\Pi(u_{EX}) = -\frac{1}{8\pi^2} \frac{a^3 b^3}{m^2 b^2 + n^2 a^2}. \quad (4.43)$$

Let $m = 3$, $n = 2$. Using p-extension on uniform meshes with 1, 4, and 16 elements, tabulate the effectivity indices as in Table 4.4.

Exercise 4.3.9 Analyze the notched rectangular plate loaded by a shear force V and bending moments M_1, M_2 as shown in Figure 4.12. The material is 6061-T6 aluminum: $E = 69.6$ GPa, $v = 0.365$, $\sigma_{\text{yld}} = 276$ MPa. The plate is loaded by a shear force V and bending moments M_1, M_2 as shown.

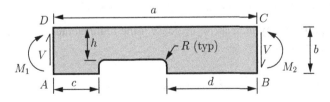

Figure 4.12 Notation for Exercise 4.3.9.

1. Define the solution domain. Define all dimensions as parameters.[4] For initial values use $a = 180$ mm, $b = 30$ mm, $c = 45$ mm, $d = 65$ mm, $h = 24$ mm, $R = 3$ mm, thickness $t_z = 4$ mm. Impose the necessary restrictions on the parameters (e.g., $c + d + 2R < a$).

2. Construct a finite element mesh manually. Observe how the mesh changes as you change the parameters.

3. Assign the thickness and the material properties.

4. Define formulas for the normal and shearing tractions on boundary segments BC and DA. Define V and M_1 as the independent load parameters.[5] Use $V = 250$ N and $M_1 = 10^4$ N mm. Hint: Define a local coordinate system, the x-axis of which is normal to BC and DA and passes through the center of the boundary segments:

$$T_n = -\frac{My}{I} \qquad T_t = -\frac{VQ}{It_z}$$

where T_n, T_t are the normal and tangential components of the traction vector and

$$I = \frac{b^3 t_z}{12} \qquad Q = Q(y) = \frac{1}{8}(b^2 - 4y^2)t_z.$$

Perform an equilibrium check after you have entered the data.

5. Impose rigid body constraints.

6. Obtain finite element solutions for $p = 1, 2, \ldots, 8$. First use the trunk space then switch to the product space.

7. Plot the first principal stress and check the dependence of the maximum value on the number of degrees of freedom.

8. Solve the problem again, but instead of manual meshing use the automatic mesh generator.

Exercise 4.3.10 The openings in the elastic rings shown in Figure 4.13 are to be closed (resp. aligned) and the two ends bonded. The objective is to determine the resulting stress distributions by finite element analysis. The thickness of the rings is 4.0 mm. Let $r_i = 135$ mm, $r_o = 150$ mm, $E = 200$ GPa, $\nu = 0.30$, $\alpha = 0.01$ rad and $\delta = 1$ mm.

1. Write down the generalized formulations based on the principle of virtual work. Utilize symmetry and antisymmetry as appropriate. Assume plane stress conditions.

2. Determine whether rigid body displacements are possible.

3. Will the strain energy increase or decrease if the finite element space is progressively enlarged? Explain.

---

[4] Note that e, t, r are among the reserved parameter names in StressCheck, so use other parameter names (e.g., th for thickness etc.)

[5] Note that M_2 is not an independent parameter. From equilibrium, $M_2 = aV + M_1$.

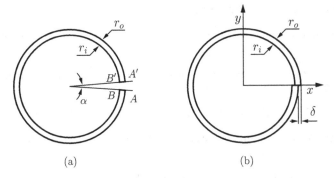

Figure 4.13 Elastic rings for Exercise 4.3.10.

4. Solve both problems using finite element software of your choice. Report the maximum and minimum principal stresses and show that the relative error is under 2.0%.

Exercise 4.3.11 A square panel with symmetrically located circular holes is subjected to constant shearing tractions as shown in Figure 4.14. Utilizing symmetry and antisymmetry, sketch the smallest domain on which this problem can be solved and specify the boundary conditions. Are rigid body constraints necessary? Explain.

4.4 Elastodynamic models

When the loads acting on a deformable body are functions of time then consideration of inertial forces may be necessary. The equilibrium equation is given by (3.48). If viscous damping is present then the equilibrium equation is

$$\sigma_{ij,j} + F_i = c_{ij}^{(d)} \frac{\partial u_j}{\partial t} + \varrho \frac{\partial^2 u_i}{\partial t^2}$$

where $c_{ij}^{(d)}$ represents the viscous damping coefficients. Time-dependent excitation and damping may be associated with the boundary conditions as well.

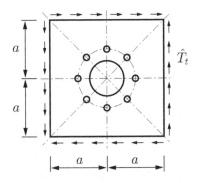

Figure 4.14 Panel for Exercise 4.3.11.

134   GENERALIZED FORMULATIONS

4.4.1 Undamped free vibration

Mathematical models of undamped free vibration of elastic bodies are based on the following assumptions:

1. The boundary conditions are homogeneous: $u_i = 0$ on $\partial\Omega_u$; $T_i = 0$ on $\partial\Omega_T$; and $d_i = 0$ on $\partial\Omega_s$. That is, $\tilde{E}(\Omega) = E^0(\Omega)$.

2. The inertial force is the only body force.

3. If spring boundary conditions are applied, the inertial forces associated with boundary springs are zero.

4. The effects of temperature change are negligible.

Given these assumptions, the generalized formulation (4.14) can be written as follows:

$$\int_\Omega C_{ijkl}\epsilon_{ij}^{(v)}\epsilon_{kl}\,dV + \int_{\partial\Omega_s} k_{ij}u_j v_i\,dS = -\int_\Omega \varrho\frac{\partial^2 u_i}{\partial t^2}v_i\,dV \qquad (4.44)$$

which holds for every t. The exact solution satisfies Equation (4.44), the homogeneous essential boundary conditions on $\partial\Omega_u$ and the initial conditions

$$u_i(x_j, 0) = f_i(x_j), \quad \left.\frac{\partial u_i}{\partial t}\right|_{(x_j,0)} = g_i(x_j) \qquad (4.45)$$

where $f_i \in E^0(\Omega)$ and $g_i \in E^0(\Omega)$.

Separation of variables

We now apply the separation of variables technique and write $u_i(x_j, t) = w_i(x_j)T(t)$ to obtain

$$T\left[\int_\Omega C_{ijkl}\epsilon_{ij}^{(v)}\epsilon_{kl}^{(w)}\,dV + \int_{\partial\Omega_s} k_{ij}w_j v_i\,dS\right] = -\frac{d^2 T}{dt^2}\int_\Omega \varrho w_i v_i\,dV$$

where

$$\epsilon_{kl}^{(w)} := \frac{1}{2}(w_{k,l} + w_{l,k}).$$

This can be written as

$$\frac{\int_\Omega C_{ijkl}\epsilon_{ij}^{(v)}\epsilon_{kl}^{(w)}\,dV + \int_{\partial\Omega_s} k_{ij}w_j v_i\,dS}{\int_\Omega \varrho w_i v_i\,dV} = -\frac{1}{T}\frac{d^2 T}{dt^2} = \omega^2. \qquad (4.46)$$

The expression on the left is a function of w_i and v_i, but not of t. The expression in the middle is a function of t. Therefore both expressions must be equal to some constant. This constant has to be a positive number for the following reason. By Assumption 1 the boundary conditions are homogeneous, therefore $\tilde{E}(\Omega) = E^0(\Omega)$. Since $v_i \in E^0(\Omega)$ is arbitrary, we may select $v_i = w_i$, in which case the expression on the left is a positive number that we set equal to ω^2. Therefore $T(t)$ satisfies the ordinary differential equation

$$\frac{d^2 T}{dt^2} + \omega^2 T = 0, \quad \text{the solution of which is} \quad T = A \cos \omega t + B \sin \omega t.$$

Similarly, w_i satisfies

$$\underbrace{\int_\Omega C_{ijkl} \epsilon_{ij}^{(v)} \epsilon_{kl}^{(w)} \, dV + \int_{\partial \Omega_s} k_{ij} w_j v_i \, dS}_{B(\mathbf{w},\mathbf{v})} - \omega^2 \underbrace{\int_\Omega \varrho w_i v_i \, dV}_{D(\mathbf{w},\mathbf{v})} = 0 \quad (4.47)$$

which we will write in shorthand as $B(\mathbf{w}, \mathbf{v}) - \omega^2 D(\mathbf{w}, \mathbf{v}) = 0$. This is a characteristic value problem. The set of numbers ω_1^2, ω_2^2, ... for which this equation has a non-trivial solution is called the spectrum. A particular value ω_k^2 for which a non-trivial solution exists is called a characteristic value or eigenvalue. The corresponding solution, denoted by $\mathbf{w}_k$, is called a characteristic function or eigenfunction, or mode of vibration. The eigenvalue ω_k, together with the corresponding eigenfunction $\mathbf{w}_k$, is called an eigenpair. The number ω_k is called the natural frequency measured in 1/s units. Commonly the cyclic frequency, defined by $f := \omega/2\pi$ (hertz), is reported.

Since the solution of Equation (4.47) lies in $E^0(\Omega)$ for every ω_k, the solution of Equation (4.44) can be written in the following form:

$$u_i(x_j, t) = \sum_{k=1}^\infty \left(A_k w_i^{(k)}(x_j) \cos \omega_k t + B_k w_i^{(k)}(x_j) \sin \omega_k t \right) \quad (4.48)$$

where the coefficients $A_1, A_2, \ldots$ and $B_1, B_2, \ldots$ are determined from the initial conditions, see Equation (4.45).

Properties of the exact solution

In the following the basic properties of the exact solution corresponding to the generalized formulation represented by Equation (4.47) are summarized. For details see Babuška and Osborn [8] and the references listed therein.

1. It can be shown that the generalized formulation, "Find $w_i \in E^0(\Omega)$ such that $B(\mathbf{w}, \mathbf{v}) - \omega^2 D(\mathbf{w}, \mathbf{v}) = 0$ for all $v \in E^0(\Omega)$," has an infinite number of solutions $(\omega_k^2, w_i^{(k)}(x_j))$ and all characteristic values are real. In the following the solutions will be numbered in ascending order of the characteristic values $\omega_1^2 \leq \omega_2^2 \leq \ldots$ and the characteristic functions will be normalized such that $D(\mathbf{w}^{(k)}, \mathbf{w}^{(k)}) = 1$.

2. When $\omega_i \neq \omega_j$ then the solutions satisfy the orthogonality relations

$$B(\mathbf{w}^{(i)}, \mathbf{w}^{(j)}) = \begin{cases} \omega_i^2 & \text{if } i = j \\ 0 & \text{if } i \neq j \end{cases} \quad (4.49)$$

and

$$D(\mathbf{w}^{(i)}, \mathbf{w}^{(j)}) = \begin{cases} 1 & \text{if } i = j \\ 0 & \text{if } i \neq j. \end{cases} \quad (4.50)$$

3. When two or more characteristic values are equal then the corresponding characteristic functions are linearly independent and any linear combination is also a characteristic function. Linear combinations can be constructed so that the orthogonality conditions are satisfied.

4. Any function $r_i(x_j) \in E^0(\Omega)$ can be written as

$$r_i(x_j) = \sum_{k=1}^{\infty} c_k w_i^{(k)}(x_j)$$

in the sense that

$$\lim_{n \to \infty} \left\| r_i - \sum_{k=1}^{n} c_k w_i^{(k)} \right\|_E = 0.$$

Making use of the orthogonality relations, the coefficients c_k can be computed from

$$B(\mathbf{r}, \mathbf{w}^{(j)}) = B\left(\sum_{k=1}^{\infty} c_k \mathbf{w}^{(k)}, \mathbf{w}^{(j)} \right) = c_j B\left(\mathbf{w}^{(j)}, \mathbf{w}^{(j)} \right) = c_j \omega_j^2$$

therefore

$$c_j = \frac{1}{\omega_j^2} B(\mathbf{r}, \mathbf{w}^{(j)}).$$

In this way the coefficients A_k and B_k in Equation (4.48) can be determined from the initial conditions (4.45).

5. The characteristic values minimize the Rayleigh quotient $R(\mathbf{u})$ defined by

$$R(\mathbf{u}) := \frac{B(\mathbf{u}, \mathbf{u})}{D(\mathbf{u}, \mathbf{u})} \quad (4.51)$$

in the following sense:

$$\omega_1^2 = \min_{\mathbf{u} \in E^0(\Omega)} R(\mathbf{u}) = R(\mathbf{u}_1)$$

$$\omega_k^2 = \min_{\mathbf{u} \in E_k^0(\Omega)} R(\mathbf{u}) = R(\mathbf{u}_k)$$

where $E_k^0(\Omega)$ is the space $E^0(\Omega)$ "purified" of the characteristic functions $\mathbf{u}_1, \mathbf{u}_2, \ldots, \mathbf{u}_{k-1}$ such that

$$E_k^0(\Omega) : \{\mathbf{u} \mid \mathbf{u} \in E^0(\Omega), \ B(\mathbf{u}, \mathbf{u}_i) = 0, \ i = 1, 2, \ldots, k-1\}.$$

6. When $\mathbf{u}$ is a rigid body displacement mode then $B(\mathbf{u}, \mathbf{u}) = 0$ but of course $D(\mathbf{u}, \mathbf{u}) \neq 0$. Therefore, when an elastic body is not fully constrained against rigid body motion there will be one zero characteristic value corresponding to each unconstrained rigid body displacement mode.

7. It is good practice to exploit symmetry and antisymmetry in the formulation of stationary problems whenever possible. In vibration problems not all vibration modes are symmetric even when the domain is symmetric. If the available computational resources permit then the solution should be performed on the entire domain. Alternatively, all combinations of symmetry and antisymmetry boundary conditions should be investigated.

Exercise 4.4.1 Consider an elastic body which is not constrained against rigid body motion, that is, $E^0(\Omega) = E(\Omega)$. Show that there are six zero characteristic values.

Properties of the finite element solution

In the finite element method, the space $E^0(\Omega)$ is replaced by a finite-dimensional space $S^0 = S^0(\Omega, \Delta, \mathbf{p}, \mathbf{Q})$ resulting in a matrix eigenvalue problem

$$([K] - \omega^2[M])\{a\} = 0 \tag{4.52}$$

where $[K]$ is the stiffness matrix, the coefficients of which are $k_{ij} = B(\boldsymbol{\varphi}_i, \boldsymbol{\varphi}_j)$ and $[M]$ is the mass matrix, the coefficients of which are $m_{ij} = D(\boldsymbol{\varphi}_i, \boldsymbol{\varphi}_j)$ where $\boldsymbol{\varphi}_i$, $i = 1, 2, \ldots, N$, are the basis function vectors. Due to the symmetry of the forms B and D, both matrices are symmetric. The finite element solutions have the following properties:

1. The computed characteristic values are always greater than or equal to the corresponding exact characteristic values. In other words, the computed characteristic values converge from "above." This is a consequence of the property of characteristic values in that they minimize the Rayleigh quotient (4.51) on the finite element space S_k^0 which is defined analogously to $E_k^0(\Omega)$.

2. The accuracy of characteristic values depends on how well the exact characteristic functions can be approximated in the finite element space. In general, the lowest characteristic functions are approximated well, hence the error in the lowest characteristic values tends to be small. The error of approximation increases as the characteristic values increase. The error in the characteristic values is related to the error of approximation of the characteristic functions measured in the energy norm. It is of the same order as the energy norm of the error of the characteristic functions squared.

3. In finite element analysis, multiple characteristic values generally appear as distinct characteristic values, because the errors of approximation of the corresponding

138   GENERALIZED FORMULATIONS

characteristic functions tend to be different. As the finite element space is enlarged, these characteristic values will converge to the exact multiple value.

4. When an elastic body is not fully constrained against rigid body motion then the exact characteristic value corresponding to a rigid body mode is zero. However, due to various tolerances set in the numerical eigensolvers, the numerically computed value is generally not exactly zero. Some eigensolvers may even fail when the stiffness matrix $[K]$ is not positive definite. The computed characteristic values corresponding to rigid body modes are generally much smaller than the eigenvalue corresponding to the first deformation mode. Also, plotting the characteristic function makes it possible to distinguish between rigid body and deformation modes.

Example 4.4.1 Consider a beam with constant cross-section, as shown in Figure 4.15(a). Let $h = 70$ mm, $b = 80$ mm, $t_w = 10.5$ mm, $t_f = 8$ mm, $r_0 = 7$ mm. The length is 1200 mm. The material properties are $E = 2.0 \times 10^5$ MPa, $\nu = 0.295$, $\rho = 7.860 \times 10^{-9}$ N s^2/mm^4. The beam is simply supported, that is, $u_y = u_z = 0$ on both ends. There are two planes of symmetry.

In this problem an elastic body is constrained against displacement in the y and z directions and against rotation about all three coordinate axes, but not against displacement in the x direction. Therefore, there will be one zero characteristic value and the corresponding characteristic function is rigid body displacement along the x-axis. The finite element mesh, consisting of 14 hexahedral finite elements, is shown in Figure 4.15

The second computed natural frequency corresponds to the first deformation mode, shown in Figure 4.15(b). It is seen that this mode is asymmetric with respect to the xz plane, which is one of the planes of symmetry of the undeformed beam.

The convergence of $\omega_2/2\pi$ and the estimated relative errors are listed in Table 4.5 with respect to increasing polynomial degree on the finite element mesh shown in Figure 4.15(b). Since the eigenvalues minimize the Rayleigh quotient defined in Equation (4.51) on the finite element spaces, convergence is monotonic when a hierarchic sequence of finite element spaces is used. Therefore estimation of the error by extrapolation is possible, using the technique described in Section 2.7.

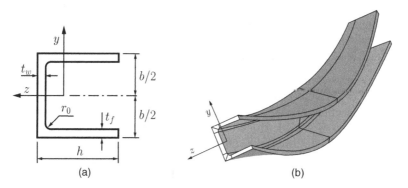

Figure 4.15 Example 4.4.1: simply supported channel beam. Finite element mesh consisting of 14 hexahedral elements. Vibration mode $\mathbf{w}^{(2)}$.

Table 4.5 Example 4.4.1: simply supported channel beam. Convergence of $\omega_2/2\pi$ (hertz).

p	N	$\omega_2/2\pi$	e_r (%)
3	492	1.17651179×10^2	3.24
4	865	1.14424638×10^2	0.41
5	1405	1.14053329×10^2	0.08
6	2154	1.13990184×10^2	0.03
7	3154	1.13971113×10^2	0.01
8	4447	1.13964393×10^2	0.00
∞	∞	1.13959652×10^2	0.00

Exercise 4.4.2 Estimate the first five natural frequencies and plot the corresponding vibration modes for the problem described in Example 4.4.1. Repeat the solution with the fillet omitted and compare the results. This example demonstrates that the omission of small geometric details does not significantly affect the results when the first few natural frequencies are of interest. In engineering practice fillets and other small features are often omitted in structural analysis in order to facilitate meshing and reduce the size of the problem.

Exercise 4.4.3 (a) Demonstrate the effects of rigid body constraints enforced on a vibrating plane elastic body by computing the eigenvalues with and without rigid body constraints. (b) Explain why rigid body constraints should not be enforced for dynamic problems through the imposition of point constraints as done for static problems. Hint: Are the six conditions represented by Equation (4.22) satisfied in a dynamic problem?

Exercise 4.4.4 A circular disk is attached to a tube. The generating section is shown in Figure 4.16. The axis of rotational symmetry is labeled AB.

Let us estimate the natural frequency of the first torsional mode. We can do this by idealizing the assembly such that the stiffness is associated with the tube and the mass is associated with the disk. In other words, the mass of the tube is neglected and the disk is considered to be perfectly rigid. Under this assumption we have the ordinary differential equation

$$\frac{GJ_t}{L_t}\phi + J_d T_d \varrho \frac{d^2\phi}{dt^2} = 0$$

where ϕ is the angle of rotation with reference to the cross-section labeled Γ_1, G is the shear modulus, ϱ is the mass density, J_t and J_d are the polar moments of inertia of the tube and the disk respectively, L_t is the length of the tube and T_d is the thickness of the disk. Then, by definition,

$$\omega^2 = \frac{GJ_t}{\varrho J_d T_d L_t} \quad \text{(rad/s)}.$$

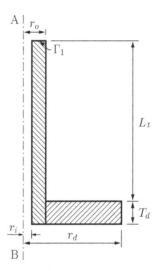

Figure 4.16 Notation for Exercise 4.4.4.

Let $r_i = 15$ mm, $r_o = 25$ mm, $L_t = 400$ mm, $r_d = 100$ mm, $T_d = 40$ mm, $E = 2.0 \times 10^5$ MPa, $\nu = 0.295$, $\varrho = 7.86 \times 10^{-9}$ Ns^2/mm^4. In this case we find $\omega = 1445$ rad/s (230 Hz).

1. Assume that the tube is fixed on the boundary Γ_1. Using the classical beam formulation, and assuming that the disk is a rigid mass, estimate the frequency of the first bending mode.

2. Construct a three-dimensional model of the tube and disk and estimate the first six natural frequencies and corresponding vibration modes by finite element analysis. Compare the estimated values of the first torsional and bending modes with the results obtained by the finite element method.

4.5 Incompressible materials

In Section 3.5 the problem of incompressible isotropic elastic materials was formulated resulting in Equations (3.76) and (3.77). The corresponding generalized formulation is based on the principle of virtual work (4.13) and the generalized statement of the incompressibility condition (3.77). Substituting the material law (3.75) into (4.13), we obtain

$$\int_\Omega \sigma_0 \epsilon_{ii}^{(v)} \, dV + \int_\Omega \frac{2E}{3}(\epsilon_{ij}\epsilon_{ij}^{(v)} - \alpha T_\Delta \epsilon_{ii}^{(v)}) \, dV = \int_\Omega F_i v_i \, dV + \int_{\partial\Omega} T_i v_i \, dS.$$

Since $\epsilon_{i,i}^{(v)} \equiv v_{i,i}$ this equation can be written as

$$\int_\Omega \sigma_0 v_{i,i} \, dV + \frac{2}{3}\int_\Omega E\epsilon_{ij}\epsilon_{ij}^{(v)} \, dV = \int_\Omega F_i v_i \, dV$$
$$+ \int_{\partial\Omega} T_i v_i \, dS + \frac{2}{3}\int_\Omega E\alpha T_\Delta v_{i,i} \, dV. \quad (4.53)$$

Let us now multiply Equation (3.77) by a scalar test function q and integrate:

$$\int_\Omega u_{i,i} q \, dV = 3 \int_\Omega \alpha T_\Delta q \, dV. \tag{4.54}$$

Equations (4.53) and (4.54) represent a generalized formulation which is qualitatively different from the one based on the principle of virtual work. The key difference is that the average normal stress, σ_0, is treated independently from the displacement components. For this reason it is called a mixed formulation.

It is customary to use the following abbreviated notation:

$$a(\mathbf{u}, \mathbf{v}) := \frac{2}{3} \int_\Omega E \epsilon_{ij} \epsilon_{ij}^{(v)} \, dV$$

$$b(\sigma_0, \mathbf{v}) := \int_\Omega \sigma_0 v_{i,i} \, dV$$

$$f_1(\mathbf{v}) := \int_\Omega F_i v_i \, dV + \int_{\partial\Omega} T_i v_i \, dS + \frac{2}{3} \int_\Omega E \alpha T_\Delta v_{i,i} \, dV$$

$$f_2(q) := 3 \int_\Omega \alpha T_\Delta q \, dV$$

and write Equations (4.53) and (4.54) in the following form:

$$a(\mathbf{u}, \mathbf{v}) + b(\sigma_0, \mathbf{v}) = f_1(\mathbf{v}) \tag{4.55}$$

$$b(\mathbf{u}, q) = f_2(q). \tag{4.56}$$

We need to define the spaces $H^1(\Omega)$ and $H^0(\Omega)$

$$H^1(\Omega) := \left\{ u_i \, \Big| \, \int_\Omega (u_{i,j} u_{i,j} + u_i u_i) \, dV \leq C < \infty \right\} \tag{4.57}$$

$$H^0(\Omega) := \left\{ u_i \, \Big| \, \int_\Omega u_i u_i \, dV \leq C < \infty \right\}, \tag{4.58}$$

the space of admissible functions

$$H^1_\star(\Omega) := \{ u_i \, | \, u_i \in H^1(\Omega), \, u_i = \hat{u}_i \text{ on } \partial\Omega_u \}$$

and the space of test functions

$$H^1_0(\Omega) := \{ u_i \, | \, u_i \in H^1(\Omega), \, u_i = 0 \text{ on } \partial\Omega_u \}.$$

The problem is to find $u_i \in H^1_\star(\Omega)$, $\sigma_0 \in H^0(\Omega)$ such that (4.55) and (4.56) are satisfied for all $v_i \in H^1_0(\Omega)$ and $q \in H^0(\Omega)$. Specific statements of the mixed formulation depend on the boundary conditions in the same way that statements of the principle of virtual work depend on the boundary conditions.

142   GENERALIZED FORMULATIONS

Remark 4.5.1 When displacement $u_i = \hat{u}_i$ is prescribed on the entire boundary then the prescribed displacement must satisfy the incompressibility constraint. Referring to (4.54), letting $q = 1$ and applying the divergence theorem we obtain

$$\int_{\partial\Omega} u_i n_i \, dS - 3 \int_\Omega \alpha T_\Delta \, dV = 0. \tag{4.59}$$

In this case σ_0 can be determined only up to an arbitrary constant.

Exercise 4.5.1 Show that when displacement $u_i = \hat{u}_i$ is prescribed on the entire boundary then σ_0 can be determined only up to an arbitrary constant.

4.5.1 The saddle point problem

In the displacement formulation it was shown that the exact solution minimizes the potential energy Π. In the mixed formulation the functions $\mathbf{u}$ and σ_0 that satisfy Equations (4.55) and (4.56) render the following functional stationary:

$$\mathcal{J}(\mathbf{u}, \sigma_0) := \frac{1}{2} a(\mathbf{u}, \mathbf{u}) - f_1(\mathbf{u}) + b(\mathbf{u}, \sigma_0) - f_2(\sigma_0), \quad (\mathbf{u}, \sigma_0) \in H^1_\star(\Omega) \times H^0(\Omega). \tag{4.60}$$

To show this, let us perturb $\mathcal{J}$ by the functions $\mathbf{v} \in H^1_0(\Omega)$ and $q \in H^0(\Omega)$:

$$\mathcal{J}(\mathbf{u} + \mu_1 \mathbf{v}, \sigma_0 + \mu_2 q) = \frac{1}{2} a(\mathbf{u}, \mathbf{u}) + \mu_1 a(\mathbf{u}, \mathbf{v}) + \mu_1^2 \frac{1}{2} a(\mathbf{v}, \mathbf{v}) - f_1(\mathbf{u}) - \mu_1 f_1(\mathbf{v})$$
$$+ b(\mathbf{u}, \sigma_0) + \mu_1 b(\mathbf{v}, \sigma_0) + \mu_2 b(\mathbf{u}, q) + \mu_1 \mu_2 b(\mathbf{v}, q) - f_2(\sigma_0)$$
$$- \mu_2 f_2(q)$$

where μ_1, μ_2 are arbitrary real numbers. By letting

$$\left.\frac{\partial \mathcal{J}}{\partial \mu_1}\right|_{\mu_1,\mu_2=0} = 0, \qquad \left.\frac{\partial \mathcal{J}}{\partial \mu_2}\right|_{\mu_1,\mu_2=0} = 0$$

Equations (4.55) and (4.56) are obtained. This is a saddle point problem, in the sense that the second derivative of $\mathcal{J}$ with respect to μ_1 is positive for all $\mathbf{v}$, but the second derivative of $\mathcal{J}$ with respect to μ_2 is zero and the mixed second derivative $b(\mathbf{v}, q)$ is indefinite.

4.5.2 Poisson's ratio locking

The mixed formulation was derived for the case of incompressible elasticity. The question may be asked: "What happens when the material is nearly incompressible, that is, ν is close to $1/2$, say $\nu = 0.4999$?" In that case, referring to Equation (4.33), λ is a large number in comparison with $2G$, and in effect minimization of the potential energy expression can be viewed as a constrained minimization problem where λ is the penalty term forcing

$$\left(e_{kk} - \frac{1 - \nu}{\nu} \alpha T_\Delta \right) \to \varepsilon(E, \nu)$$

where ε is a small number. This constraint in effect reduces the number of degrees of freedom, slowing convergence and making it difficult or even impractical to control the error of approximation. This phenomenon is called Poisson's ratio locking. The mixed formulation provides effective means for handling problems of elasticity when the material is nearly incompressible. Alternatively, the generalized formulation based on the principle of virtual work can be used in conjunction with high polynomial degrees. In that case, computation of the normal stresses requires special procedures [80], the discussion of which is beyond the scope of this book.

4.5.3 Solvability

In analyzing a finite element method, the following questions have to be addressed: "(a) Does the generalized solution exist and is it unique; and (b) does the finite element solution exist and is it unique?" A brief discussion follows.

Existence of the generalized solution

In Section 4.1.2 we have stated sufficient conditions that guarantee the existence of a solution $u \in \tilde{E}(\Omega)$ of the displacement formulation. We have shown that the solution is unique in the energy space. Because of the relationship between the bilinear form and the norm of $E(\Omega)$ (i.e., $B(u, u) = 2\|u\|_E^2$) the analysis was simple. It was essential that we were interested only in solutions that lie in $\tilde{E}(\Omega)$.

In the case of the mixed formulation the situation is more complicated because $a(u, u) + b(q, q)$ is not a norm. Nevertheless, it is possible to show that the solution (u_{EX}, q_{EX}), $u_{EX} \in H^1(\Omega)$, $q_{EX} \in H^0(\Omega)$ exists and is unique.

Existence of the finite element solution

In the displacement formulation convergence in the energy norm is assured if a sequence of functions $u_n \in S_n \subset \tilde{E}(\Omega)$ exits such that $\|u_{EX} - u_n\|_E \to 0$ as $n \to \infty$. Furthermore, by Theorem 2.4.2, $\|u_{EX} - u_{FE}\|_E \leq \|u_{EX} - u_n\|_E$.

In the case of the mixed formulation the situation is more complicated because it is possible to have sequences of functions $u_n \in S_n^{(u)} \subset H^1(\Omega)$ and $q_n \in S_n^{(q)} \subset H^0(\Omega)$ such that

$$\|u_{EX} - u_n\|_{H^1(\Omega)} + \|q_{EX} - q_n\|_{H^0(\Omega)} \to 0 \text{ as } n \to \infty$$

but the corresponding sequence of finite element solutions may diverge. In order to assure convergence of the finite element solutions of the mixed formulation, it is necessary to have a relationship between the finite element spaces $S_n^{(u)} \subset H^1(\Omega)$ and $S_n^{(q)} \subset H^0(\Omega)$. Analysis of the mixed formulation is beyond the scope of this book.

4.6 Chapter summary

A close analogy exists between the one-dimensional formulation and higher dimensional formulations, but there is an important difference: functions that lie in the energy space are continuous in one dimension but are not necessarily continuous in two and three dimensions. Consequently, concentrated forces and point constraints are admissible in one dimension but inadmissible in two and three dimensions except that point constraints may be used for the enforcement of rigid body constraints subject to the conditions that (a) the appropriate equilibrium conditions are satisfied and (b) in the locations of point constraints the exact solution is known to be continuous.

A clear understanding of the generalized formulation and its properties is an essential prerequisite to finite element analysis. A mathematical model is meaningful only when the corresponding exact solution is uniquely defined and the data are consistent with the formulation.

5

Finite element spaces

Finite element spaces are sets of continuous functions characterized by the finite element mesh Δ, a polynomial space defined on standard elements and the functions used for mapping the standard elements onto the elements of the mesh. We have seen a simple example of this in Section 2.5.3 where finite element spaces in one dimension were described. There the standard element was the interval $I = (-1, +1)$, the standard space, denoted by S^p, was a polynomial space of degree p, and the mapping function was the linear function given by Equation (2.60). Finite element spaces in two dimensions are denoted by

$$S = S(\Omega, \Delta, \mathbf{p}, \mathbf{Q})$$
$$:= \{\mathbf{u} \mid \mathbf{u} \in E(\Omega),\ \mathbf{u}(Q_x^{(k)}(\xi, \eta),\ Q_y^{(k)}(\xi, \eta)) \in S^{p_k}(\Omega_{\text{st}}),\ k = 1, 2, \ldots, M(\Delta)\} \quad (5.1)$$

where $\mathbf{p}$ and $\mathbf{Q}$ represent, respectively, the arrays of the assigned polynomial degrees and the mapping functions. The expression

$$\mathbf{u}(Q_x^{(k)}(\xi, \eta),\ Q_y^{(k)}(\xi, \eta)) \in S^{p_k}(\Omega_{\text{st}})$$

indicates that the basis functions defined on element Ω_k are mapped from the shape functions of a polynomial space defined on standard triangular and quadrilateral elements.

In this chapter the standard elements, the corresponding shape functions (i.e., standard basis function) and mapping functions are described for two- and three-dimensional problems.

5.1 Standard elements in two dimensions

Two-dimensional finite element meshes are composed of triangular and quadrilateral elements. The standard quadrilateral element will be denoted by $\Omega_{st}^{(q)}$ and the standard triangular element

146 FINITE ELEMENT SPACES

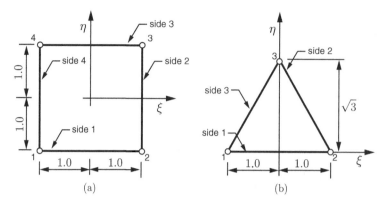

Figure 5.1 Standard quadrilateral and triangular elements $\Omega_{st}^{(q)}$ and $\Omega_{st}^{(t)}$.

by $\Omega_{st}^{(t)}$. The definition of standard elements is arbitrary; however, certain conveniences in mapping and assembly are achieved when the standard elements are defined as shown in Figure 5.1. Note that the sides of the elements have the same length (2) as in one dimension.

5.2 Standard polynomial spaces

The standard polynomial spaces in two and three dimensions are generalizations of the standard polynomial space $S^p(I_{st})$ defined in Section 2.5.2. Whereas the polynomial degree could be characterized by a single number (p) in one dimension, in higher dimensions more than one interpretation is possible.

5.2.1 Trunk spaces

Trunk spaces, also called "serendipity" spaces, are polynomial spaces spanned by the set of monomials $\xi^i \eta^j$, $i, j = 0, 1, 2, \ldots, p$, subject to the restriction $i + j = 0, 1, 2, \ldots, p$. In the case of quadrilateral elements these are supplemented by one or two monomials of degree $p + 1$:

1. Triangles. The dimension of the space is $n(p) = (p+1)(p+2)/2$. For example, the space $S^6(\Omega_{st}^{(t)})$ is spanned by the 28 monomial terms indicated in Figure 5.2. All polynomials of degree less than or equal to p are included in $S^p(\Omega_{st}^{(t)})$.

2. Quadrilaterals. Monomials of degree less than or equal to p are supplemented by $\xi\eta$ for $p = 1$ and by $\xi^p\eta$, $\xi\eta^p$ for $p \geq 2$. For example, the space $S^4(\Omega_{st}^{(q)})$ is spanned by the 17 monomial terms indicated in Figure 5.2. The dimension of space $S^p(\Omega_{st}^{(q)})$ is

$$n(p) = \begin{cases} 4p & \text{for } p \leq 3 \\ 4p + (p-2)(p-3)/2 & \text{for } p \geq 4. \end{cases} \quad (5.2)$$

All polynomials of degree less than or equal to p are included in $S^p(\Omega_{st}^{(q)})$.

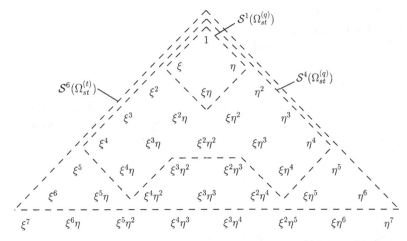

Figure 5.2 Trunk space. Illustration of spanning sets for $\mathcal{S}^1(\Omega_{st}^{(q)})$, $\mathcal{S}^4(\Omega_{st}^{(q)})$ and $\mathcal{S}^6(\Omega_{st}^{(t)})$.

5.2.2 Product spaces

In two dimensions product spaces are spanned by the monomials $1, \xi, \xi^2, \ldots, \xi^p$, $1, \eta, \eta^2, \ldots, \eta^q$ and their products. Thus the dimension of product spaces is $n(p,q) = (p+1)(q+1)$. Product spaces on triangles will be denoted by $\mathcal{S}^{p,q}(\Omega_{st}^{(t)})$ and on quadrilaterals by $\mathcal{S}^{p,q}(\Omega_{st}^{(q)})$. The spanning set of monomials for the space $\mathcal{S}^{4,2}(\Omega_{st}^{(q)})$ is illustrated in Figure 5.3.

5.3 Shape functions

As in the one-dimensional case, we will discuss two types of shape functions: shape functions based on Lagrange polynomials and hierarchic shape functions based on the integrals of

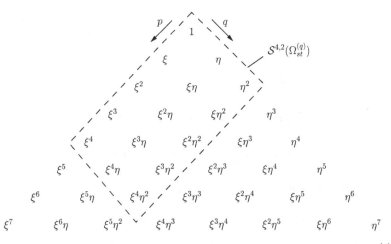

Figure 5.3 Product space. Illustration of spanning set for the space $\mathcal{S}^{4,2}(\Omega_{st}^{(q)})$.

Legendre polynomials. We will use the notation $N_i(\xi, \eta)$ $(i = 1, 2, \ldots, n)$ for both. The shape functions in two dimensions along the sides of each element are the same as the shape functions defined for the one-dimensional standard element I_{st}.

5.3.1 Lagrange shape functions

Elements with shape functions that span $\mathcal{S}^p(\Omega_{st}^{(t)})$ and $\mathcal{S}^p(\Omega_{st}^{(q)})$ with $p = 1$ and $p = 2$ are widely used in engineering practice. Each of the Lagrange shape functions is unity in one of the node points and zero in the other node points. Therefore the approximating function u can be written as

$$u(\xi, \eta) = \sum_{i=1}^{n} u_i N_i(\xi, \eta)$$

where u_i is the value of u in the ith node point.

Quadrilateral elements

The shape functions of four-node quadrilateral elements span the space $\mathcal{S}^1(\Omega_{st}^{(q)})$:

$$N_1(\xi, \eta) := \frac{1}{4}(1 - \xi)(1 - \eta) \tag{5.3}$$

$$N_2(\xi, \eta) := \frac{1}{4}(1 + \xi)(1 - \eta) \tag{5.4}$$

$$N_3(\xi, \eta) := \frac{1}{4}(1 + \xi)(1 + \eta) \tag{5.5}$$

$$N_4(\xi, \eta) := \frac{1}{4}(1 - \xi)(1 + \eta). \tag{5.6}$$

The shape functions of eight-node quadrilateral elements span the space $\mathcal{S}^2(\Omega_{st}^{(q)})$. The shape functions corresponding to the vertex nodes are

$$N_1(\xi, \eta) := \frac{1}{4}(1 - \xi)(1 - \eta)(-\xi - \eta - 1) \tag{5.7}$$

$$N_2(\xi, \eta) := \frac{1}{4}(1 + \xi)(1 - \eta)(\xi - \eta - 1) \tag{5.8}$$

$$N_3(\xi, \eta) := \frac{1}{4}(1 + \xi)(1 + \eta)(\xi + \eta - 1) \tag{5.9}$$

$$N_4(\xi, \eta) := \frac{1}{4}(1 - \xi)(1 + \eta)(-\xi + \eta - 1) \tag{5.10}$$

and the shape functions corresponding to the mid-side nodes are

$$N_5(\xi, \eta) := \frac{1}{2}(1 - \xi^2)(1 - \eta) \tag{5.11}$$

$$N_6(\xi, \eta) := \frac{1}{2}(1 + \xi)(1 - \eta^2) \tag{5.12}$$

$$N_7(\xi, \eta) := \frac{1}{2}(1 - \xi^2)(1 + \eta) \tag{5.13}$$

$$N_8(\xi, \eta) := \frac{1}{2}(1 - \xi)(1 - \eta^2). \tag{5.14}$$

Note that if we denote the coordinates of the vertices and the mid-points of the sides by (ξ_i, η_i), $i = 1, 2, \ldots, 8$, then $N_i(\xi_j, \eta_j) = \delta_{ij}$.

The shape functions of the nine-node quadrilateral element span $\mathcal{S}^{2,2}(\Omega_{st}^{(q)})$ (i.e., the product space). In addition to the four nodes located in the vertices and the four nodes located in the mid-points of the sides there is a node in the center of the element. Construction of these shape functions is left to the reader in the following exercise.

Exercise 5.3.1 Write down the shape functions for the nine-node quadrilateral element. Sketch the shape function associated with the node in the center of the element and one of the vertex shape functions and one of the side shape functions.

Triangular elements

The shape functions for triangular elements are usually written in terms of the triangular coordinates, defined as follows:

$$L_1 := \frac{1}{2}\left(1 - \xi - \frac{\eta}{\sqrt{3}}\right) \tag{5.15}$$

$$L_2 := \frac{1}{2}\left(1 + \xi - \frac{\eta}{\sqrt{3}}\right) \tag{5.16}$$

$$L_3 := \frac{\eta}{\sqrt{3}}. \tag{5.17}$$

Note that L_i is unity at node i and zero on the side opposite to node i. Also, $L_1 + L_2 + L_3 = 1$. The space $\mathcal{S}^1(\Omega_{st}^{(t)})$ is spanned by the following shape functions:

$$N_i := L_i, \quad i = 1, 2, 3. \tag{5.18}$$

These elements are called "three-node triangles." For the six-node triangles the shape functions are

$$N_1 := L_1(2L_1 - 1) \tag{5.19}$$

$$N_2 := L_2(2L_2 - 1) \tag{5.20}$$

$$N_3 := L_3(2L_3 - 1) \tag{5.21}$$

$$N_4 := 4L_1L_2 \tag{5.22}$$

$$N_5 := 4L_2L_3 \tag{5.23}$$

$$N_6 := 4L_3L_1 \tag{5.24}$$

which span $\mathcal{S}^2(\Omega_{st}^{(t)})$.

5.3.2 Hierarchic shape functions

Hierarchic shape functions based on the integrals of Legendre polynomials are described for the nodes, sides and vertices of quadrilateral and triangular elements. The shape functions associated with nodes and sides are the same for the product and trunk spaces. Only the number of internal shape functions is different.

Quadrilateral elements

The nodal shape functions are the same as those for the four-node quadrilateral, given by Equations (5.3) through (5.6).

The side shape functions are constructed by multiplying the shape functions $N_3, N_4, \ldots$, defined for the one-dimensional element (see Figure 2.9), by linear blending functions. We define

$$\phi_k(s) := \sqrt{\frac{2k-1}{2}} \int_{-1}^{s} P_{k-1}(t)\, dt, \quad k = 2, 3, \ldots. \tag{5.25}$$

Note that the index k represents the polynomial degree. The shape functions of degree $p \geq 2$ are defined for the four sides as follows:

$$\text{side 1:} \quad N_k^{(1)}(\xi, \eta) := \frac{1}{2}(1 - \eta)\phi_k(\xi) \tag{5.26}$$

$$\text{side 2:} \quad N_k^{(2)}(\xi, \eta) := \frac{1}{2}(1 + \xi)\phi_k(\eta) \tag{5.27}$$

$$\text{side 3:} \quad N_k^{(3)}(\xi, \eta) := \frac{1}{2}(1 + \eta)\phi_k(-\xi) \tag{5.28}$$

$$\text{side 4:} \quad N_k^{(4)}(\xi, \eta) := \frac{1}{2}(1 - \xi)\phi_k(-\eta) \tag{5.29}$$

where $k = 2, 3, \ldots, p$. Thus there are $4(p-1)$ side shape functions. The argument of ϕ_k is negative for sides 3 and 4 because the positive orientation of the sides is counterclockwise. This will affect shape functions of odd degrees only.

The internal shape functions are zero on the sides. For the trunk space there are $(p-2)(p-3)/2$ internal shape functions ($p \geq 4$) constructed from the products of ϕ_k:

$$N_p^{(k,l)}(\xi, \eta) := \phi_k(\xi)\phi_l(\eta), \quad k, l = 2, 3, \ldots, p, \quad k + l = 4, 5, \ldots, p. \tag{5.30}$$

The shape functions are assigned unique sequential numbers, as shown for example in Figure 5.4. For the product space there are $(p-1)(q-1)$ internal shape functions, defined for $p, q \geq 2$ by

$$N_{pq}^{(k,l)}(\xi, \eta) := \phi_k(\xi)\phi_l(\eta), \quad k = 2, 3, \ldots, p, \ l = 2, 3, \ldots, q, \ k + l \leq p + q. \tag{5.31}$$

Triangular elements

The nodal shape functions are the same as those for the three-node triangles, given by Equation (5.18). The side shape functions are constructed as follows: Define

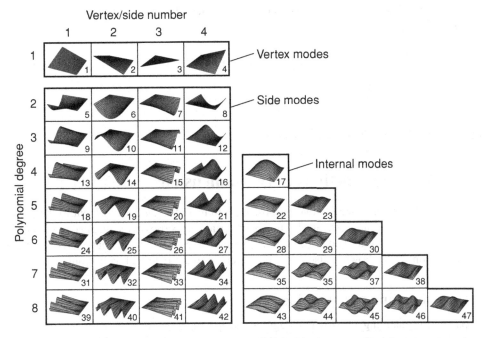

Figure 5.4 Hierarchic shape functions for quadrilateral elements. Trunk space, $p = 1$ to $p = 8$. First published in [80]. Reprinted with permission of John Wiley & Sons, Inc. ©1991.

$$\tilde{\phi}_k(s) := 4\frac{\phi_k(s)}{1 - s^2}, \quad k = 2, 3, \ldots \tag{5.32}$$

where $\phi_k(s)$ is the function defined by Equation (5.25). For example,

$$\tilde{\phi}_2(s) = -\sqrt{6}, \quad \tilde{\phi}_3(s) = -\sqrt{10}s, \quad \tilde{\phi}_4(s) = -\sqrt{\frac{7}{8}}(5s^2 - 1), \text{ etc.}$$

Using

$$s = \begin{cases} L_2 - L_1 & \text{for side 1} \\ L_3 - L_2 & \text{for side 2} \\ L_1 - L_3 & \text{for side 3} \end{cases} \tag{5.33}$$

the definition of the side shape functions is

$$\text{side 1:} \quad N_k^{(1)}(L_1, L_2, L_3) := L_1 L_2 \tilde{\phi}_k(L_2 - L_1) \tag{5.34}$$

$$\text{side 2:} \quad N_k^{(2)}(L_1, L_2, L_3) := L_2 L_3 \tilde{\phi}_k(L_3 - L_2) \tag{5.35}$$

$$\text{side 3:} \quad N_k^{(3)}(L_1, L_2, L_3) := L_3 L_1 \tilde{\phi}_k(L_1 - L_3) \tag{5.36}$$

where $k = 2, 3, \ldots, p$. Thus there are $3(p - 1)$ side shape functions.

For the trunk space there are $(p-1)(p-2)/2$ internal shape functions ($p \geq 3$) defined as follows:

$$N_p^{(k,l)} := L_1 L_2 L_3 P_k(L_2 - L_1) P_l(2L_3 - 1), \quad k, l = 0, 1, 2, \ldots, p-3 \qquad (5.37)$$

where $k + l \leq p - 3$ and P_k is the kth Legendre polynomial.

Exercise 5.3.2 Sketch the shape function $N_3^{(2)}(L_1, L_2, L_3)$.

5.4 Mapping functions in two dimensions

Mapping functions are used for two purposes: (a) to map the standard element to the corresponding elements of the mesh; and (b) to impart certain properties to the mapped polynomials so that they will be better suited for approximating some known local characteristics of the exact solution. The following discussion is concerned with the first objective only.

5.4.1 Isoparametric mapping

Isoparametric mapping utilizes the Lagrange shape functions described in Section 5.3.1. The most commonly used isoparametric mapping procedures are the linear and quadratic mappings.

Quadrilateral elements

Linear mapping of quadrilateral elements from the standard quadrilateral element shown in Figure 5.1(a) to the kth element is defined by

$$x = Q_x^{(k)}(\xi, \eta) := \sum_{i=1}^{4} N_i(\xi, \eta) X_i \qquad (5.38)$$

$$y = Q_y^{(k)}(\xi, \eta) := \sum_{i=1}^{4} N_i(\xi, \eta) Y_i \qquad (5.39)$$

where (X_i, Y_i) are the coordinates of vertex i of the kth element numbered in counterclockwise order and N_i are the shape functions defined by Equations (5.3) through (5.6).

Quadratic mapping of quadrilateral elements from the standard quadrilateral element is defined by

$$x = Q_x^{(k)}(\xi, \eta) := \sum_{i=1}^{8} N_i(\xi, \eta) X_i \qquad (5.40)$$

$$y = Q_y^{(k)}(\xi, \eta) := \sum_{i=1}^{8} N_i(\xi, \eta) Y_i \qquad (5.41)$$

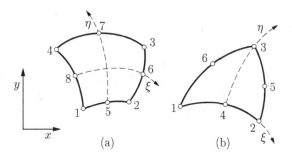

Figure 5.5 Isoparametric quadrilateral and triangular elements.

where (X_i, Y_i) are the coordinates of the four vertices numbered in counterclockwise order and the mid-points of the four sides also numbered in counterclockwise order. The side between nodes 1 and 2 is the first side. N_i are the shape functions defined by Equations (5.7) through (5.14). Quadratic isoparametric mapping of a quadrilateral element and typical numbering of the node points are illustrated in Figure 5.5(a).

Triangular elements

Linear mapping of triangles from the standard triangular element shown in Figure 5.1(b) to the kth element is defined by

$$x = Q_x^{(k)}(L_1, L_2, L_3) := \sum_{i=1}^{3} L_i X_i \qquad (5.42)$$

$$y = Q_y^{(k)}(L_1, L_2, L_3) := \sum_{i=1}^{3} L_i Y_i. \qquad (5.43)$$

Quadratic isoparametric mapping of triangular elements from the standard quadrilateral element is given by

$$x = Q_x^{(k)}(L_1, L_2, L_3) := \sum_{i=1}^{6} N_i(L_1, L_2, L_3) X_i \qquad (5.44)$$

$$y = Q_y^{(k)}(L_1, L_2, L_3) := \sum_{i=1}^{6} N_i(L_1, L_2, L_3) Y_i \qquad (5.45)$$

where N_i are the shape functions defined by Equations (5.19) through (5.24). The mapping of a triangular element and typical numbering of the node points are illustrated in Figure 5.5(b).

The term isoparametric mapping is meant to convey the idea that the same shape functions are used for providing topological descriptions for elements as for the element-level approximations. If the mapping is of lower (resp. higher) polynomial degree than the approximating functions then it is said to be subparametric (resp. superparametric).

Remark 5.4.1 The mapped shape functions are polynomials only in the special cases when the standard triangle is mapped into a straight-side triangle and when the standard quadrilateral element is mapped into a parallelogram. In general the mapped shape functions are not polynomials. Mapped shape functions are called pull-back polynomials. The accuracy of finite element approximation is governed by the properties of the pull-back polynomials. Depending on the exact solution, approximation by pull-back polynomials can be better or worse than approximation by polynomials. In some special cases the mapping is designed to improve the approximation.

Exercise 5.4.1 Show that quadratic parametric mapping applied to straight-side triangular and quadrilateral elements is identical to linear mapping.

5.4.2 Mapping by the blending function method

To illustrate the method, let us consider a simple case where only one side (side 2) of a quadrilateral element is curved, as shown in Figure 5.6. The curve $x = x_2(\eta)$, $y = y_2(\eta)$ is given in parametric form with $-1 \leq \eta \leq 1$. We can now write

$$x = \frac{1}{4}(1-\xi)(1-\eta)X_1 + \frac{1}{4}(1+\xi)(1-\eta)X_2 + \frac{1}{4}(1+\xi)(1+\eta)X_3$$
$$+ \frac{1}{4}(1-\xi)(1+\eta)X_4 + \left(x_2(\eta) - \frac{1-\eta}{2}X_2 - \frac{1+\eta}{2}X_3\right)\frac{1+\xi}{2}. \quad (5.46)$$

Observe that the first four terms in this expression are the linear mapping terms given by Equation (5.38). The fifth term is the product of two functions. One function, the bracketed expression, represents the difference between $x_2(\eta)$ and the x coordinates of the chord that connects points (X_2, Y_2) and (X_3, Y_3). The other function is the linear blending function $(1+\xi)/2$ which is unity along side 2 and zero along side 4. Therefore we can write

$$x = \frac{1}{4}(1-\xi)(1-\eta)X_1 + \frac{1}{4}(1-\xi)(1+\eta)X_4 + x_2(\eta)\frac{1+\xi}{2}. \quad (5.47)$$

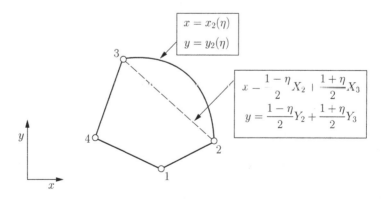

Figure 5.6 Quadrilateral element with one curved side.

Similarly,

$$y = \frac{1}{4}(1-\xi)(1-\eta)Y_1 + \frac{1}{4}(1-\xi)(1+\eta)Y_4 + y_2(\eta)\frac{1+\xi}{2}. \qquad (5.48)$$

In the general case all sides may be curved. We write the curved sides in parametric form:

$$x = x_i(s), \quad y = y_i(s), \quad -1 \le s \le +1 \quad \text{where } s = \begin{cases} \xi & \text{on sides 1 and 3} \\ \eta & \text{on sides 2 and 4} \end{cases}$$

and the subscripts represent the side numbers of the standard element. In this case the mapping functions are

$$x = \frac{1}{2}(1-\eta)x_1(\xi) + \frac{1}{2}(1+\xi)x_2(\eta) + \frac{1}{2}(1+\eta)x_3(\xi) + \frac{1}{2}(1-\xi)x_4(\eta)$$

$$- \frac{1}{4}(1-\xi)(1-\eta)X_1 - \frac{1}{4}(1+\xi)(1-\eta)X_2 - \frac{1}{4}(1+\xi)(1+\eta)X_3$$

$$- \frac{1}{4}(1-\xi)(1+\eta)X_4 \qquad (5.49)$$

$$y = \frac{1}{2}(1-\eta)y_1(\xi) + \frac{1}{2}(1+\xi)y_2(\eta) + \frac{1}{2}(1+\eta)y_3(\xi) + \frac{1}{2}(1-\xi)y_4(\eta)$$

$$- \frac{1}{4}(1-\xi)(1-\eta)Y_1 - \frac{1}{4}(1+\xi)(1-\eta)Y_2 - \frac{1}{4}(1+\xi)(1+\eta)Y_3$$

$$- \frac{1}{4}(1-\xi)(1+\eta)Y_4. \qquad (5.50)$$

The inverse mapping, that is, $\xi = Q_\xi^{(k)}(x, y)$, $\eta = Q_\eta^{(k)}(x, y)$, cannot be given explicitly in general, but (ξ, η) can be computed very efficiently for any given (x, y) by means of the Newton–Raphson method or a similar procedure.

Exercise 5.4.2 Refer to Figure 5.7(a). Show that the mapping of the quadrilateral element by the blending function method is

$$x = r_i \cos(\theta_m + \eta\theta_d)\frac{1-\xi}{2} + r_o \cos(\theta_m + \eta\theta_d)\frac{1+\xi}{2}$$

$$y = r_i \sin(\theta_m + \eta\theta_d)\frac{1-\xi}{2} + r_o \sin(\theta_m + \eta\theta_d)\frac{1+\xi}{2}$$

where

$$\theta_m := \frac{\theta_1 + \theta_2}{2}, \quad \theta_d := \frac{\theta_2 - \theta_1}{2}.$$

Exercise 5.4.3 Refer to Figure 5.7(b). A quadrilateral element is bounded by two circles. The centers of the circles are offset as shown. Write down the mapping by the blending

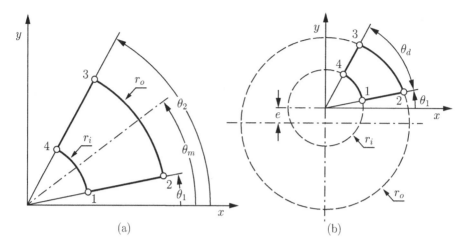

Figure 5.7 Quadrilateral elements bounded by circular segments.

function method in terms of the given parameters. Hint: Using the cosine law, the radius of the arc between node 2 and node 3 is

$$r_{2-3} = -e \sin\theta + \sqrt{r_o^2 - e^2 \cos^2\theta}$$

where θ is the angle measured from the x-axis.

5.4.3 Mapping of high-order elements

High-order elements are usually mapped by the blending function method with the bounding curves approximated by polynomial functions, similar to isoparametric mapping. The reasons for this are that (a) the boundary curves are generally not available in analytical form and (b) standard treatment of all bounding curves is preferable from the point of view of implementation. For example, the boundaries of a domain are typically represented by a collection of splines in computer-aided design (CAD) software products. In the interest of generality of implementation, the bounding curves are interpolated using the Lagrangian basis functions defined in Equation (2.64). The quality of the approximation depends on the choice of the interpolation points. Specifically, the interpolation points must be such that, for the given polynomial degree of interpolation, the interpolation function is close to the best possible approximation of the boundary curve by polynomials in maximum norm. The abscissas of the Lobatto points[1] are close to the optimal interpolation points. For details we refer to [19].

5.4.4 Rigid body rotations

In two-dimensional elasticity, infinitesimal rigid body rotation is represented by the displacement vector $\vec{u} = C\{-y\ x\}^T$. Having introduced the mappings $x = Q_x^{(k)}(\xi, \eta)$, $y = Q_y^{(k)}(\xi, \eta)$, infinitesimal rigid body rotations are represented exactly by element k only when $Q_x^{(k)}(\xi, \eta) \in$

---

[1] See Appendix B.

$S^{p_k}(\Omega_{st})$ and $Q_y^{(k)}(\xi, \eta) \in S^{p_k}(\Omega_{st})$. Iso- and subparametric mappings satisfy this condition, hence rigid body rotation imposed on an element will not induce strains. Superparametric mappings and mapping by the blending function method when the sides are not polynomials, or are polynomials of degree higher than p_k, do not satisfy this condition. It has been argued that for this reason only iso- and subparametric mappings should be employed.

This argument is flawed, however. One should view this question in the following light. Errors are introduced when rigid body rotations are not represented exactly and also by the approximation of boundary curves. With the blending function method analytic curves, such as circles, are represented exactly, but the rigid body rotation terms are approximated. With iso- and subparametric mappings the boundaries are approximated but the rigid body rotation terms are represented exactly. In either case, the errors of approximation will go to zero as the number of degrees of freedom is increased whether by mesh refinement or increasing the polynomial degrees. This is illustrated by the following exercises.

Exercise 5.4.4 Refer to Figure 5.7(a). Let $\theta_1 = 0$, $\theta_2 = 60°$, $r_i = 1.0$ and $r_o = 2.0$. Use $E = 200$ GPa, $\nu = 0.3$, plane strain. Impose nodal displacements consistent with rigid body rotation about the origin: $\mathbf{u} = C\{y - x\}$. For example, let $u_x^{(1)} = 0$, $u_y^{(1)} = Cr_i$, $u_y^{(2)} = Cr_o$ where the superscripts indicate the node numbers and C is the angle of rotation (in radians) about the positive z-axis. Let $C = 0.1$ and compute the maximum equivalent strain[2] for $p = 1, 2, \ldots, 6$. Very rapid convergence to zero will be seen.[3]

Exercise 5.4.5 Repeat Exercise 5.4.4 using uniform mesh refinement and p fixed at $p = 1$ and $p = 2$. Plot the maximum equivalent strain vs. the number of degrees of freedom.

Remark 5.4.2 The constant function is in the finite element space $S^{p_k}(\Omega_{st})$ independently of the mapping. Therefore rigid body displacements are represented exactly.

5.5 Elements in three dimensions

Three-dimensional finite element meshes are composed of hexahedral, tetrahedral and pentahedral elements; less frequently other types of elements, such as pyramid elements, are used. The standard hexahedral element, denoted by $\Omega_{st}^{(h)}$, is the set of points $-1 \leq \xi, \eta, \zeta \leq +1$. The standard tetrahedral element, denoted by $\Omega_{st}^{(th)}$, and the standard pentahedral element, denoted by $\Omega_{st}^{(p)}$, are shown in Figure 5.8. Note that the edges of the elements have the length 2.0, as in one and two dimensions.

The shape functions are analogous to those in one and two dimensions. For example, the eight-node hexahedral element has vertex shape functions such as

$$N_1 = \frac{1}{8}(1 - \xi)(1 - \eta)(1 - \zeta).$$

---

[2] The equivalent strain is proportional to the RMS root-mean-square (RMS) of the differences of principal strains and therefore it is an indicator of the maximum shearing strain.

[3] Curves are approximated by polynomials of degree 5 in StressCheck. This is a default value that can be changed by setting a parameter. When the default value is used, the mapping is superparametric for $p \leq 4$, isoparametric at $p = 5$ and subparametric for $p \geq 6$.

158   FINITE ELEMENT SPACES

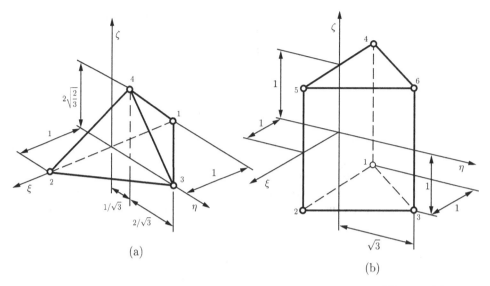

Figure 5.8 The standard tetrahedral and pentahedral elements $\Omega_{st}^{(th)}$ and $\Omega_{st}^{(p)}$.

The 20-node hexahedron is a generalization of the 8-node quadrilateral element to three dimensions. Similarly, the 4-node and 10-node tetrahedra are generalizations of the 3-node and 6-node triangles.

The hierarchic shape functions are also analogous to the shape functions defined in one and two dimensions. A detailed description of shape functions for hexahedral elements can be found in [26]. The shape functions associated with the edges and faces are the same along the edges and on the faces as in two dimensions. For example, the edge shape function of the pentahedral element, associated with the edge between nodes 1 and 2, corresponding to $p = 2$ is

$$N_7 = L_1 L_2 \tilde{\phi}_2(L_2 - L_1) \frac{1-\zeta}{2}$$

where $\tilde{\phi}_2$ is defined by Equation (5.32). The face shape function of the pentahedral element, associated with the face defined by nodes 1, 2 and 3, at $p = 3$ is

$$N_{13} = L_1 L_2 L_3 \frac{1-\zeta}{2}.$$

The internal shape functions of the pentahedral elements are the products of the internal shape functions defined for the triangular elements in Equation (5.37) and the function $\phi_k(\eta)$ defined by Equation (5.25).

5.6 Integration and differentiation

In Section 2.5.4, the coefficients of the stiffness matrix, Gram matrix and the right hand side vector were computed on the standard element. In two and three dimensions the corresponding procedures are analogous; however, with the exception of some important special cases, the mappings are generally nonlinear.

The mapping functions

$$x = Q_x^{(k)}(\xi, \eta, \zeta), \quad y = Q_y^{(k)}(\xi, \eta, \zeta), \quad z = Q_z^{(k)}(\xi, \eta, \zeta) \quad (5.51)$$

map a standard element Ω_{st} onto the kth element Ω_k. In the following we will drop the superscript when it is clear that we are referring to the mapping of the kth element. A mapping is said to be proper when the following three conditions are met: (a) the mapping functions Q_x, Q_y, Q_z are single-valued functions of ξ, η, ζ and possess continuous first derivatives; (b) the Jacobian determinant $|J|$ (defined below) does not vanish anywhere in Ω_{st}; and (c) $|J|$ is positive at every point of Ω_{st}. The mapping functions used in finite element analysis (FEA) must meet these criteria.

5.6.1 Volume and area integrals

Volume integrals on the kth element are computed on the corresponding standard element. The volume integral of a scalar function $F(x, y, z)$ on Ω_k is

$$\int_{\Omega_k} F(x, y, z)\, dx\, dy\, dz = \int_{\Omega_{st}} \mathcal{F}(\xi, \eta, \zeta) |J|\, d\xi\, d\eta\, d\zeta \quad (5.52)$$

where $\mathcal{F}(\xi, \eta, \zeta) := F(Q_x(\xi, \eta, \zeta), Q_y(\xi, \eta, \zeta), Q_z(\xi, \eta, \zeta))$ and $|J|$ is the determinant of the Jacobian matrix,[4] called the Jacobian determinant. The Jacobian determinant in Equation (5.52) arises from the definition of the differential volume. Let us denote the position vector of an arbitrary point P in the element $\bar{\Omega}_k$ by $\vec{r}$:

$$\vec{r} := x\vec{e}_x + y\vec{e}_y + z\vec{e}_z \quad (5.53)$$

where $\vec{e}_x, \vec{e}_y, \vec{e}_z$ are the orthonormal basis vectors of a right-handed Cartesian coordinate system. By definition, the differential volume is understood to be the scalar triple product:

$$dV := \left(\frac{\partial \vec{r}}{\partial x} dx \times \frac{\partial \vec{r}}{\partial y} dy\right) \cdot \frac{\partial \vec{r}}{\partial z} dz = (\vec{e}_x \times \vec{e}_y) \cdot \vec{e}_z\, dx\, dy\, dz = dx\, dy\, dz.$$

Given the change of variables of Equation (5.51), the analogous expression is

$$dV = \left(\frac{\partial \vec{r}}{\partial \xi} d\xi \times \frac{\partial \vec{r}}{\partial \eta} d\eta\right) \cdot \frac{\partial \vec{r}}{\partial \zeta} d\zeta = \begin{vmatrix} \dfrac{\partial x}{\partial \xi} & \dfrac{\partial y}{\partial \xi} & \dfrac{\partial z}{\partial \xi} \\ \dfrac{\partial x}{\partial \eta} & \dfrac{\partial y}{\partial \eta} & \dfrac{\partial z}{\partial \eta} \\ \dfrac{\partial x}{\partial \zeta} & \dfrac{\partial y}{\partial \zeta} & \dfrac{\partial z}{\partial \zeta} \end{vmatrix} d\xi\, d\eta\, d\zeta \equiv |J|\, d\xi\, d\eta\, d\zeta \quad (5.54)$$

---

[4] Carl Gustav Jacob Jacobi (1804–1851).

which was to be shown. The vectors $\partial \vec{r}/\partial \xi$, $\partial \vec{r}/\partial \eta$, $\partial \vec{r}/\partial \zeta$ are a set of right-handed basis vectors, that is, their scalar triple product yields a positive number. If the Jacobian determinant is negative then the right-handed coordinate system is transformed into a left-handed one, in which case the mapping is improper.

In two dimensions the mapping is

$$x = Q_x(\xi, \eta), \quad y = Q_y(\xi, \eta), \quad z = Q_z(\zeta) = \frac{t_z}{2}\zeta$$

where t_z is the thickness, see Figure 3.3. When the thickness is constant the integration in the transverse (ζ) direction can be performed explicitly and only an area integral needs to be evaluated:

$$dV = t_z\, dA = t_z \begin{vmatrix} \dfrac{\partial x}{\partial \xi} & \dfrac{\partial y}{\partial \xi} \\ \dfrac{\partial x}{\partial \eta} & \dfrac{\partial y}{\partial \eta} \end{vmatrix} d\xi\, d\eta. \qquad (5.55)$$

In the finite element method the integrations are performed by numerical quadrature, details of which are given in Appendix B. The minimum number of quadrature points depends on the polynomial degree of the shape functions: when the mapping is linear and the material properties are constant then the coefficients of the stiffness matrix should be exact (up to numerical round-off errors), otherwise the stiffness matrix may become singular. When the material properties vary or the mapping is nonlinear then the number of integration points should be increased so that the errors in the stiffness coefficients are small. Similar considerations apply to the load vector.

Exercise 5.6.1 Show that the Jacobian matrix of straight-side triangular elements, that is, triangular elements mapped by Equations (5.42) and (5.43), is independent of L_1, L_2 and L_3 and show that the area of a triangle in terms of its vertex coordinates (X_i, Y_i), $i = 1, 2, 3$, is

$$A = \frac{X_1(Y_2 - Y_3) + X_2(Y_3 - Y_1) + X_3(Y_1 - Y_2)}{2}.$$

Exercise 5.6.2 Show that for straight-side quadrilaterals the Jacobian determinant is constant only if the quadrilateral element is a parallelogram.

5.6.2 Surface and contour integrals

Given the mapping functions, each face is parameterized by imposing the appropriate restriction on the mapping function. For example, if integration is to be performed on the face of a hexahedron that corresponds to $\zeta = 1$ then, referring to Equation (5.51), the parametric form of the surface becomes

$$x = Q_x(\xi, \eta, 1), \quad y = Q_y(\xi, \eta, 1), \quad z = Q_z(\xi, \eta, 1) \qquad (5.56)$$

and, using the definition of $\vec{r}$ given in (5.53), the surface integral of a scalar function $F(x, y, z)$ is

$$\iint_{(\partial\Omega_k)_{\zeta=1}} F(x, y, z)\, dS = \int_{-1}^{+1}\int_{-1}^{+1} \mathcal{F}(\xi, \eta, 1) \left|\frac{\partial \vec{r}}{\partial \xi} \times \frac{\partial \vec{r}}{\partial \eta}\right| d\xi\, d\eta$$

where $\mathcal{F}$ is obtained from F by replacing x, y, z with the mapping functions Q_x, Q_y, Q_z. The treatment of the other faces is analogous.

In two dimensions the contour integral of a scalar function $F(x, y)$ on the side of a quadrilateral element corresponding to $\eta = 1$ is

$$\int_{(\partial\Omega_k)_{\eta=1}} F(x, y)\, ds = \int_{-1}^{+1} \mathcal{F}(\xi, 1) \left|\frac{d\vec{r}}{d\xi}\right| d\xi.$$

The other sides are treated analogously. The positive sense of the contour integral is counter-clockwise.

5.6.3 Differentiation

The inverse of the Jacobian matrix plays an important role in the computation of stiffness matrices and in post-processing operations. In particular, the approximating functions and hence the solution are known in terms of the shape functions defined on standard elements. Therefore differentiation with respect to x, y and z has to be expressed in terms of differentiation with respect to ξ, η and ζ. Using the chain rule we have

$$\begin{Bmatrix} \dfrac{\partial}{\partial \xi} \\ \dfrac{\partial}{\partial \eta} \\ \dfrac{\partial}{\partial \zeta} \end{Bmatrix} = \begin{bmatrix} \dfrac{\partial x}{\partial \xi} & \dfrac{\partial y}{\partial \xi} & \dfrac{\partial z}{\partial \xi} \\ \dfrac{\partial x}{\partial \eta} & \dfrac{\partial y}{\partial \eta} & \dfrac{\partial z}{\partial \eta} \\ \dfrac{\partial x}{\partial \zeta} & \dfrac{\partial y}{\partial \zeta} & \dfrac{\partial z}{\partial \zeta} \end{bmatrix} \begin{Bmatrix} \dfrac{\partial}{\partial x} \\ \dfrac{\partial}{\partial y} \\ \dfrac{\partial}{\partial z} \end{Bmatrix}. \tag{5.57}$$

On multiplying by the inverse of the Jacobian matrix, we obtain the expression used for computing the derivatives of the shape functions defined on standard elements:

$$\begin{Bmatrix} \dfrac{\partial}{\partial x} \\ \dfrac{\partial}{\partial y} \\ \dfrac{\partial}{\partial z} \end{Bmatrix} = \begin{bmatrix} \dfrac{\partial x}{\partial \xi} & \dfrac{\partial y}{\partial \xi} & \dfrac{\partial z}{\partial \xi} \\ \dfrac{\partial x}{\partial \eta} & \dfrac{\partial y}{\partial \eta} & \dfrac{\partial z}{\partial \eta} \\ \dfrac{\partial x}{\partial \zeta} & \dfrac{\partial y}{\partial \zeta} & \dfrac{\partial z}{\partial \zeta} \end{bmatrix}^{-1} \begin{Bmatrix} \dfrac{\partial}{\partial \xi} \\ \dfrac{\partial}{\partial \eta} \\ \dfrac{\partial}{\partial \zeta} \end{Bmatrix}. \tag{5.58}$$

Computation of the first derivatives from a finite element solution at a given point is discussed in Section 7.1.

5.7 Stiffness matrices and load vectors

The algorithms for the computation of stiffness matrices and load vectors for three-dimensional elasticity are outlined in the following. Their counterparts for two-dimensional elasticity and heat conduction are analogous. The algorithms are based on Equation (4.21); however, the integrals are evaluated element by element:

$$\int_{\Omega_k} ([D]\{v\})^T [E][D]\{u\} \, dV = \int_{\Omega_k} \{v\}^T \{F\} \, dV + \int_{\partial\Omega_k \cap \partial\Omega_T} \{v\}^T \{T\} \, dS$$
$$+ \int_{\Omega_k} ([D]\{v\})^T [E]\{\alpha\} T_\Delta \, dV \quad (5.59)$$

where the differential operator matrix $[D]$ and the material stiffness matrix $[E]$ are as defined in Exercise 4.3.1. The kth element is denoted by Ω_k. The second term on the right hand side represents the virtual work of tractions acting on boundary segment $\partial\Omega_T$. This term is present only when one or more of the boundary surfaces of the element lies on $\partial\Omega_T$. For the sake of simplicity in presentation we assume that the number of degrees of freedom is the same for all three fields on Ω_k. We denote the number of degrees of freedom per field by n and define the $3 \times 3n$ matrix $[N]$ as follows:

$$[N] = \begin{bmatrix} N_1 & N_2 & \cdots & N_n & 0 & 0 & \cdots & 0 & 0 & 0 & \cdots & 0 \\ 0 & 0 & \cdots & 0 & N_1 & N_2 & \cdots & N_n & 0 & 0 & \cdots & 0 \\ 0 & 0 & \cdots & 0 & 0 & 0 & \cdots & 0 & N_1 & N_2 & \cdots & N_n \end{bmatrix}.$$

The jth column of $[N]$, denoted by $\{N_j\}$, is the jth shape function vector. We write the trial and test functions as linear combinations of the shape function vectors:

$$\{u\} = \sum_{j=1}^{3n} a_j \{N_j\} \quad \text{and} \quad \{v\} = \sum_{i=1}^{3n} b_i \{N_i\}. \quad (5.60)$$

5.7.1 Stiffness matrices

The elements of the stiffness matrix k_{ij} can be written in the form

$$k_{ij}^{(k)} = \int_{\Omega_k} ([D]\{N_i\})^T [E][D]\{N_j\} \, dV. \quad (5.61)$$

We take advantage of the fact that two elements of $\{N_i\}$ are zero. The position of the zero elements depends on the value of the index i. For example, when $1 \leq i \leq n$ then

$$[D]\{N_i\} = \begin{Bmatrix} \partial/\partial x \\ 0 \\ 0 \\ \partial/\partial y \\ 0 \\ \partial/\partial z \end{Bmatrix} N_i = \underbrace{\begin{bmatrix} 1 & 0 & 0 \\ 0 & 0 & 0 \\ 0 & 0 & 0 \\ 0 & 1 & 0 \\ 0 & 0 & 0 \\ 0 & 0 & 1 \end{bmatrix}}_{[M_1]} \begin{Bmatrix} \partial/\partial x \\ \partial/\partial y \\ \partial/\partial z \end{Bmatrix} N_i = [M_1][J_k]^{-1} \begin{Bmatrix} \partial/\partial \xi \\ \partial/\partial \eta \\ \partial/\partial \zeta \end{Bmatrix} N_i$$

where $[J_k]^{-1}$ is the inverse of the Jacobian matrix corresponding to element k, see Equation (5.58) and $[M_1]$ is a logical matrix. When the index i changes, only $[M_1]$ has to be replaced. Specifically, when $(n+1) \leq i \leq 2n$ then $[M_1]$ is replaced by $[M_2]$; when $(2n+1) \leq i \leq 3n$ then $[M_1]$ is replaced by $[M_3]$. These are defined as follows:

$$[M_2] := \begin{bmatrix} 0 & 0 & 0 \\ 0 & 1 & 0 \\ 0 & 0 & 0 \\ 1 & 0 & 0 \\ 0 & 0 & 1 \\ 0 & 0 & 0 \end{bmatrix} \quad [M_3] := \begin{bmatrix} 0 & 0 & 0 \\ 0 & 0 & 0 \\ 0 & 0 & 1 \\ 0 & 0 & 0 \\ 0 & 1 & 0 \\ 1 & 0 & 0 \end{bmatrix}.$$

We define $\{\mathcal{D}\} := \{\partial/\partial \xi \; \partial/\partial \eta \; \partial/\partial \zeta\}^T$ and write Equation (5.61) in a form suitable for evaluation by numerical integration, which is described in Appendix B:

$$k_{ij}^{(k)} = \int_{\Omega_{st}} \left([M_\alpha][J_k]^{-1}\{\mathcal{D}\}N_i\right)^T [E][M_\beta][J_k]^{-1}\{\mathcal{D}\}N_j |J_k| \, d\xi d\eta d\zeta. \tag{5.62}$$

The domain of integration is the appropriate standard element, that is, hexahedral, tetrahedral or pentahedral element. The indices α and β take on the values 1, 2, 3 depending on range of the indices i and j. Therefore the element stiffness matrix $[K^{(k)}]$ consists of six blocks $[K_{\alpha\beta}^{(k)}]$:

$$[K^{(k)}] = \begin{bmatrix} [K_{11}^{(k)}] & [K_{12}^{(k)}] & [K_{13}^{(k)}] \\ & [K_{22}^{(k)}] & [K_{23}^{(k)}] \\ \text{sym.} & & [K_{33}^{(k)}] \end{bmatrix}. \tag{5.63}$$

Exercise 5.7.1 Assume that the mapping functions are given for the kth element. For example, in two dimensions

$$x = \alpha Q_x^{(k)}(\xi, \eta), \quad y = \alpha Q_y^{(k)}(\xi, \eta)$$

where $\alpha > 0$ is some real number. Assume further that the element stiffness matrix has been computed for $\alpha = 1$. How will the elements of the stiffness matrix change as functions of α in one, two and three dimensions? Assume that in two dimensions the thickness is independent of α.

Exercise 5.7.2 Refer to Equations (3.67) through (3.71). Develop an expression for the computation of the terms of the stiffness matrix, analogous to $k_{ij}^{(k)}$ given by Equation (5.62), for axisymmetric elastostatic models.

5.7.2 Load vectors

The computation of element-level load vectors corresponding to volume forces, surface tractions and thermal loading is based on the corresponding terms on the right hand side of Equation (5.59).

Volume forces

Computation of the load vector corresponding to volume force $\{F\}$ acting on element k is a straightforward application of Equation (5.52):

$$r_i^{(k)} = \int_{\Omega_{st}} \{N_i\}^T \{F\} |J_k| d\xi d\eta d\zeta \qquad i = 1, 2, \ldots, 3n. \qquad (5.64)$$

Surface tractions

Evaluation of the load vector terms corresponding to surface tractions depends on the side of the element on which the tractions are acting. For example, let us assume that traction vectors are acting on a hexahedral element on the face $\zeta = 1$. In this case the ith term of the load vector is

$$r_i^{(k)} = \int_{-1}^{+1} \int_{-1}^{+1} \{N_i\}^T \{T\} \left| \frac{\partial \vec{r}}{\partial \xi} \times \frac{\partial \vec{r}}{\partial \eta} \right|_{\zeta=1} d\xi d\eta$$

where the range of i is the set of indices of shape functions associated with the face $\zeta = 1$.

Thermal loading

The differential operator $[D]$ appears in the functional that represents thermal loading. Therefore the expression for $r_i^{(k)}$ depends on the index i:

$$r_i^{(k)} = \int_{\Omega_{st}} \left([M_\beta][J_k]^{-1}[\mathcal{D}]\{N_i\} \right)^T [E]\{\alpha\} T_\Delta |J_k| d\xi d\eta d\zeta \qquad i = 1, 2, \ldots, 3n$$

where $\beta = 1$ when $1 \leq i \leq n$, $\beta = 2$ when $(n+1) \leq i \leq 2n$ and $\beta = 3$ when $(2n+1) \leq i \leq 3n$. The matrices $[M_\beta]$ and the operator $\{\mathcal{D}\}$ are defined in Section 5.7.1.

5.8 Chapter summary

A finite element space is characterized by a finite element mesh and the polynomial degrees and mapping functions assigned to the elements of the mesh. The polynomial degrees identify a polynomial space defined on a standard element. The polynomial space is spanned by basis functions, called shape functions. Two kinds of shape functions, called Lagrange and hierarchic shape functions, were described for quadrilateral and triangular elements. The finite element space is spanned by the mapped shape functions subject to the requisite continuity requirements, discussed in Section 2.5.3.

Unless the mappings of all elements are polynomial functions of degree equal to or less than the polynomial degree of elements, rigid body rotation will not be represented exactly by the finite element solution. Nevertheless, rapid convergence to the correct solution will occur as the finite element space is progressively enlarged by h-, p-, or hp-extension. The mapping functions used in FEA must be such that the Jacobian determinant is positive at every point within the element.

6

Regularity and rates of convergence

We have seen in Section 2.4.2 (Theorem 2.4.2) that the finite element solution minimizes the error in the energy norm on the space $\tilde{S} \subset S(\Omega, \Delta, \mathbf{p}, \mathbf{Q})$. The error depends on the exact solution of the generalized formulation, the choice of the finite element mesh Δ, the polynomial degree(s) assigned to the elements $\mathbf{p}$ and the choice of the mapping functions $\mathbf{Q}$. In this chapter we address the question of how a priori information concerning the exact solution should be used in (a) defining the space S and (b) interpreting the results. A posteriori error estimation and adaptive techniques are briefly discussed at the end of the chapter.

6.1 Regularity

The degree of difficulty associated with approximating a function with polynomials depends on the size of its derivatives. We have seen an example of this in Section 2.6.5 where the error estimate (2.97) for approximation with piecewise polynomials of degree 1 was shown to depend on the absolute value of the second derivative of the exact solution.

If all derivatives of a function are bounded on domain $\bar{\Omega}$ then the function is analytic on $\bar{\Omega}$. For example, on $\bar{\Omega} = [0, 1]$ the function $u_1 = \sin m\pi x$ ($m = 1, 2, \ldots$) is an analytic function but $u_2 = \sqrt{x}$ and $u_3 = x \log x$ are not. However, u_2 and u_3 are analytic on $\bar{\Omega} = [a, b]$ where $0 < a < b < 1$. On comparing two functions, the function whose derivatives grow slower on Ω is the smoother function. In our example the absolute value of the kth derivative of $\sin m\pi x$ is bounded by $(m\pi)^k$ on $\Omega = (0, 1)$. Therefore if $n > m$ then $\sin m\pi x$ is smoother than $\sin n\pi x$, which is smoother than $\sqrt{x}$ for any n. The terms regularity and smoothness are used interchangeably.

The regularity of the exact solutions of various stationary problems is characterized by the following input data:

(a) The domain Ω.

(b) The boundary conditions, namely, Dirichlet, Neumann and Robin boundary conditions, or, equivalently, the displacements, tractions and spring boundary conditions.

(c) The material properties.

(d) The source term, that is, the heat source Q in Equation (3.11) and the volume forces F_i in Equation (3.51).

The input data are said to be admissible if the exact solution of the corresponding generalized formulation lies in the energy space. In the following we describe admissible data for two-dimensional stationary heat conduction and elasticity problems. In three dimensions the main points are similar but a detailed description would be more complicated.

The data are characterized by piecewise analytic functions[1] on domains and subdomains bounded by piecewise analytic curves.[2] A piecewise analytic curve is the union of a set of analytic arcs. The ends of the analytic arcs are called singular points. We assume that if two analytic arcs have a common point then the two arcs are either coincident or the common point is an isolated point.

Understanding the behavior of the exact solution in the neighborhood of singular points is important because the solution is analytic at every point that is not a singular point and hence all derivatives are bounded outside of the neighborhood of singular points. In the neighborhood of a singular point some derivatives may be bounded but not all derivatives are bounded [46a].

The goals of computation in heat conduction, elasticity and similar problems usually include the determination of functionals that are related to the first derivatives of the solution. For example, stress is related to strain by Hooke's law in the linear theory of elasticity. In many mathematical models the first derivatives of the exact solution are not finite in certain points; however, the first derivatives computed from the finite element solution are always finite. In such cases, reporting the maximum value of temperature gradients, strains and related data, such as fluxes or stresses, computed from the finite element solution as approximations to the corresponding functionals determined from the exact solution, would be erroneous and misleading. Therefore if the results of computation are to be interpreted correctly when the data of interest are functions of the derivatives then the question of whether the derivatives of the exact solution are bounded must be answered. This question is related to the regularity of the exact solution.

The following examples illustrate singular points associated with domains, boundary conditions, material properties and source functions. Singular points are indicated by open circles in the figures.

Example 6.1.1 The domain shown in Figure 6.1(a) consists of two curves Γ_1 and Γ_2. Curve Γ_1 is analytic whereas Γ_2 is the union of four analytic arcs: $\Gamma_2 = \cup_{j=1}^{4} \Gamma_{2j}$, hence it is piecewise analytic. In this case there are four geometric singular points.

The boundary of the domain shown in Figure 6.1(b) consists of one curve Γ_1 which is the union of six analytic arcs. There are seven geometric singular points. When $\alpha = 2\pi$ the arcs Γ_{15} and Γ_{16} are coincident and form a crack.

---

[1] The definition of analytic function is given in Appendix A, Section A.7.1.
[2] The definition of analytic curve is given in Appendix A, Section A.7.2.

REGULARITY 169

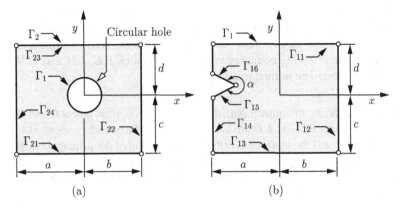

Figure 6.1 Typical geometric singular points associated with planar domains.

Example 6.1.2 Various commonly occurring boundary conditions are illustrated schematically in Figure 6.2(a). The prescribed boundary conditions and the associated coefficients (e.g., spring constants, radiation coefficients, etc.) are analytic functions prescribed on analytic arcs.

Referring to Figure 6.2(a), the boundary is a piecewise analytic curve, comprising 15 analytic arcs. Arc AB represents prescribed temperature or displacement boundary conditions;

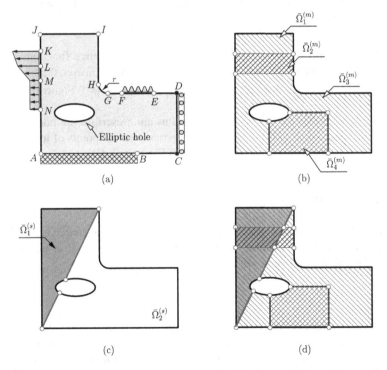

Figure 6.2 Typical singular points associated with (a) boundary conditions, (b) material interfaces, (c) source terms, (d) a combination of material interfaces and source terms.

arcs $BC, DE, FG, GH, HI, IJ, JK, ABNA$ and the boundary of the elliptic hole represent zero-flux or zero-traction boundary conditions; boundary segment CD represents symmetry boundary condition (zero flux or zero normal displacement, zero shearing traction); arc EF represents Robin or spring boundary conditions; and arcs KL, LM and MN represent prescribed flux or prescribed normal tractions.

Remark 6.1.1 Referring to Figure 6.2(a), the solution is analytic at points C and D, provided that boundary segments BC and ED are at right angles to the boundary segment CD and the material is isotropic, or if the material is orthotropic then one of the principal material axes is aligned with the boundary segment CD. When boundary segment CD is a line of symmetry then we can visualize the entire domain on which points C and D are not endpoints of analytic arcs and hence not singular points.

Example 6.1.3 The material properties are analytic functions on a set of closed subdomains $\bar{\Omega}_i \subset \Omega$ that collectively cover the entire domain. The boundaries of $\Omega_i^{(m)}$, denoted by $\partial \Omega_i^{(m)}$, are called material interfaces. Any point in $\partial \Omega$ is a point of a material interface.

The material properties, represented (for example) by $\kappa(x, y)$ and $c(x, y)$ in Equation (4.4), are defined separately on each closed subdomain $\bar{\Omega}_i$. In the points that lie on material interfaces $x \in \partial \Omega_i \cap \partial \Omega_j, i \neq j$, there may be a discontinuity in material properties, or their derivatives. In that case the coefficients are not analytic in those points; nevertheless $\kappa^{(i)}(x, y) \in \bar{\Omega}_i$ and $\kappa^{(j)}(x, y) \in \bar{\Omega}_j$ are analytic functions.

Figure 6.2(b) is a schematic representation of a domain that is the union of a set of subdomains on which three different material properties, represented by the different shading patterns, were defined.

Example 6.1.4 Analogously to the material properties, the source functions are analytic functions defined on a set of subdomains $\bar{\Omega}_i^{(s)}, i = 1, 2, \ldots, M_s$; however, the source functions are not analytic on the source interfaces $(x, y) \in \partial \Omega_i^{(s)} \cap \partial \Omega_j^{(s)}, i \neq j$. Figure 6.2(c) illustrates a case where $M_s = 2$.

When the subdomains on which source functions are prescribed are not coincident with the subdomains associated with the material properties then the points of intersection of the boundaries are singular points. This is illustrated in Figure 6.2(d).

6.2 Classification

The solution domain, boundary conditions, material properties and source terms characterize the regularity of the exact solution of the mathematical problem. It is useful to classify mathematical problems into three broad categories, as follows.

Category A

A mathematical problem is said to be in Category A when there are no singular points in $\bar{\Omega}$. In this case the exact solution u_{EX} of the problem is continuous on $\bar{\Omega}$ and analytic on every $\bar{\Omega}_i^{(m)}$ and $\bar{\Omega}_i^{(s)}$, but does not have to be analytic on $\bar{\Omega}$. An example is shown in Figure 4.1 where the domain is the union of three subdomains to which the material properties described in Example 4.1.1 are assigned. The exact solution is not analytic on the material interfaces, nevertheless the problem is in Category A.

A precise definition of problems in Category A is based on Equation (A.15). We define the set of points Ω_c that do not lie on interfaces

$$\Omega_c := \Omega - \cup_{i \neq j} \partial\Omega_i^{(m)} \cap \partial\Omega_j^{(m)} - \cup_{i \neq j} \partial\Omega_i^{(s)} \cap \partial\Omega_j^{(s)} - \cup_{i \neq j} \partial\Omega_i^{(m)} \cap \partial\Omega_j^{(s)}.$$

If there exist constants K and R, independent of $(x, y) \in \Omega_c$, such that for any s we have

$$\left| \frac{\partial^s u_{EX}}{\partial x^k \partial y^{s-k}} \right| \leq K R^s s! \qquad k = 0, 1, \ldots, s, \quad s = 1, 2, \ldots \tag{6.1}$$

then the exact solution u_{EX} can be approximated by a Taylor series in the neighborhood of any point in Ω_c and, furthermore, the Taylor series converges exponentially. Estimation of the size of the derivatives is important because the error term in a Taylor series approximation of degree p depends on the size of the derivative of degree $p + 1$. This is generally true for approximation by polynomials.

Example 6.2.1 Consider the problem $\Delta u = 0$ on the domain

$$\Omega = \{x, y \,|\, (x - 1 - d)^2 + y^2 < 1\}, \quad d > 0$$

shown in Figure 6.3. The boundary condition is

$$u(x, y) = \Re(z^{-1}) = \frac{x}{x^2 + y^2} \quad \text{on } \partial\Omega. \tag{6.2}$$

Since $u(x, y)$ satisfies $\Delta u = 0$, the exact solution is $u(x, y)$. The derivatives of u are bounded by

$$\left| \frac{\partial^s u}{\partial x^k \partial y^{s-k}} \right| \leq K\, s! \left(\frac{1}{d} \right)^{s+1}. \tag{6.3}$$

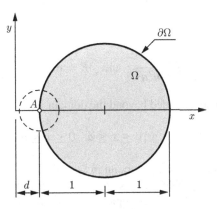

Figure 6.3 Definition of the domain for Example 6.2.1.

Note that the exact solution is analytic on $\bar{\Omega}$ and can be extended beyond $\bar{\Omega}$ through Taylor series expansion. For example, on a circle centered at point A, the radius of which is less than d, the Taylor series converges.

It is seen from Equation (6.2) that the size of the derivatives grows very rapidly for small values of d. This influences the design of the finite element mesh, which has to be much finer in the neighborhood of point A than elsewhere. This feature will be discussed further in Section 6.4.2.

Remark 6.2.1 We have seen in Example 6.2.1 that the exact solution of a problem in Category A can have very low regularity. It is demonstrated in Example 6.2.2 that in exceptional cases the exact solution of a problem in Category B can be an analytic function.

Category B

A problem is said to be in Category B when there are a finite number of singular points in $\bar{\Omega}$. Figures 6.1 and 6.2 represent various Category B problems. The neighborhood of a singular point (x_i, y_i) is denoted by $\omega_i^{(\epsilon)}$ and defined by

$$\omega_i^{(\epsilon)} = \{(x, y) \mid (x - x_i)^2 + (y - y_i)^2 \leq \epsilon^2\}.$$

The solutions of Category B problems satisfy Equation (6.1) on $\Omega_c - \cup \bar{\omega}_i^{(\epsilon)}, i = 1, 2, \ldots, m$. The behavior of solutions in ω_i will be discussed in Section 6.3.

Example 6.2.2 The problem in Exercise 4.3.8 is in Category B, nevertheless its exact solution given by Equation (4.42)

$$u_{EX}(x, y) = \frac{1}{\pi^2} \frac{a^2 b^2}{m^2 b^2 + n^2 a^2} \sin(m\pi x/a) \sin(n\pi y/b)$$

satisfies Equation (6.1) in all points of $\bar{\Omega}$:

$$\left| \frac{\partial^s u_{EX}}{\partial x^k \partial y^{s-k}} \right| \leq \frac{1}{\pi^2} \frac{a^2 b^2}{m^2 b^2 + n^2 a^2} \left(\frac{m\pi}{a} \right)^k \left(\frac{n\pi}{b} \right)^{s-k} \leq K R^s$$

where

$$K = \frac{1}{\pi^2} \frac{a^2 b^2}{m^2 b^2 + n^2 a^2} \quad \text{and} \quad R = \max \left(\frac{m\pi}{a}, \frac{n\pi}{b} \right).$$

Note that the solution can be extended beyond the domain

$$\bar{\Omega} = \{x, y \mid 0 \leq x \leq a, \ 0 \leq y \leq b\}$$

by Taylor series approximation about any point in $\bar{\Omega}$.

The solution u_{EX} of the problem in Exercise 4.3.8 corresponds to the source function $f = \sin(m\pi x/a) \sin(n\pi y/b)$. If, for example, $f = 1$ were given then the corresponding exact solution would have singularities in the corner points of Ω. This example illustrates that when a problem is in Category B its solution may still satisfy Equation (6.1); however, these are exceptional cases.

Category C

When the exact solution is not in Category A or Category B then it is in Category C.

Remark 6.2.2 We have described admissible data for two-dimensional stationary heat conduction and elasticity problems. We have noted that in three dimensions the data are similar but a description of the categories would be more complicated. This is because in three dimensions there are not only singular points but also singular arcs.

6.3 The neighborhood of singular points

We denote the distance from singular point (x_i, y_i) by

$$r = \sqrt{(x - x_i)^2 + (y - y_i)^2}.$$

The regularity of a function $u = u(x, y)$ is characterized by the size of its derivatives. Specifically,

$$\left| \frac{\partial^s u}{\partial x^k \partial y^{s-k}} \right| \leq K \Phi(s, r), \quad r > 0 \tag{6.4}$$

where K is independent of r and s. A very important special case is when $\Phi(r)$ is of the form

$$\Phi(s, r) \leq r^{\lambda - s} s!, \quad \lambda > 0, \quad s = 1, 2, \ldots \tag{6.5}$$

where λ is a fractional number, called the degree of singularity. In the following we will discuss such special cases in connection with the Laplace equation and the equations of elasticity.

6.3.1 The Laplace equation

Let us consider solutions of the Laplace equation in the neighborhood of a corner point, such as point B in Figure 6.4(a),

$$\Delta u \equiv \frac{\partial^2 u}{\partial r^2} + \frac{1}{r} \frac{\partial u}{\partial r} + \frac{1}{r^2} \frac{\partial^2 u}{\partial \theta^2} = 0 \tag{6.6}$$

where r, θ are polar coordinates. In particular, let us seek solutions in the form $u = r^\lambda F(\theta)$ with $\lambda \neq 0$. Such solutions are typically associated with geometric singularities, boundary conditions and intersections of material interfaces. Substituting into Equation (6.6), we get

$$F'' + \lambda^2 F = 0$$

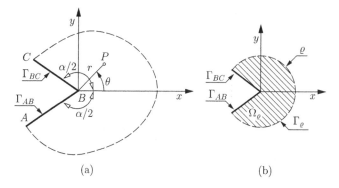

Figure 6.4 Reentrant corner: notation.

the general solution of which is

$$F = a \cos \lambda\theta + b \sin \lambda\theta \qquad (6.7)$$

where a, b are arbitrary constants. Therefore the solution is of the form

$$u = r^\lambda (a \cos \lambda\theta + b \sin \lambda\theta). \qquad (6.8)$$

Let us now consider the problem $u = 0$ on both boundary segments Γ_{AB} and Γ_{BC} (i.e., $\theta = \pm \alpha/2$):

$$a \cos \lambda\alpha/2 + b \sin \lambda\alpha/2 = 0$$
$$a \cos \lambda\alpha/2 - b \sin \lambda\alpha/2 = 0.$$

Adding and subtracting the two equations, we find

$$\cos \lambda\alpha/2 = 0 \quad \text{therefore} \quad \lambda\alpha/2 = \pm(2m-1)\pi/2, \quad m = 1, 2, \ldots \qquad (6.9)$$
$$\sin \lambda\alpha/2 = 0 \quad \text{therefore} \quad \lambda\alpha/2 = \pm n\pi, \quad n = 1, 2, \ldots. \qquad (6.10)$$

Note that the values of λ that satisfy Equations (6.9) and (6.10) can be either positive or negative. However, the function u, given by Equation (6.8), lies in the energy space only when $\lambda \geq 0$. Therefore, having excluded $\lambda = 0$ from consideration, we will be concerned with $\lambda > 0$ only. Denoting

$$\bar{\lambda}_m := \frac{(2m-1)\pi}{\alpha} \qquad \lambda_n^\star := \frac{2n\pi}{\alpha}$$

where $m, n = 1, 2, \ldots$, the solution can be written in the following form:

$$u = \sum_{m=1}^\infty a_m r^{\bar{\lambda}_m} \cos \bar{\lambda}_m \theta + \sum_{n=1}^\infty b_n r^{\lambda_n^\star} \sin \lambda_n^\star \theta, \quad r \leq r_c \qquad (6.11)$$

where r_c is the radius of convergence of the infinite series.

Observe that the first sum in Equation (6.11) is a symmetric function with respect to the *x*-axis, the second is an antisymmetric function. If the solution is symmetric (with respect to the *x*-axis) then $b_n = 0$; if it is antisymmetric then $a_m = 0$. Let us consider the symmetric part of the solution and assume that $a_1 \neq 0$ and $\bar{\lambda}_{\min} = \bar{\lambda}_1 = \pi/\alpha$ is not an integer. Then

$$\left|\frac{\partial^s u}{\partial x^k \partial y^{s-k}}\right| \leq K r^{\bar{\lambda}_1 - s} s! = K r^{\pi/\alpha - s} s!. \tag{6.12}$$

Therefore the degree of singularity is $\bar{\lambda}_1 = \pi/\alpha$. Note that when $\alpha = \pi/k$ where $k = 1, 2, \ldots$, the powers of r are integers and u is an analytic function. For other values of α the solution u is not analytic at the corner point and when $\alpha > \pi$ then the first derivative of u with respect to r is infinity at the corner point, provided that $a_1 \neq 0$.

Remark 6.3.1 The expressions (6.11) and (6.12) were derived under the assumption that $-\Delta u = 0$ and $u = 0$ on the straight boundary segments that intersect at the singular point. In the general case that includes $-\Delta u = f(x, y)$, where $f(x, y)$ is a smooth function, and curved boundaries, we have

$$\left|\frac{\partial^s u}{\partial x^k \partial y^{s-k}}\right| \leq K(\varepsilon) r^{\bar{\lambda}_1 - \varepsilon - s} s!, \quad k = 0, 1, 2, \ldots, s, \quad s = 1, 2, \ldots \tag{6.13}$$

where $\varepsilon > 0$ and $K(\varepsilon) \to \infty$ as $\varepsilon \to 0$.

Exercise 6.3.1 Consider the solution of $\Delta u = 0$ in the neighborhood of corner point B shown in Figure 6.4(a), and let $u = 0$ on Γ_{BC} and the normal derivative $\partial u/\partial n = 0$ on Γ_{AB}. Show that u can be written as

$$u = \sum_{n=1}^{\infty} a_n r^{\lambda_n} (\cos \lambda_n \theta + (-1)^n \sin \lambda_n \theta) \quad \text{where } \lambda_n = \frac{(2n-1)\pi}{2\alpha}.$$

Hint: The condition $\partial u/\partial n = 0$ is equivalent to $\partial u/\partial \theta = 0$.

Exercise 6.3.2 Construct the series expansion analogous to Equation (6.11) for the problem $\Delta u = 0$ in the neighborhood of corner point B shown in Figure 6.4(a), given that $\partial u/\partial n = 0$ on Γ_{AB} and Γ_{BC}.

Exercise 6.3.3 Show that $u(r, \theta)$, defined by Equation (6.8), is not in the energy space when $\lambda < 0$.

Exercise 6.3.4 Consider functions $u = r^\lambda F(\theta, \phi)$ where r, θ, ϕ are spherical coordinates centered on a corner point and F is a smooth function. Show that $\partial u/\partial r$ is square integrable in three dimensions for $\lambda > -1/2$.

6.3.2 The Navier equations

In the following discussion it is assumed that the material is isotropic and elastic, hence the material properties are characterized by the two material constants E and ν, and the volume forces are zero. We examine the solution in the neighborhood of corner points when the intersecting edges are stress free. The treatment of other cases is analogous.

The stress fields in planar elasticity can be derived from the Airy stress function denoted by $U(r, \theta)$. The Airy stress function satisfies the biharmonic equation

$$\left(\frac{\partial^2}{\partial r^2} + \frac{1}{r}\frac{\partial}{\partial r} + \frac{1}{r^2}\frac{\partial^2}{\partial \theta^2}\right)\left(\frac{\partial^2}{\partial r^2} + \frac{1}{r}\frac{\partial}{\partial r} + \frac{1}{r^2}\frac{\partial^2}{\partial \theta^2}\right) U = 0. \tag{6.14}$$

The components of the stress tensor in polar coordinates are related to U by the following formulas (see, for example, [87]):

$$\sigma_r = \frac{1}{r}\frac{\partial U}{\partial r} + \frac{1}{r^2}\frac{\partial^2 U}{\partial \theta^2}, \quad \sigma_\theta = \frac{\partial^2 U}{\partial r^2}, \quad \tau_{r\theta} = -\frac{\partial}{\partial r}\left(\frac{1}{r}\frac{\partial U}{\partial \theta}\right). \tag{6.15}$$

The Cartesian components of the stress tensor are

$$\sigma_x = \frac{\partial^2 U}{\partial y^2}, \quad \sigma_y = \frac{\partial^2 U}{\partial x^2}, \quad \tau_{xy} = -\frac{\partial^2 U}{\partial x \partial y}. \tag{6.16}$$

The stress function can be written in complex variable form[3]

$$U = \Re(\bar{z}\varphi(z) + \chi(z)) \tag{6.17}$$

where $\varphi(z)$ and $\chi(z)$, called complex potentials, are analytic functions of the complex variable z. The overbar indicates the complex conjugate. The stress components in polar coordinates are related to $\varphi(z)$ and $\chi(z)$ as follows:

$$\sigma_r + \sigma_\theta = 2\left(\varphi'(z) + \overline{\varphi'(z)}\right) = 4\Re(\varphi'(z)) \tag{6.18}$$

$$\sigma_\theta - \sigma_r + 2i\tau_{r\theta} = 2z\bar{z}^{-1}\left(z\varphi''(z) + \chi''(z)\right). \tag{6.19}$$

The stress components in Cartesian coordinates are related to $\varphi(z)$ and $\chi(z)$ as follows:

$$\sigma_x + \sigma_y = 2\left(\varphi'(z) + \overline{\varphi'(z)}\right) = 4\Re(\varphi'(z)) \tag{6.20}$$

$$\sigma_y - \sigma_x + 2i\tau_{xy} = 2\left(\bar{z}\varphi''(z) + \chi''(z)\right). \tag{6.21}$$

The components of the displacement vector in polar coordinates (up to rigid body displacement and rotation) are related to $\varphi(z)$ and $\chi(z)$ as follows:

$$2G(u_r + iu_\theta) = z^{-1/2}\bar{z}^{1/2}\left(\kappa\varphi(z) - z\overline{\varphi'(z)} - \overline{\chi'(z)}\right) \tag{6.22}$$

and in Cartesian coordinates

$$2G(u_x + iu_y) = \kappa\varphi(z) - z\overline{\varphi'(z)} - \overline{\chi'(z)} \tag{6.23}$$

---

[3] This formula is attributed to Edouard Jean-Baptiste Goursat (1858–1936).

where κ is given by Equation (4.39). This is known as the Kolosov–Muskhelishvili method.[4] Details and derivations are available in books on elasticity, such as [45], [75], [87].

We are interested in solutions corresponding to

$$\varphi(z) = (a_1 - ia_2)z^\lambda, \quad \chi(z) = (a_3 - ia_4)z^{\lambda+1}, \quad \lambda \geq 0, \lambda \neq 1 \quad (6.24)$$

where a_i ($i = 1, 2, 3, 4$) and λ are real numbers. The corresponding stress function is

$$U = r^{\lambda+1}(a_1 \cos(\lambda - 1)\theta + a_2 \sin(\lambda - 1)\theta + a_3 \cos(\lambda + 1)\theta + a_4 \sin(\lambda + 1)\theta). \quad (6.25)$$

In the case of $\lambda = 1$

$$\varphi(z) = a_1 z - ia_2 z \log z, \quad \chi(z) = (a_3 - ia_4)z^2, \quad (6.26)$$

and the corresponding stress function is

$$U = r^2(a_1 + a_2\theta + a_3 \cos 2\theta + a_4 \sin 2\theta). \quad (6.27)$$

Stress-free edges

We refer to Figure 6.4 and assume that the boundary segments Γ_{AB} and Γ_{BC} are stress free, that is, $\sigma_\theta = \tau_{r\theta} = 0$ at $\theta = \pm \alpha/2$. Using Equations (6.25) and (6.15) we find

$$\sigma_\theta = r^{\lambda-1}\lambda(\lambda + 1)[a_1 \cos(\lambda - 1)\theta + a_2 \sin(\lambda - 1)\theta$$
$$+ a_3 \cos(\lambda + 1)\theta + a_4 \sin(\lambda + 1)\theta]$$
$$\tau_{r\theta} = r^{\lambda-1}\lambda(\lambda - 1)[a_1 \sin(\lambda - 1)\theta - a_2 \cos(\lambda - 1)\theta)]$$
$$+ r^{\lambda-1}\lambda(\lambda + 1)[a_3 \sin(\lambda + 1)\theta - a_4 \cos(\lambda + 1)\theta].$$

On setting $\sigma_\theta = \tau_{r\theta} = 0$ at $\theta = \pm\alpha/2$, following straightforward algebraic manipulation, we obtain

$$\begin{bmatrix} \cos(\lambda - 1)\alpha/2 & \cos(\lambda + 1)\alpha/2 \\ -\Lambda \sin(\lambda - 1)\alpha/2 & \sin(\lambda + 1)\alpha/2 \end{bmatrix} \begin{Bmatrix} a_1 \\ a_3 \end{Bmatrix} = 0 \quad (6.28)$$

and

$$\begin{bmatrix} \sin(\lambda - 1)\alpha/2 & \sin(\lambda + 1)\alpha/2 \\ -\Lambda \cos(\lambda - 1)\alpha/2 & \cos(\lambda + 1)\alpha/2 \end{bmatrix} \begin{Bmatrix} a_2 \\ a_4 \end{Bmatrix} = 0 \quad (6.29)$$

where

$$\Lambda := \frac{1-\lambda}{1+\lambda}.$$

---

[4] Gury Vasilievich Kolosov (1867–1936), Nikolai Ivanovich Muskhelishvili (1891–1976).

Note that a_1 and a_3 (resp. a_2 and a_4) are coefficients of symmetric (resp. antisymmetric) functions in Equation (6.25). Therefore, analogously to Equation (6.11), $U(r, \theta)$ can be written in terms of sums of symmetric and antisymmetric functions. The symmetric functions associated with the eigenvalues of Equation (6.28) are usually called Mode I eigenfunctions and the antisymmetric functions associated with the eigenvalues of Equation (6.29) are called Mode II eigenfunctions.

Non-trivial solutions exist if either the determinant of Equation (6.28) or the determinant of Equation (6.29) vanishes. This will occur if either

$$\cos(\lambda - 1)\alpha/2 \sin(\lambda + 1)\alpha/2 + \Lambda \sin(\lambda - 1)\alpha/2 \cos(\lambda + 1)\alpha/2 = 0,$$

which can be simplified to

$$\sin \lambda \alpha + \lambda \sin \alpha = 0, \quad \lambda \neq 0, \pm 1, \qquad (6.30)$$

or

$$\sin(\lambda - 1)\alpha/2 \cos(\lambda + 1)\alpha/2 + \Lambda \cos(\lambda - 1)\alpha/2 \sin(\lambda + 1)\alpha/2 = 0,$$

which can be simplified to

$$\sin \lambda \alpha - \lambda \sin \alpha = 0, \quad \lambda \neq 0, \pm 1. \qquad (6.31)$$

We denote

$$Q(\lambda \alpha) := \frac{\sin \lambda \alpha}{\lambda \alpha} \quad \text{and} \quad q(\alpha) := \frac{\sin \alpha}{\alpha} \qquad (6.32)$$

and discuss the eigenvalues corresponding to the symmetric and antisymmetric eigenfunctions separately in the following. For more detailed treatment of the subject we refer to [39], [93].

Eigenvalues corresponding to symmetric (Mode I) eigenfunctions Equation (6.30) can be written as

$$Q(\lambda \alpha) + q(\alpha) = 0. \qquad (6.33)$$

The function $Q(\lambda \alpha)$ is plotted on the interval $0 < \lambda \alpha < 4\pi$ in Figure 6.5. The problem is to find the roots of Equation (6.33) for a given α. In the interval $0 < \alpha < \alpha_A$ where $\alpha_A = 2.553\,591$ ($146.31°$) the line $-q(\alpha)$ has no points in common with $Q(\lambda \alpha)$, therefore there are no real roots. At $\alpha = \alpha_A$ the line $-q(\alpha_A)$ is tangent to $Q(\lambda \alpha)$ at point A. Therefore there is a double root at this angle. There are double roots at $\alpha = \alpha_B^{(1)} = 3.625\,739$ ($207.74°$), $\alpha = \alpha_B^{(2)} = 5.499\,379$ ($315.09°$) and also at $\alpha = \alpha_C = 2.875\,839$ ($164.77°$). It is seen that in the interval $\alpha_A < \alpha < \alpha_B$ there are at least two real and simple roots. At $\alpha = \pi$ and $\alpha = 2\pi$ there are infinitely many real roots. Furthermore, at $\alpha = \pi$ all roots are integers. Point D corresponds to $\alpha = 3\pi/2$ ($270°$) where $\lambda \alpha = 2.565\,819$, hence $\lambda = 0.544\,484$. Point E corresponds to $\alpha = 2\pi$ where $\lambda \alpha = \pi$, hence $\lambda = 1/2$. Point E also corresponds to $\alpha = \pi$, which is a special case, discussed next.

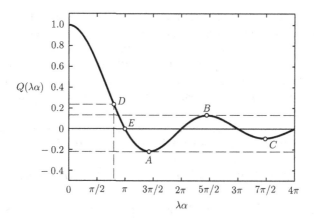

Figure 6.5 The function $Q(\lambda\alpha)$ on the interval $0 < \lambda\alpha < 4\pi$.

In formulating Equation (6.33) we excluded $\lambda = 1$ from consideration. When $\lambda = 1$ then U is given by Equation (6.27). Considering the symmetric terms only and using Equation (6.15), we have

$$\sigma_\theta = 2(a_1 + a_3 \cos 2\theta), \quad \tau_{r\theta} = 2a_3 \sin 2\theta.$$

Letting $\sigma_\theta(\pm\alpha/2) = \tau_{r\theta}(\pm(\alpha/2) = 0$, we find that a non-trivial solution exists only if

$$\det \begin{bmatrix} 1 & \cos\alpha \\ 0 & \sin\alpha \end{bmatrix} = 0.$$

Therefore $\alpha = n\pi$ ($n = 1, 2, \ldots$). Point E in Figure 6.5 represents $\alpha = \pi$.

Remark 6.3.2 To find the complex roots of Equation (6.33) we write $\lambda = \xi + i\eta$. Therefore Equation (6.33) becomes

$$\frac{\sin(\xi\alpha + i\eta\alpha)}{\xi\alpha + i\eta\alpha} = q(\alpha) \tag{6.34}$$

which is equivalent to the following system of two equations:

$$\sin\xi\alpha \cosh\eta\alpha = \xi q(\alpha) \tag{6.35}$$

$$\cos\xi\alpha \sinh\eta\alpha = \eta q(\alpha). \tag{6.36}$$

For a detailed description we refer to [89].

Eigenvalues corresponding to antisymmetric (Mode II) eigenfunctions Equation (6.31) can be written as

$$Q(\lambda\alpha) - q(\alpha) = 0. \tag{6.37}$$

180 REGULARITY AND RATES OF CONVERGENCE

Note that $\lambda = 1$ trivially satisfies Equation (6.37) for all α and recall that we have excluded $\lambda = 1$ from consideration when we formulated Equation (6.37). There are no real roots in the interval $0 < \alpha < \alpha_B$ where $\alpha_B = 2.777\,068$ (159.11°). There are at least two real roots in the interval $\alpha_B < \alpha < \alpha_C$ where $\alpha_C^{(1)} = 3.463\,416$ (198.44°) and $\alpha_C^{(2)} = 5.732\,235$ (328.43°). As in the case of Mode I, there are infinitely many real roots at $\alpha = \pi$ and $\alpha = 2\pi$, and at $\alpha = \pi$ all roots are integers. There is only one real root in the interval $\alpha_C < \alpha < \alpha_A$ where $\alpha_A = 4.493\,409$ (257.45°).

The angle α_A is a special angle which corresponds to $\lambda = 1$. To show this, we consider the antisymmetric terms in Equation (6.27). Using Equation (6.15), we have

$$\sigma_\theta = 2(a_2\theta + a_4 \sin 2\theta), \quad \tau_{r\theta} = -(a_2 + 2a_4 \cos 2\theta).$$

Letting $\sigma_\theta(\pm\alpha/2) = \tau_{r\theta}(\pm\alpha/2) = 0$, we find that a non-trivial solution exists when

$$\det \begin{bmatrix} \alpha/2 & \sin\alpha \\ 1 & 2\cos\alpha \end{bmatrix} = 0.$$

Therefore $\alpha = \tan\alpha$. In the interval $0 < \alpha < 2\pi$ there is one root: $\alpha = \alpha_A = 4.493\,409$ (257.45°). This corresponds to point A in Figure 6.5.

Cracks An important special case is when the singular point is a crack tip, that is, $\alpha = 2\pi$. The Airy stress function corresponding to the symmetric part of the asymptotic expansion is

$$U_s = \sum_{i=1}^{\infty} a_i \Re(\bar{z}z^{\lambda_i} + Q_i z^{\lambda_i+1}), \quad \lambda_i = i/2 \qquad (6.38)$$

where $Q_i = (2-i)/(2+i)$ when i is odd and $Q_i = -1$ when i is even. The Airy stress function corresponding to the antisymmetric part of the asymptotic expansion is

$$U_a = \sum_{i=1}^{\infty} b_i \Im(\bar{z}z^{\lambda_i} + Q_i z^{\lambda_i+1}), \quad \lambda_i = i/2 \qquad (6.39)$$

where $\Im(\cdot)$ represents the imaginary part of $(\cdot)$ and $Q_i = -1$ when i is odd and $Q_i = (2-i)/(2+i)$ when i is even. This can be verified by comparing eq (6.39) with the antisymmetric terms in Equation (6.25). Using Equations (6.38), (6.39) and (6.15), (6.16), the stress distribution in the neighborhood of a crack tip can be determined up to the coefficients a_i, b_i ($i = 1, 2, \ldots, \infty$). Procedures for the determination of these coefficients from finite element solutions are outlined in Chapter 7 and Appendix D.

Exercise 6.3.5 Show that: (a) the stress components corresponding to the second term of Equation (6.38) are $\sigma_x = 4a_2$, $\sigma_y = \tau_{xy} = 0$; and (b) the stress components corresponding to the second term of Equation (6.39) are $\sigma_x = \sigma_y = \tau_{xy} = 0$. Note that the stress $\sigma_x = 4a_2$ (constant) is usually denoted by T and is called the T-stress. According to some models, the T-stress influences crack growth. See, for example, [55].

THE NEIGHBORHOOD OF SINGULAR POINTS

Exercise 6.3.6 Let $\alpha = 3\pi/2$ and assume that the solution is symmetric.

1. Verify that the lowest positive eigenvalue of Equation (6.30) is $\lambda_1 = 0.544\,483\,737$.

2. Refer to Equation (6.28) and show that the Airy stress function corresponding to λ_1 can be written as

$$U = a_1 r^{\lambda_1+1}(\cos(\lambda_1 - 1)\theta + Q_1 \cos(\lambda_1 + 1)\theta) \tag{6.40}$$

where a_1 is an arbitrary real number and $Q_1 = 0.543\,075\,579$. Verify that Equation (6.40) is equivalent to

$$U = a_1 \Re(\bar{z} z^{\lambda_1} + Q_1 z^{\lambda_1+1}). \tag{6.41}$$

Exercise 6.3.7 Let $\alpha = 3\pi/2$ and assume that the solution is symmetric. Using Equations (6.41) and (6.16) show that the stress components corresponding to λ_1 are

$$\sigma_x = a_1 \lambda_1 r^{\lambda_1-1}[(2 - Q_1(\lambda_1 + 1))\cos(\lambda_1 - 1)\theta - (\lambda_1 - 1)\cos(\lambda_1 - 3)\theta]$$
$$\sigma_y = a_1 \lambda_1 r^{\lambda_1-1}[(2 + Q_1(\lambda_1 + 1))\cos(\lambda_1 - 1)\theta + (\lambda_1 - 1)\cos(\lambda_1 - 3)\theta]$$
$$\tau_{xy} = a_1 \lambda_1 r^{\lambda_1-1}[(\lambda_1 - 1)\sin(\lambda_1 - 3)\theta + Q_1(\lambda_1 + 1)\sin(\lambda_1 - 1)\theta].$$

The angle $\alpha = 3\pi/2$ is representative of frequently occurring geometric details and this stress field will be used for illustrating the convergence characteristics of the finite element method.

Exercise 6.3.8 Let $\alpha = 3\pi/2$ and assume that the solution is symmetric. Show that the displacement components corresponding to λ_1, up to rigid body displacement and rotation terms, are

$$u_x = \frac{a_1}{2G} r^{\lambda_1}[(\kappa - Q_1(\lambda_1 + 1))\cos\lambda_1\theta - \lambda_1 \cos(\lambda_1 - 2)\theta]$$
$$u_y = \frac{a_1}{2G} r^{\lambda_1}[(\kappa + Q_1(\lambda_1 + 1))\sin\lambda_1\theta + \lambda_1 \sin(\lambda_1 - 2)\theta]$$

where κ is defined by Equation (4.39). Hint: Use Equations (6.23) and (6.41).

Exercise 6.3.9 Assume that λ is real. Show that the corresponding stress field is in the energy space only if $\lambda > 0$. Hint: A displacement field is in the energy space if the stress components are square integrable:

$$\int_\Omega (\sigma_r^2 + \sigma_\theta^2 + \tau_{r\theta}^2)\, r\, dr d\theta \leq C < \infty.$$

Complex eigenvalues

We have seen that in planar elasticity, in contrast to the Laplace equation, λ can be complex and it can be either a simple or a multiple root. If λ is complex then its conjugate is also a root. In the case of multiple roots special treatment is necessary which is not discussed here. We refer to [58] for details.

Consider the Airy stress function in the form $U = r^{\lambda+1} F(\theta)$. If λ is complex we write $\lambda = \xi + i\eta$ and $F = f + ig$. Therefore

$$U = r^{(\xi+1+i\eta)}(f + ig) = r^{(\xi+1)} e^{(\ln r^{i\eta})}(f + ig).$$

Writing

$$e^{(\ln r^{i\eta})} = e^{(i\eta \ln r)} = \cos(\eta \ln r) + i \sin(\eta \ln r)$$

we get

$$U = r^{\xi+1}[(f \cos(\eta \ln r) - \eta \sin(\eta \ln r)) + i(f \sin(\eta \ln r) - \eta \cos(\eta \ln r))]. \tag{6.42}$$

Both the real and imaginary parts of U are solutions of the biharmonic equation (6.14). Since $\ln r \to -\infty$ as $r \to 0$, the sinusoidal terms oscillate with a wavelength approaching zero. The singularity is characterized by the estimate of the derivatives, see Equation (6.13), with λ_1 replaced by $\xi_1 = \Re(\lambda_1)$:

$$\left| \frac{\partial \mathbf{u}^s}{\partial x^k \partial y^{s-k}} \right| \leq K(\varepsilon) r^{\xi_1 - s - \varepsilon} s!. \tag{6.43}$$

Note that this estimate is independent of the imaginary part.

The lowest values of $\Re(\lambda)$

The lowest values of $\Re(\lambda)$ are listed in Table 6.1 for three kinds of homogeneous boundary conditions prescribed on the corner edges: (a) the free–free condition:[5] on both edges $\sigma_\theta = \tau_{r\theta} = 0$; (b) the fixed–free condition (plane stress, $\nu = 0.3$): on one edge $u_r = u_\theta = 0$, on the

Table 6.1 Lowest positive values of $\Re(\lambda)$ at corner points for three kinds of homogeneous boundary conditions.

	Free–free		Fixed–free	Fixed–fixed	
α	$\Re(\lambda_1^{(s)})$	$\Re(\lambda_1^{(a)})$	$\Re(\lambda_1)$	$\Re(\lambda_1^{(s)})$	$\Re(\lambda_1^{(a)})$
45°	5.390 53	9.562 71	1.304 34	5.573 28	2.608 31
90°	2.739 59	4.808 25	0.758 35	2.825 79	1.490 46
135°	1.885 37	3.242 81	0.693 39	1.573 23	1.160 88
180°	1.000 00	2.000 00	0.500 00	1.000 00	1.000 00
225°	0.673 58	1.302 09	0.405 94	0.735 54	0.877 23
270°	0.544 48	0.908 53	0.340 32	0.604 04	0.744 46
315°	0.505 01	0.659 70	0.287 84	0.537 93	0.609 45
360°	0.500 00	0.500 00	0.250 00	0.500 00	0.500 00

---

[5] In the case of free–free boundary conditions the eigenvalues are independent of Poisson's ratio ν.

other edge $\sigma_\theta = \tau_{r\theta} = 0$; (c) the fixed–fixed condition (plane stress, $\nu = 0.3$): $u_r = u_\theta = 0$ on both edges.

In the free–free and fixed–fixed cases the characteristic functions are either symmetric or antisymmetric. These are indicated in Table 6.1 by the superscripts s and a, respectively. At $\alpha = 180°$ the characteristic values are integers.

Note that if λ is a solution of Equation (6.30) or Equation (6.31) then $-\lambda$ is also a solution; however, the corresponding stress field has finite strain energy only when the real part of λ is greater than zero.

6.3.3 Material interfaces

Some singular points associated with material interfaces are illustrated in Figure 6.2(b). The solution in the neighborhood of these singular points is also characterized by eigenvalues and eigenfunctions which are typically of the form $u = r^\lambda F(\theta)$ in the case of the Laplace equation and $u_i = r^\lambda F_i(\theta)$ ($i = 1, 2$) in the case of the two-dimensional Navier equations.[6] However, unlike in the case of homogeneous materials, $F(\theta)$ and $F_i(\theta)$ are piecewise analytic functions: at angles corresponding to material interfaces the derivatives are discontinuous. It is possible to determine the eigenvalues and eigenfunctions by classical methods; however, the problem is now more complicated than in the homogeneous case. It is possible (and much more convenient) to determine the eigenpairs numerically using a method known as the Steklov method.[7]

The Steklov method is based on the observation that on a circular contour of radius ϱ, shown in Figure 6.4(b), the normal derivative of the function $u = r^\lambda F(\theta)$ is

$$\left(\frac{\partial u}{\partial n}\right)_{\Gamma_\varrho} = \left(\frac{\partial u}{\partial r}\right)_{r=\varrho} = \lambda \varrho^{\lambda-1} F(\theta) = \frac{\lambda}{\varrho} u. \qquad (6.44)$$

Consider now problems where two or more isotropic materials are bonded and the material interfaces are planar.[8] The generalized form of Equation (3.18), subject to the assumptions that $\bar{Q} = 0$, steady state conditions exist, and either $u = 0$ or $\partial u/\partial n = 0$ on Γ_{AB} and Γ_{BC}, is

$$\int_{\Omega_\varrho} \text{grad}\, v [\kappa] \text{grad}\, u\, dxdy = \frac{\lambda}{\varrho} \int_{\Gamma_\varrho} uv\, ds \quad \text{for all } v \in E^0(\Omega_\varrho) \qquad (6.45)$$

where $[\kappa(x, y)]$ is a positive-definite matrix of material properties which are discontinuous at the material interfaces[9] and Ω_ϱ is defined in Figure 6.4(b). Equation (6.45) is a characteristic value problem; the characteristic values are real. On solving Equation (6.45) by the finite element method, the characteristic functions are approximations of $F(\theta)$ by the (mapped) piecewise polynomial functions. The size of the numerical characteristic value problem can be reduced to the number of degrees of freedom associated with the contour Γ_ϱ.

---

[6] In the three-dimensional Navier equations, $u_i = r^\lambda F_i(\theta, \phi)$ at vertices where (r, θ, ϕ) are spherical coordinates and $u_i = r^\lambda F_i(\theta, z)$ along edges, where (r, θ, z) are cylindrical coordinates, the axial coordinate z being coincident with the edge.
[7] Vladimir Andreevich Steklov (1864–1926).
[8] The treatment of anisotropic materials and curved interfaces is not considered here.
[9] The elements of the matrix $[\kappa]$ are usually piecewise constant functions.

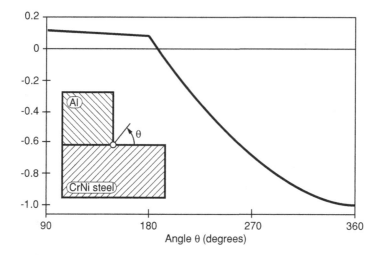

Figure 6.6 Example 6.3.1: the characteristic function $F_1(\theta)$.

The method is applicable to the Navier equations as well. In the case of the Navier equations the characteristic values may be complex [58], [85], [94].

Example 6.3.1 A chrome–nickel steel plate with coefficient of thermal conductivity $k_{cns} = 16.3$ W/(m K) is bonded to a pure aluminum plate with $k_{al} = 202$ W/(m K). The edges are offset, forming a corner similar to that shown in Figure 6.2(c). Assuming that the edges intersecting at the reentrant corner are perfectly insulated ($q_n = 0$), the first characteristic value is $\lambda_1 = 0.5238$. The corresponding characteristic function $F_1(\theta)$ is shown in Figure 6.6. This function is determined up to an arbitrary multiplier.

Exercise 6.3.10 Determine the second eigenvalue and the corresponding eigenfunction for the problem of Example 6.3.1. Partial answer: $\lambda_2 = 1.4762$.[10]

Exercise 6.3.11 Referring to Figure 6.7, assume that boundary segments IB and BC are traction free.[10] The material properties for Material 1 (resp. Material 2) are $E_1 = 200$ GPa, $\nu_1 = 0.3$ (resp. $E_2 = qE_1$, $\nu_2 = \nu_1$). Assuming plane strain conditions and zero volume

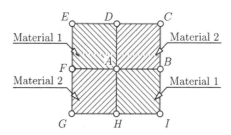

Figure 6.7 Exercise 6.3.11: notation.

---

[10] Solution of this exercise requires procedures that are not supported by StressCheck

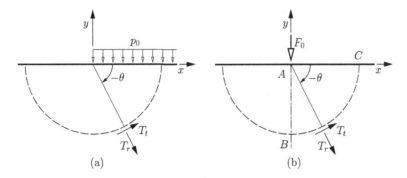

Figure 6.8 (a) Loading by a step function. (b) Loading by a concentrated force.

forces, determine the degrees of singularity at points A and B for $q = 0.5, q = 0.1, q = 0.01$ and $q = 0.001$.

This exercise shows that $\lambda_{\min}$ decreases as the ratio E_2/E_1 decreases. Partial solution: for $q = 0.1$, $\lambda_{\min} = 0.396\,08$ at point A, $\lambda_{\min} = 0.801\,53$ at point B.

6.3.4 Forcing functions acting on boundaries

The regularity of the solution is influenced by the forcing function also. Consider, for example, constant normal traction acting along the positive x-axis, as shown in Figure 6.8(a). The point where the normal traction changes from zero to $-p_0$ is a singular point. In this case the stress components are finite but not single valued at the singular point. The stress function is

$$U = -\frac{p_0 r^2}{2\pi}\left(\pi + \theta - \frac{1}{2}\sin 2\theta\right) \quad -\pi \leq \theta \leq 0$$

and the stress components are

$$\sigma_r = \frac{1}{r}\frac{\partial U}{\partial r} + \frac{1}{r^2}\frac{\partial^2 U}{\partial \theta^2} = -\frac{p_0}{\pi}\left(\pi + \theta + \frac{1}{2}\sin 2\theta\right) \tag{6.46}$$

$$\sigma_\theta = \frac{\partial^2 U}{\partial r^2} = -\frac{p_0}{\pi}\left(\pi + \theta - \frac{1}{2}\sin 2\theta\right) \tag{6.47}$$

$$\tau_{r\theta} = -\frac{\partial}{\partial r}\left(\frac{1}{r}\frac{\partial U}{\partial \theta}\right) = \frac{p_0}{2\pi}(1 - \cos 2\theta). \tag{6.48}$$

Observe that along the x-axis ($\theta = 0$) $\tau_{r\theta} = 0$ at the origin, but along the y-axis ($\theta = -\pi/2$) $\tau_{r\theta} = p_0/\pi$. At the origin it is multi-valued. Similarly, the sum of the normal stresses $\sigma_r + \sigma_\theta = -2p_0(1 + \theta/\pi)$ ranges from $-2p_0$ to 0 at the origin, depending on the direction from which the origin is approached. It can be shown that the derivatives of the displacement function are bounded analogously to Equation (6.13):

$$\left|\frac{\partial u^s}{\partial x^k \partial y^{s-k}}\right| \leq K(\varepsilon) r^{\lambda-\varepsilon-s} s!, \quad k = 0, 1, 2, \ldots, s, \quad s = 1, 2, \ldots \tag{6.49}$$

with $\lambda = \lambda_1 = 1$ in the example shown in Figure 6.8(a).

Remark 6.3.3 The regularity estimate (6.49) holds also when the load is applied through the imposition of spring displacement when the spring displacement is a step function [23].

Concentrated force

The problem of a concentrated force acting on an elastic body is often a source of conceptual errors and therefore merits special discussion. Referring to Figure 6.8(b), the linear functional corresponding to the concentrated force F_0 is

$$F(\mathbf{v}) = -F_0 v_y(0, 0). \tag{6.50}$$

The derivation of this expression is analogous to the one-dimensional case, see Equation (2.81) in Section 2.5.5. However, unlike in the one-dimensional case, in two and higher dimensions $\mathbf{v}$ is not necessarily bounded, see Section 4.1.1. Therefore $F(\mathbf{v})$ does not satisfy condition 3 in Section A.3 and hence concentrated forces are inadmissible data in mathematical models based on the principle of virtual work.

An alternative way to show the inadmissibility of concentrated forces is to consider the Airy stress function corresponding to a concentrated force:

$$U = -\frac{F_0}{\pi} r\theta \cos\theta \quad -\pi \leq \theta \leq 0$$

which can be found in textbooks on elasticity, such as Timoshenko and Goodier [87]. It is left to the reader to verify that

$$\sigma_r = \frac{2F_0}{r\pi} \sin\theta \tag{6.51}$$

and $\sigma_\theta = \tau_{r\theta} = 0$. On a solution domain that includes the origin, this stress field is not square integrable, hence the solution does not lie in the energy space. Therefore sequences of solutions corresponding to finite element spaces constructed by the h-, p- or hp-methods do not have a limit in the energy space.

Nevertheless, concentrated forces are widely used in finite element analysis with reasonably good results. This apparent contradiction is related to the fact that for a fixed mesh and polynomial degree the load vector corresponding to a concentrated force can be understood as a load vector computed for some statically equivalent distributed traction, corresponding to which an exact solution exists in the energy space. This is illustrated by Example 6.3.2.

Although the statically equivalent tractions are not uniquely defined, the differences in stress distributions corresponding to statically equivalent tractions decay with distance, see Remark 6.3.4. This is related to Saint-Venant's principle.[11] An illustration is given in Example 6.3.3.

Concentrated forces are typically used for reasons of convenience in modeling. For example, tractions transmitted by bodies in elastic contact, such as, for example, contact between the shanks of fasteners and the fastener holes, are often represented by statically equivalent concentrated forces. The rationale is that the details of the distribution of tractions in the vicinity of points where the concentrated forces are applied do not have significant influence on the data of interest. Two assumptions are implied: (a) the data of interest are far from the points of application of concentrated forces; and (b) there is a finite length or area, such as the diameter of a fastener hole or the area of mechanical contact, and the approximation is understood to be an approximation of distributed tractions acting over that finite length or area. Should the mesh be refined such that the elements are of the size of the finite length or area, then the concentrated forces must be replaced with the appropriate tractions.

Remark 6.3.4 The rate of decay of the differences in stress distribution corresponding to statically equivalent tractions is related to Green's function. It can be shown that in the scalar elliptic problem and in elasticity the rate of decay is $(d/\varrho)^2$ where d represents the diameter of the area on which the statically equivalent tractions are applied and ϱ is the distance. In the finite element method d is approximately the diameter of the largest element that contains the point of application of the concentrated force.

Example 6.3.2 Consider the eight-node quadrilateral element shown in Figure 6.9. For the sake of simplicity the element is defined such that the mapping is $x = \xi$, $y = \eta$. Therefore the shape functions defined by Equations (5.7) through (5.14) are also the basis functions.

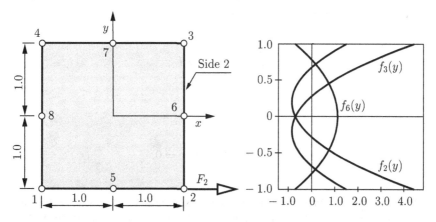

Figure 6.9 Eight-node quadrilateral element. Polynomial tractions $f_2(y)$, $f_3(y)$, $f_6(y)$ that are equivalent to unit forces applied at nodes 2, 3, 6 respectively.

---

[11] Adhémar Jean Claude Barré de Saint-Venant (1797–1886).

Let us assume that a concentrated force F_2 is applied at node 2. In this case the only non-zero load vector term is $r_2 = F_2$ which follows from the definition of v_x on the element:

$$v_x(x, y) = \sum_{i=1}^{8} v_x^{(i)} N_i(x, y)$$

where $v_x^{(i)}$ is the value of $v(x)$ at node i. Since at node 2 all basis functions are zero with the exception of N_2, we have

$$F(\mathbf{v}) = F_2 v(1, -1) = F_2 v_x^{(2)} = r_2 v_x^{(2)}.$$

The load vector corresponding to the traction $T_x(y) = F_2 f_2(y)$, acting on side 2, that satisfies the conditions

$$r_2 = \int_{-1}^{1} T_x(y) N_2(1, y)\, dy = F_2 \tag{6.52}$$

$$r_3 = \int_{-1}^{1} T_x(y) N_3(1, y)\, dy = 0 \tag{6.53}$$

$$r_6 = \int_{-1}^{1} T_x(y) N_6(1, y)\, dy = 0 \tag{6.54}$$

is the same as the load vector corresponding to the concentrated force F_2 acting on node 2. For statical equivalency the resultant of T_x must be F_2:

$$\int_{-1}^{1} T_x(y)\, dy = F_2; \tag{6.55}$$

and the moment about node 2 must be zero:

$$\int_{-1}^{1} T_x(y)(1 + y)\, dy = 0. \tag{6.56}$$

Noting that $N_2(1, y) + N_3(1, y) + N_6(1, y) = 1$, on summing Equations (6.52) through (6.54) we see that (6.55) is satisfied. Equation (6.56) is satisfied by (6.53), (6.54) because $2N_3(1, y) + N_6(1, y) = (1 + y)$.

Let us construct, for example, $f_2(y)$ as a linear combination of $N_2(1, y), N_3(1, y), N_6(1, y)$. It is left to the reader to verify that

$$f_2(y) = \frac{9}{2} N_2(1, y) + \frac{3}{2} N_3(1, y) - \frac{3}{4} N_6(1, y) \tag{6.57}$$

satisfies Equations (6.52) through (6.56). This function and the analogous functions corresponding to unit forces applied in nodes 3 and 6

$$f_3(y) = \frac{3}{2}N_2(1, y) + \frac{9}{2}N_3(1, y) - \frac{3}{4}N_6(1, y) \qquad (6.58)$$

$$f_6(y) = -\frac{3}{4}N_2(1, y) - \frac{3}{4}N_3(1, y) + \frac{9}{8}N_6(1, y) \qquad (6.59)$$

are plotted in Figure 6.9.

These results demonstrate that it is possible to construct smooth tractions that are statically equivalent to the applied concentrated force and the load vector is the same as the load vector corresponding to the concentrated force. Therefore the finite element solution can be understood as an approximation to the exact solution of a problem with statically equivalent smooth tractions. These tractions are not uniquely defined and depend on the element and its polynomial degree.

Remark 6.3.5 It can be shown that sequences of finite element solutions $\mathbf{u}_{(\Delta,p)}$ corresponding to $F(\mathbf{v})$ defined in Equation (6.50) converge to the exact solution of the concentrated force problem at every point inside the domain.

Example 6.3.3 Let us solve the problem shown Figure 6.8(b). Since AB is a line of symmetry, our solution domain is the circular sector ABC and the magnitude of the concentrated force is $F_0/2$. On the boundary BC the radial normal stress given by Equation (6.51) is applied. In order to prevent rigid body motion in the direction of the y-axis, one point constraint is specified.

The first of three geometrically graded finite element meshes used in this example is shown in Figure 6.10(a). This mesh consists of four finite elements. The size of elements 1 and 2 is approximately r_0, the size of elements 3 and 4 is αr_0. A detail of the second mesh, consisting of six-elements, is shown in Figure 6.10(b). Here the size of elements that have a node at the point of application of the force F_0 is $\alpha^2 r_0$. The third mesh, consisting of eight

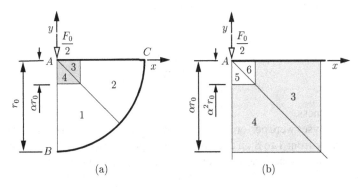

Figure 6.10 The concentrated force problem for Example 6.3.3: (a) four-element mesh; (b) detail of the six-element mesh.

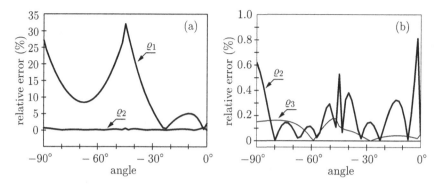

Figure 6.11 Example 6.3.3: the relative error in the radial stress σ_r along circular arcs centered on point of application of the concentrated force, measured in maximum norm. Eight-element mesh, $p = 8$, $\alpha = 0.15$, trunk space.

elements, is not shown; however, the refinement is analogous: the size of the elements that have a node at point A is $\alpha^3 r_0$.

Since the exact stress distribution is known, see Equation (6.51), we can compute the relative error in maximum norm at any point other than point A. For the eight-element mesh we construct three circles centered on point A. The circles pass through the mid-points of the nodes that lie on AB and CD, that is, the radii are

$$\varrho_1 = \frac{\alpha^3}{2} r_0, \quad \varrho_2 = \frac{\alpha^2}{2}(1+\alpha) r_0, \quad \varrho_3 = \frac{\alpha}{2}(1+\alpha) r_0.$$

The relative error in maximum norm along the three circles is shown in Figure 6.11 for $\alpha = 0.15$. It can be seen that the error is large along the smallest circle (radius ϱ_1) which passes through the elements that have a vertex at point A, but rapidly decreases with respect to increasing radius.

Example 6.3.4 To illustrate the strong mesh dependence of the solution at the point of application of concentrated forces, we compute the normalized displacement component at point A defined by

$$\bar{u}_y^A := \frac{u_y^A E t}{F_0} \tag{6.60}$$

where t is the thickness.

Two sets of analyses were performed. In the first set the polynomial degree of elements was increased uniformly from 1 to 8 on each of the three geometrically graded meshes described above using the trunk space. The normalized displacement component $\bar{u}_y^A$ is plotted vs. the number of degrees of freedom N in Figure 6.12(a). It can be seen that the absolute value of $\bar{u}_y^A$ monotonically increases with N. The reason for this is that the exact solution is not in the energy space. Note that the strain energy is proportional to $F_0 u_y^A$ and, since F_0 is fixed, u_y^A

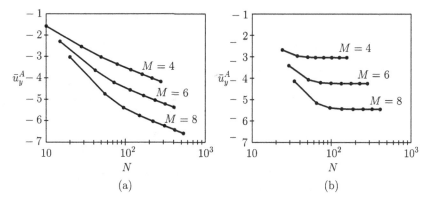

Figure 6.12 The normalized displacement component $\bar{u}_y^A$ vs. the number of degrees of freedom N. (a) Uniform p-distribution, p ranging from 1 to 8. (b) On the smallest elements p is fixed at 3, uniformly increased from 1 to 8 on the other elements. The number of elements is denoted by M.

tends to infinity as N is increased. In fact, in the neighborhood of point A, u_y^A is proportional to $\log r$ where r is the distance from point A.

In the second set of analyses p was fixed at 3 on the elements that have a node at point A and increased uniformly from 1 to 8 on the other elements. $\bar{u}_y^A$ is plotted vs. N on a semi-log scale in Figure 6.12(b). In this case $\bar{u}_y^A$ converges but its limit value depends on the choice of mesh.

Example 6.3.5 In Example 6.3.3 we saw that the displacement in the point of application of a concentrated force, corresponding to the finite element solution, is strongly mesh dependent and tends to infinity with respect to increasing N. In this example we demonstrate that the concentrated force corresponding to a displacement imposed on a node tends to zero as N is increased.

Let us consider once again the domain ABC shown in Figure 6.10. This time we prescribe $u_y^A = -\Delta$ at point A; symmetry boundary conditions on AB; and zero normal displacement on arc BC. We are interested in computing the concentrated force F_y^A from a sequence of finite element solutions.

We construct a sequence of geometrically graded meshes of 4, 6, ..., 16 elements. The first two meshes are shown in Figure 6.10. We fix the polynomial degree at $p = 8$, and compute the strain energy U corresponding to each mesh. The force F_y^A is determined from U using the relationship $U = F_y^A \Delta/2$. The results of computation are shown in Figure 6.13 where the non-dimensional force, defined by

$$\bar{F}_y^A := \frac{F_y^A}{Et\Delta}, \tag{6.61}$$

is plotted vs. $\log N$.

Remark 6.3.6 It is common practice in finite element analysis to idealize fastened connections as rigid or elastic links; that is, point constraints that restrict the relative or absolute

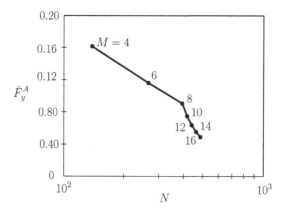

Figure 6.13 Example 6.3.5: mesh dependence of the force corresponding to displacement imposed on a node point. Geometrically graded meshes, $p = 8$.

value of one or more displacement components at node points. The expectation is that the stiffness of the connection and the distribution of nodal forces[12] will be a reasonable approximation of the fastened connection that the model is supposed to represent. In view of the foregoing discussion and the results of Examples 6.3.4 and 6.3.5, the stiffness and force distribution will be artifacts of the discretization. The results may appear to be plausible, but the appearance of plausibility is caused by large errors in discretization masking the conceptual error in the formulation. Engineering decisions should not be based on the expectation that two large errors will cancel one another.

Exercise 6.3.12 Show that the traction $T_x = f_2(y)$, where $f_2(y)$ is given by Equation (6.57), satisfies Equations (6.52) through (6.56).

Exercise 6.3.13 Solve the problem shown in Figure 6.8(a) on a semicircular domain, using the finite element method. Specify $T_r = \sigma_r$, $T_t = \tau_{r\theta}$ on the circular segment where σ_r (resp. $\tau_{r\theta}$) is given by Equation (6.46) (resp. Equation (6.48)). Compare the computed values σ_y with the exact value along the edges of the elements that have a vertex at the origin. Hint:

$$\sigma_y = \frac{\sigma_r + \sigma_\theta}{2} - \frac{\sigma_r - \sigma_\theta}{2} \cos 2\theta - \tau_{r\theta} \sin 2\theta.$$

Exercise 6.3.14 Construct three graded meshes as in Example 6.3.3. In point A prescribe the displacement $u_y^A = \Delta$. Solve the problem shown in Figure 6.8(b) on a quarter-circular solution domain ABC, utilizing symmetry.

6.3.5 Strong and weak singular points

In many engineering applications the goals of computation involve determination of the first derivatives of the solution. For example, in elasticity the first derivatives define the strain

---

[12] The definition and computation of nodal forces are discussed in Chapter 7.

and stress states, in heat transfer the temperature gradient and in porous flow problems the piezometric gradient.

Note that when $\lambda_1 < 1$ in Equation (6.49) the first derivative of $\mathbf{u}_{EX}$ with respect to r is infinity. Hence, according to the linear theory of elasticity, the strain and stress are infinity at singular points where $\lambda < 1$. Similarly, in heat transfer and viscous flow problems the gradients are infinity. For problems where $\lambda < 1$ it is not meaningful to try to compute the maximum value of the first derivatives. The computed values of the first derivatives will be finite but will converge to infinity in h-, p- and hp-extensions. Therefore at a singular point the exact solution u_{EX} will be called strongly singular if $\lambda < 1$, weakly singular otherwise. This applies to singular points and edges in three dimensions also.

Exercise 6.3.15 Refer to Exercise 6.3.11 and Figure 6.7. Show that points A and B are strongly singular for any $q \neq 1$.

6.4 Rates of convergence

In this section some key theoretical results that establish relationships between the error in the energy norm and the number of degrees of freedom associated with sequences of finite element spaces $S_1, S_2, \ldots \subset E(\Omega)$ are presented. Recall that the error in the energy norm depends on the choice of finite element spaces and the finite element spaces are characterized by the mesh Δ, the polynomial degree of elements $\mathbf{p}$ and the mapping functions $\mathbf{Q}$. The sequences of finite element spaces are hierarchic if $S_1 \subset S_2 \subset S_3 \subset \ldots \subset E(\Omega)$.

A priori estimates of the error in the energy norm are available for solutions of problems in Categories A, B and C. Convergence is either algebraic or exponential. The algebraic estimate is of the form

$$\|u_{EX} - u_{FE}\|_{E(\Omega)} \leq \frac{k}{N^\beta} \qquad (6.62)$$

and the exponential estimate is of the form

$$\|u_{EX} - u_{FE}\|_{E(\Omega)} \leq \frac{k}{\exp(\gamma N^\theta)} \qquad (6.63)$$

where N is the number of degrees of freedom. These estimates should be understood to mean that there exist some positive constants k, β (resp. γ and θ) that depend on u_{EX}, such that the error will be bounded by the algebraic (resp. exponential) estimate as N is increased. For sufficiently large N the "less than or equal" sign ($\leq$) often can be replaced with "approximately equal" ($\approx$). In those cases the estimate is called the asymptotic estimate and the rate of convergence is called the asymptotic rate of convergence.

In the following we assume that the finite element meshes are regular: that is, in two dimensions any two elements may have a vertex in common, an entire side in common or no points in common. For example, the mesh in Figure 6.14(a) satisfies this restriction but the mesh in Figure 6.14(b) does not. The node points highlighted by open circles in Figure 6.14(b) are called irregular nodes. If a mesh has irregular node points then the mesh is irregular.

Analogous restrictions apply in three dimensions where in a regular mesh any two elements may have a vertex, an entire edge or an entire face in common, or no points in common. If these conditions are not satisfied then the mesh is irregular. We note that some implementations

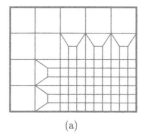
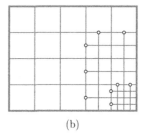

(a) (b)

Figure 6.14 (a) Example of a regular mesh. (b) Example of an irregular mesh. The open circles represent irregular node points.

permit the use of irregular meshes. This simplifies the meshing of complicated domains very substantially. With certain restrictions, it is possible to enforce continuity on the basis functions such that the basis functions lie in the energy space $E(\Omega)$.

We assume that the elements do not become too distorted, in the sense that none of the interior angles approach zero or 180° as the mesh is refined.

We further assume that in two-dimensional problems that are in Categories A and B all analytic arcs are covered by element boundaries and all singular points are node points. Analogous assumptions apply in three dimensions.

In the early implementations of the finite element method the polynomial degrees were restricted to $p = 1$ or $p = 2$ only. Sequences of finite element spaces were produced by mesh refinement, that is, by reduction of the diameter of the largest element, usually denoted by h. Subsequently this limitation was removed, allowing the creation of sequences of finite element spaces by increasing the polynomial degree of elements, usually denoted by p, while keeping the mesh fixed. To distinguish between the two approaches, the terms "h-version" or "h-method" and "p-version" or "p-method" gained currency. We will consider three strategies for constructing sequences of finite element spaces:

1. The h-method. The polynomial degree of elements is fixed, typically at some low number, such as $p = 1$ or $p = 2$, and the number of elements is increased such that the diameter of the largest element, denoted by h, is progressively reduced. The pattern of mesh refinement may be quasiuniform or non-uniform. A sequence of meshes is quasiuniform when the ratio of the largest and smallest diameters of the elements is bounded, see Equation (2.70). Examples of non-uniform refinement are shown in Figure 6.14. Quasiuniform meshes are used for Category A problems. Non-uniform mesh refinement is used for Category B problems with strong refinement in the neighborhoods of singular points. The use of quasiuniform meshes for Category B problems generally leads to poor performance.

2. The p-method. The mesh is fixed and the polynomial degree of elements, denoted by **p**, is increased either uniformly or non-uniformly. The sequences of finite element spaces are hierarchic, that is, $S_1 \subset S_2 \subset S_3 \ldots$.

3. The hp-method. The mesh is refined and the polynomial degree of elements is concurrently increased. The hp-method is used for Category B problems. With proper refinement and p-distribution, exponential rates of convergence are obtained. The first mathematical analysis of the p-method was published in 1981 [4a].

Table 6.2 Asymptotic rates of convergence in two dimensions.

Category	Type of extension		
	h	p	hp
A	Algebraic $\beta = p/2$	Exponential $\theta \geq 1/2$	Exponential $\theta \geq 1/2$
B	Algebraic, Note 1 $\beta = \frac{1}{2}\min(p, \lambda)$	Algebraic $\beta = \lambda$	Exponential $\theta \geq 1/3$
C	Algebraic $\beta > 0$	Algebraic $\beta > 0$	Note 2

A fourth strategy, not considered here, introduces basis functions, other than the mapped polynomial basis functions described in Chapter 5, to represent some local characteristics of the exact solution. This is variously known as the space enrichment method, partition of unity method and meshless method.

The asymptotic rates of convergence for two-dimensional problems are summarized in Table 6.2 and for three-dimensional problems in Table 6.3. It is assumed that (a) if quadrilateral or hexahedral elements are used then the standard polynomial space is the product space and (b) the element boundaries are aligned with material and source interfaces.

Note that with reference to Tables 6.2 and 6.3:

1. Uniform or quasiuniform mesh refinement is assumed. In the case of optimal or nearly optimal mesh refinement, $\beta_{\max} = p/2$. This is demonstrated in Example 6.4.2.

2. When u_{EX} has a recognizable structure then it is possible to achieve faster than algebraic rates of convergence with hp-adaptive methods.

3. In three dimensions u_{EX} cannot be characterized by a single parameter. Nevertheless, the rate of p-convergence is at least twice the rate of h-convergence when uniform or quasiuniform mesh refinement is used.

Table 6.3 Asymptotic rates of convergence in three dimensions.

Category	Type of extension		
	h	p	hp
A	Algebraic $\beta = p/3$	Exponential $\theta \geq 1/3$	Exponential $\theta \geq 1/3$
B		Note 3	Exponential $\theta \geq 1/5$
C	Algebraic $\beta > 0$	Algebraic $\beta > 0$	Note 2

Remark 6.4.1 The rates of convergence in Table 6.2 were proven under the assumption that the finite element meshes are triangular or, if quadrilateral, then the product space is used. If trunk spaces are used and the elements are not parallelograms (i.e., the mapping is not linear) then the asymptotic rate of convergence of the h-version is only one-half ($p/4$ rather than $p/2$) for Categories A and B. This is visible only when the finite element solutions are highly accurate, however.

6.4.1 The choice of finite element spaces

The a priori estimates provide an important conceptual framework for the construction of finite element spaces.

Problems in Category A

Referring to Tables 6.2 and 6.3, it is seen that for problems in Category A exponential rates of convergence are possible when p- and hp-extensions are used. The optimal mesh consists of the smallest number of elements required to partition the solution domain into triangular and quadrilateral elements in two dimensions, and tetrahedral, pentahedral and hexahedral elements in three dimensions. Whenever possible, quadrilateral elements should be used in two dimensions, hexahedral elements in three dimensions.

When h-extensions are used the optimal rate of convergence is algebraic with $\beta = p/2$. Uniform or nearly uniform meshes should be used.

Example 6.4.1 The exact solution of a long cylindrical tube, made of an elastic material and subjected to constant internal and/or external pressure, is characterized by the radial displacement function $u_r(r)$:

$$u_r = \frac{1}{E}\left(-C_1\frac{1+\nu}{r} + 2C_2(1-\nu-2\nu^2)r\right) \qquad r_i \leq r \leq r_o, \quad r_i > 0$$

where E, ν are the modulus of elasticity and Poisson's ratio, C_1, C_2 are constants to be determined from the boundary conditions, r_i, r_o are the insid and outside radius, respectively (see, for example, [29] page 244). Assuming that the axial component of the displacement is zero at the ends, plane strain conditions exist. In this case the domain and loading are analytic functions and the problem is in Category A.

If the tube is made of two or more materials, so that the material boundaries are concentric circles or smooth lines, then the problem is still in Category A; however, the material interfaces must be covered by element boundaries. The reason for this is that the solution is not analytic on the material interfaces.

Problems in Category B

The behavior of the solution in the neighborhood of singular points in two dimensions is characterized by the estimate (6.13). Ideally, the finite element space should be designed such that the interpolation error is approximately the same for all elements in the neighborhoods of singular points. This goal is achieved if the finite elements are laid out such that the sizes of elements decrease in geometric progression toward the singular point. The optimal grading

RATES OF CONVERGENCE 197

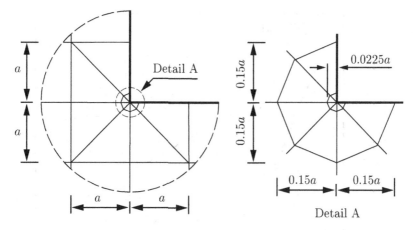

Figure 6.15 Example of a geometric mesh (detail).

is characterized by the common factor $q = (\sqrt{2} - 1)^2 \approx 0.17$, which is independent of the degree of singularity $\lambda_{\min}$ [32]. In practice $q = 0.15$ can be used.

For example, in a mesh consisting of three elements on $I = (0, \ell)$, the node points should be located at $x_1 = 0$, $x_2 = q^2\ell$, $x_3 = q\ell$, $x_4 = \ell$. These are called geometric meshes. An example of a geometric mesh in two dimensions is given in Figure 6.15.

The ideal distribution of polynomial degrees is that the lowest polynomial degree is associated with the smallest element and the polynomial degrees increase linearly away from the singular points. This is because the errors in the vicinity of singular points depend primarily on the size of the elements, whereas errors associated with elements farther from singular points, where the solution is smooth, depend mainly on the polynomial degree of elements. In practice a uniform p-distribution is used, which yields very nearly optimal results in the sense that exponential rates of convergence are realized and the work penalty associated with using a uniform rather than an optimal polynomial degree distribution is not substantial.

Example 6.4.2 Let us consider an L-shaped domain loaded by tractions corresponding to the lowest eigenvalue determined in Exercise 6.3.6. Plane strain and Poisson's ratio of 0.3 are assumed. The reentrant edges are stress free, the other boundaries are loaded by tractions computed from the stress field given in Exercise 6.3.7. Since the exact solution is known, it is possible to compute the exact value of the potential energy from Equation (4.40):

$$\Pi(\mathbf{u}_{EX}) = -\frac{1}{2} \oint [u_x(\sigma_x n_x + \tau_{xy} n_y) + u_y(\tau_{xy} n_x + \sigma_y n_y)] \, ds$$

$$= -4.154\,544\,23 \frac{A_1 a^{2\lambda_1}}{E}$$

where u_x, u_y are the displacement components given in Exercise 6.3.8; n_x, n_y are the components of the unit normal to the boundary; and a is the dimension shown in the inset in Figure 6.16. The errors in the energy norm were computed using Equation (2.93). The error curves for h-extension using uniform meshes ($p = 2$), p-extension, trunk space, on a uniform mesh,

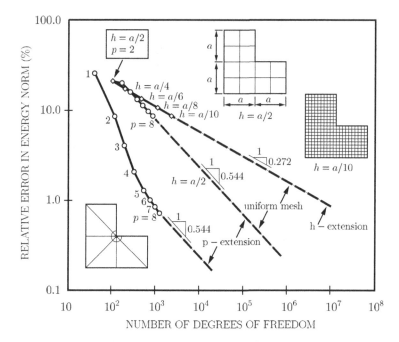

Figure 6.16 Example 6.4.2: error curves for h- and p-extensions. First published in [80]. Reprinted with permission of John Wiley & Sons, Inc. © 1991.

and p-extension, trunk space, on an 18-element geometric mesh constructed with the grading factor $q = 0.15$, are shown in Figure 6.16.

It can be seen that the asymptotic rates of convergence are exactly as predicted by the estimate (6.62). However, when p-extension is used on a geometric mesh the pre-asymptotic rate is exponential, or nearly so. This can be explained by observing that the geometric mesh shown in Figure 6.15 is over-refined for low polynomial degrees, hence the dominant source of error is that part of the domain where the exact solution is smooth and thus the rate of convergence is exponential, as predicted by the estimate (6.63). Convergence slows to the algebraic rate for small errors, where the dominant source of error is the immediate vicinity of the singular point.

Error curves for h-extension using radical meshes ($p = 2$) and hp-extension are shown in Figure 6.17. For purposes of comparison, the error curve for p-extension on an 18-element geometric mesh, shown in Figure 6.16, is shown in Figure 6.17 also.

The nodes of the radical meshes were located at distances d_k from the singular point defined by

$$d_k = a \left(\frac{k}{M}\right)^\theta, \quad k = 1, 2, \ldots, M, \quad \theta = \frac{p+1}{\lambda_1} = \frac{3}{0.544} = 5.51$$

where a is the dimension shown in Figure 6.17 and $M = 3, 4, \ldots, 11$ is the number of layers of elements. This choice of θ is based on the consideration that an ideal mesh is one in which the

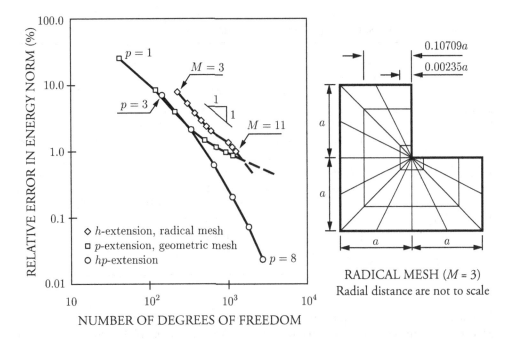

Figure 6.17 Example 6.4.2: error curves for h-, p- and hp-extensions and the radical mesh corresponding to $M = 3$ used in h-extension (not to scale).

error contribution of each element is approximately the same. For a detailed discussion of the one-dimensional case we refer to [70]. Note that grading in the radial direction is not to scale in Figure 6.17. This is because the grading is so strong that only the first layer would be visible on a mesh drawn to scale. The circumferential distribution of elements is nearly uniform, as shown in Figure 6.17. The circumferential distribution was kept constant for $M = 3$ to $M = 8$ then doubled for $M = 9$ to $M = 11$. This causes the convergence path to have a small "kink." It can be seen that the optimal rate of algebraic convergence $\beta = p/2 = 1.0$ is very nearly realized.

The error curve for the hp-extension was obtained using geometric meshes with the grading factor $q = 0.15$. The polynomial degree was increased uniformly from $p = 3$ to $p = 8$ (trunk space). Starting with the 18-element mesh shown in Figure 6.16, for each increase in polynomial degree a new layer of elements was added in the neighborhood of the singular point. It can be seen that the rate of convergence is stronger than algebraic; in fact it is exponential. When high accuracy is desired then hp-extension has to be used.

The estimated relative error in the energy norm was computed by the method described in Section 2.7 from data generated by p-extension on the 18-element geometric mesh shown in Figure 6.16. The estimated and true errors and the effectivity indices are listed in Table 6.4. The parameter β is that in Equation (6.62).

The extrapolated value of the potential energy was computed from the last three points on the convergence path, corresponding to $p = 6$ to $p = 8$. These points are located where the error curve is transitioning from the pre-asymptotic range to the asymptotic range. For this reason the relative errors are under-estimated; however, the effectivity index remains reasonably close to unity.

Table 6.4 Example 6.4.2: L-shaped domain, geometric mesh, 18 elements, trunk space. Plane strain, $\nu = 0.3$. Estimated and true relative errors in energy norm. Effectivity index (θ).

p	N	$\dfrac{\Pi(\mathbf{u}_{FE})E}{A_1^2 a^{2\lambda_1} t_z}$	β Est.	β True	$(e_r)_E$ (%) Est.	$(e_r)_E$ (%) True	θ
1	41	−3.886 332	—	—	25.42	25.41	1.00
2	119	−4.124 867	1.03	1.03	8.44	8.46	1.00
3	209	−4.148 121	1.37	1.36	3.91	3.93	0.99
4	335	−4.152 651	1.33	1.30	2.09	2.14	0.98
5	497	−4.153 636	0.99	0.94	1.42	1.48	0.96
6	695	−4.153 975	0.78	0.68	1.09	1.17	0.93
7	929	−4.154 139	0.69	0.60	0.89	0.99	0.89
8	1199	−4.154 238	0.69	0.56	0.75	0.86	0.87
∞	∞	−4.154 470		0.54		0	

Remark 6.4.2 This model problem is representative of the neighborhoods of corners (resp. edges) in two-dimensional (resp. three-dimensional) solution domains. Although the physical objects being modeled are typically filleted, the fillets are frequently omitted in order to facilitate meshing. Therefore the problem being solved by the finite element method is often without fillets. If the fillets are omitted then the mathematical model cannot be used for approximating the maximum stress unless special post-processing procedures are employed.

Remark 6.4.3 In Figure 6.17 the convergence curve corresponding to the radical mesh ($p = 2$) was obtained with a sequence refinement characterized by the number of layers of elements around the singular point denoted by M. Assuming that $a = 1.0$ m, the size of the smallest elements at $M = 11$ is 1.8×10^{-6} m which is smaller than the typical grain size of metals. Therefore the idealization of material properties as perfectly homogeneous cannot be justified on this scale. Nevertheless, the goal of numerical analysis is to find a *verified* approximation to the exact solution of the mathematical model. The problem of definition and *validation* of a mathematical model is separate from its numerical approximation which is the problem under consideration here. It would not be correct to say for example that the smallest element should not me smaller than the grain size.

Exercise 6.4.1 Refer to Exercise 6.3.1 and Figure 6.4(b). Let $\varrho = 1.0$ and $\alpha = 7\pi/4$. On Γ_{AB} and Γ_{BC} use the same boundary conditions as in Exercise 6.3.1. On Γ'_ϱ prescribe the normal flux

$$q_n^{(k)} = -\lambda_k \varrho^{\lambda_k - 1}(\cos \lambda_k \theta + (-1)^k \sin \lambda_k \theta) \quad \text{where } \lambda_k = \frac{(2k-1)\pi}{2\alpha}.$$

1. Solve this problem for $k = 1, 2, 3$ and compare the estimated rates of convergence with the theoretical rates given in Table 6.2. Discuss the results.

2. Prescribe the Dirichlet conditions

$$u^{(k)} = \varrho^{\lambda_k}(\cos \lambda_k \theta + (-1)^k \sin \lambda_k \theta)$$

on Γ_ϱ and repeat the solutions for $k = 1, 2, 3$. Compare the results with the previously obtained data. Comment on the differences.

3. Show that the exact value of the strain energy is given by

$$U_k = \frac{\lambda_k \varrho^{2\lambda_k}}{2} \int_{-\alpha/2}^{+\alpha/2} (\cos \lambda_k \theta + (-1)^k \sin \lambda_k \theta)^2 \, d\theta$$

hence

$$U_1 = 0.785\,398\,163\,397 \varrho^{2\lambda_1}$$
$$U_2 = 2.356\,194\,490\,19 \varrho^{2\lambda_2}$$
$$U_3 = 3.926\,990\,816\,99 \varrho^{2\lambda_3}.$$

4. Using Theorem 2.6.3 and the relationship between the potential energy and the strain energy of the exact solution, compute the effectivity indices.

Exercise 6.4.2 Using Table 6.2, estimate the asymptotic rate of convergence in the energy norm for the problem shown in Figure 6.8(a) for h- and p-extensions. Construct a coarse mesh on a semicircular domain and use p-extension, then let $p = 2$ and refine the mesh uniformly. As in Exercise 6.3.13, specify $T_r = \sigma_r$, $T_t = \tau_{r\theta}$ on the semicircular boundary. Use plane strain and $\nu = 0.3$. Compare the observed rates of convergence with the predicted ones. Hint: Make use of Theorem 2.6.3. The exact value of the potential energy is $-0.375\,722\,08\, p_0^2 R^2 t/E$ where R is the radius of the semicircular domain and t is the thickness.

Problems in Category C

When the exact solution does not have a recognizable structure, as in highly heterogeneous media, then h-extensions with uniform or nearly uniform mesh refinement should be used.

6.4.2 Uses of a priori information

Understanding the relationship between input data and the regularity of the exact solution is very important in finite element analysis for two reasons. First, in constructing an initial finite element mesh, whether manually or by automatic mesh generators, the presence of singular points, singular arcs and boundary layers should be taken into account. Second, if the goal of the computation is to determine the maximal stress, strain, flux, etc., then the degree of singularity λ (see Equation (6.13)) must be greater than one in the region where the maximum is sought.

The accuracy of the finite element solution depends on the smoothness of the exact solution and the finite element space $\tilde{S}(\Omega, \Delta, \mathbf{p}, \mathbf{Q})$. All implementations of finite element analysis impose limitations on the maximum polynomial degree and the choice of mapping functions. Therefore an important task is to construct a finite element mesh such that the desired accuracy can be achieved at $p \leq p_{\max}$ where $p_{\max}$ is the maximum p-level allowed by the implementation. In the following we focus on the goal to obtain a finite element solution that has a reasonably small error in the energy norm. The problem is formulated as follows: given a tolerance of relative error in the energy norm τ, design a mesh such that the relative error in the energy norm is within the allowed tolerance at some value of p that is less than or equal to $p_{\max}$.

Problems in Category A

For problems in Category A the rules of mesh construction are as follows:

1. The material and source term interfaces coincide with the boundaries of elements.

2. The mesh is such that:

   (a) The maximum ratio of the length of any two sides of an element is less than about 6.5.

   (b) On curved sides the ratio of the length of an element side and the radius of curvature of the side is less than about 0.5.

   (c) The vertex angles α_i satisfy the condition $0.1\pi < \alpha_i < 0.9\pi$.

   (d) The prescribed boundary conditions are interpolated by the trace of the finite element space such that the interpolation error is approximately equal among the elements that lie on the boundary at $p \approx p_{\max}/2$ and, furthermore, the interpolation error in maximum norm is less than about $\tau/10$.

When the mesh satisfies rules 1 through 2(c) then the rate of convergence of the p-version is exponential for problems in Category A, as indicated in Table 6.1, hence rapid convergence is achieved. This should not be interpreted to mean that the computational effort is necessarily small, because the constant in Equation (6.63) can be very large and entry into the asymptotic range may occur only at much higher values of p than the maximum allowed by the implementation ($p_{\max}$). Rule 2(d) is made necessary by the fact that p cannot exceed $p_{\max}$. In the following example we illustrate an application of rule 2(d) and outline the underlying considerations.

Example 6.4.3 In the following we consider the model problem $\Delta u = 0$ on $\Omega = \{x, y \mid -1 < x < 1, \ 0 < d < y < 1+d\}$ with the Dirichlet boundary condition $u = g$ prescribed on $\partial\Omega$. We define $g = r^m \cos m\theta$ where r and θ are polar coordinates centered on the origin. We fix $m = -2$, utilize symmetry, and solve the problem on the square domain $\Omega_s = \{x, y \mid 0 < x < 1, \ 0 < d < y < 1+d\}$ shown in Figure 6.18(a).

In the finite element method g is replaced by its interpolant $\bar{g}$ which depends on the mesh Δ, the polynomial degrees $\mathbf{p}$ and the mapping functions of those elements that have points on $\partial\Omega$. Our goal is to create a mesh in such a way that (a) the interpolation error $\|g - \bar{g}\|_{\max}$ is approximately the same for all elements along the boundary and (b) the relative error in the energy norm, estimated by the method outlined in the following, is within a given error tolerance at $p \approx p_{\max}/2$. Specifically, we would like to construct a mesh such that the relative error in the energy norm is less than 2.5% at $p = 4$. The reason for wanting to satisfy the error tolerance requirement at $p \approx p_{\max}/2$ is that only a rough estimate of the relative error

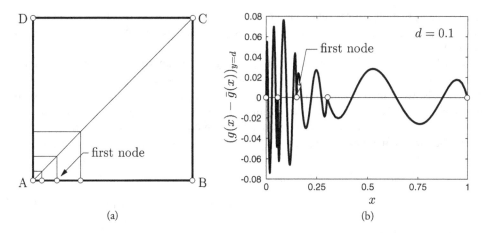

Figure 6.18 Example 6.4.3: (a) the domain Ω_s and the finite element mesh; (b) the error $g - \bar{g}$ along side AB for $d = 0.1$ and $p = 4$ with three internal nodes located on the side AB.

in the energy norm can be obtained on the basis of information available prior to performing the finite element analysis and we would like to make it reasonably certain that the desired error tolerance can be satisfied for $p \leq p_{\max}$.

A procedure for ensuring that the interpolation error $\|g - \bar{g}\|_{\max}$ is approximately the same for all elements along the boundary is that we define a node along one of the boundary segments, for example, segment AB shown in Figure 6.18(a), and change its position until $\|g - \bar{g}\|_{\max}$ is the same on either side. We then fix the first node and locate one node on each side of the first node to satisfy the same criterion. Next we select the two segments where $\|g - \bar{g}\|_{\max}$ is the largest and repeat the procedure.

The dependence of $\|g - \bar{g}\|_{\max}$ on the location of the first node along side AB is shown in Figure 6.19 for $d = 0.1$ and $d = 0.01$. It can be seen that for $d = 0.1$ the first node is located at approximately $x = 0.15$ and for $d = 0.01$ it is located at $x = 0.02$. Dividing the segments on either side of the first node results in two new node locations: for $d = 0.1$ the new node locations are $x = 0.058$ and $x = 0.3$. The error $g - \bar{g}$, following the second refinement, is shown in Figure 6.18(b).

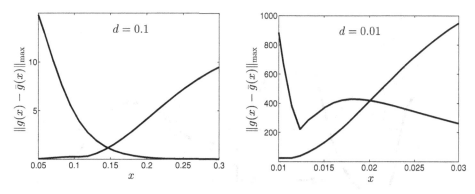

Figure 6.19 Example 6.4.3: dependence of the absolute error $\|g - \bar{g}\|_{\max}$ on the location of the first node along boundary segment AB for $p = 4$ and (a) $d = 0.1$, (b) $d = 0.01$.

The interpolation error is much larger on side AB than on sides BC and CD, therefore as long as $\|g - \bar{g}\|_{\max}$ is largest on side AB it is not necessary to locate nodes on the sides BC and CD.

In addition to locating nodes such that $\|g - \bar{g}\|_{\max}$ is approximately the same for all elements along the boundary, we need a criterion for deciding how many nodes to create. In other words, we need a rough estimate of the relative error in the energy norm, based on the information available to us prior to performing a finite element analysis. We denote the exact solution corresponding to the Dirichlet boundary condition $\bar{g}(s)$ prescribed on $\partial\Omega$ by $\bar{u}_{EX}$. Using the triangle inequality, we get

$$\|u_{EX} - u_{FE}\|_{E(\Omega)} = \|u_{EX} - \bar{u}_{EX} + \bar{u}_{EX} - u_{FE}\|_{E(\Omega)}$$
$$\leq \|u_{EX} - \bar{u}_{EX}\|_{E(\Omega)} + \|\bar{u}_{EX} - u_{FE}\|_{E(\Omega)}. \quad (6.64)$$

Furthermore, the function $u_{EX} - \bar{u}_{EX}$ satisfies

$$\Delta(u_{EX} - \bar{u}_{EX}) = 0 \quad \text{with } g(s) - \bar{g}(s) \text{ prescribed on } \partial\Omega.$$

The function $g(s) - \bar{g}(s)$ can be determined for any partition of $\partial\Omega$. We make the following assumption:

$$\|\bar{u}_{EX} - u_{FE}\|_{E(\Omega)} \leq C \|u_{EX} - \bar{u}_{EX}\|_{E(\Omega)} \quad (6.65)$$

where the constant C depends on Δ, $\mathbf{p}$ and $\mathbf{Q}$. We assume that C is not very large (of the order of 10). Using this assumption, we need to estimate $\|u_{EX} - \bar{u}_{EX}\|_{E(\Omega)}$. This estimate is based on the following consideration: for any partition of $\partial\Omega$ we can compute $u_{EX} - \bar{u}_{EX}$. This is illustrated in Figure 6.20.

For each interval $s_k < s < s_{k+1}$ we approximate $u_{EX} - \bar{u}_{EX}$ by a sinusoidal function and construct a function $w_k(s, t)$ such that $\Delta w_k = 0$:

$$w_k = \beta_k \sin\left(\pi \frac{s - s_k}{\ell_k}\right) \exp\left(-\frac{\pi t}{\ell_k}\right). \quad (6.66)$$

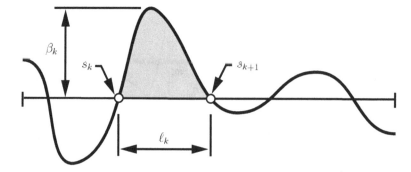

Figure 6.20 The function $u_{EX} - \bar{u}_{EX}$ on a boundary segment.

We now compute $\|w_k\|_{E(\Omega)}$ on the domain Ω_k defined by

$$\Omega_k := \{s, t \mid s_k < s < s_{k+1}, \ 0 < t < \infty\}$$

and find

$$\|w_k\|_{E(\Omega_k)}^2 = \frac{\pi}{4}\beta_k^2. \tag{6.67}$$

On summing over all intervals, a rough estimate of the absolute error in energy norm is obtained:

$$\|u_{EX} - \bar{u}_{EX}\|_{E(\Omega)}^2 \approx \frac{\pi}{4}\sum_k \beta_k^2. \tag{6.68}$$

Specifically for this example we have $\|u_{EX} - \bar{u}_{EX}\|_{E(\Omega)}^2 \approx 2.90 \times 10^{-2}$. Therefore $\|u_{EX} - \bar{u}_{EX}\|_{E(\Omega)} \approx 0.17$. Assuming $C \approx 10$ in Equation (6.65), we estimate $\|u_{EX} - u_{FE}\|_{E(\Omega)}$ to be approximately 1.7. Since the potential energy of the exact solution is known, $\pi(u_{EX}) = 2944.823\,67$, it is possible to compute the exact value of $\|u_{EX} - u_{FE}\|_{E(\Omega)} = 1.2$. Therefore the estimated value of the absolute error in the energy norm is reasonably close to its exact value.

Estimation of the relative error requires an estimate of $\|u\|_{E(\Omega)}$. We denote the mean value of u_{EX} on $\partial\Omega$ by u_m. For each interval $s_k < s < s_{k+1}$ we approximate $u_{EX} - u_m$ by a sinusoidal function as in Equation (6.66) and proceed as before. Note that subtraction of u_m does not affect the estimate of $\|u\|_{E(\Omega)}$. The estimate of $\|u\|_{E(\Omega)}$ is generally less accurate than the estimate given by Equation (6.68). In this example $u_m = -0.348$ and $\|u\|_{E(\Omega)}$ is estimated to be 89.4. Therefore the estimated relative error at $p = 4$ is

$$\frac{\|u_{EX} - u_{FE}\|_{E(\Omega)}}{\|u\|_{E(\Omega)}} = \frac{1.7}{89.4} = 0.019 \quad (1.9\%)$$

and the true relative error at $p = 4$ is

$$\frac{\|u_{EX} - u_{FE}\|_{E(\Omega)}}{\|u\|_{E(\Omega)}} = \frac{1.2061}{54.2663} = 0.022 \quad (2.2\%)$$

where $\|u\|_{E(\Omega)} = \sqrt{\pi(u_{EX})}$. The results indicate that reasonably accurate estimates can be provided for the relative error in the energy norm based on the input data. This provides a basis for the design of the finite element mesh.

The computed potential energy $\pi(u_{FE})$, the absolute error in the energy norm $\|e\|_{E(\Omega)} := \|u_{EX} - u_{FE}\|_{E(\Omega)}$ and the relative error in the energy norm $(e_r)_E$ (%) are given in Table 6.5 for $p = 1, 2, \ldots, 8$. It can be seen that the mesh is adequate for reducing the relative error in the energy norm to under 1% at $p = 5$.

In the case of Neumann boundary conditions the normal derivative $\partial u_{EX}/\partial n$ is specified. This is approximated by $\partial \bar{u}_{EX}/\partial n$ and the difference $\partial u_{EX}/\partial n - \partial \bar{u}_{EX}/\partial n$ can be represented

Table 6.5 The computed potential energy $\pi(u_{FE})$, the absolute error in the energy norm $\|e\|_{E(\Omega)} := \|u_{EX} - u_{FE}\|_{E(\Omega)}$ and the relative error in the energy norm $(e_r)_E$ (%) in Example 6.4.3.

p	N	$\pi(u_{FE})$	$\|e\|_{E(\Omega)}$	$(e_r)_E$ (%)
1	6	3206.750 69	16.184 159 5	29.82
2	20	2968.026 30	4.816 910 9	8.88
3	36	2958.070 24	3.639 583 9	6.71
4	60	2946.278 42	1.206 130 4	2.22
5	92	2945.083 89	0.510 118 0	0.94
6	132	2944.855 22	0.177 624 1	0.33
7	180	2944.826 48	0.053 012 2	0.10
8	236	2944.823 92	0.015 820 6	0.03
∞	∞	2944.823 67	—	—

by Figure 6.20. Referring to Equation (6.66) we note that

$$\frac{\partial w_k}{\partial t} = -\frac{\pi}{\ell_k} w_k.$$

Therefore when Neumann boundary conditions are specified then β_k must be scaled by $-\ell_k/\pi$ and therefore the estimate (6.68) becomes

$$\|u_{EX} - \bar{u}_{EX}\|^2_{E(\Omega)} \approx \frac{1}{4\pi} \sum_k \beta_k^2 \ell_k^2. \quad (6.69)$$

Solving the problem of this example with Neumann boundary conditions, the optimal node locations along side AB are $x_1 = 0.0462$, $x_2 = 0.110$, $x_3 = 0.266$ and $\|u_{EX} - \bar{u}_{EX}\|^2_{E(\Omega)} \approx 0.0879$.

Problems in Category B

The number of layers required for a given error tolerance (e.g., 5% percent error in the energy norm) is related to the magnitude of the coefficient of the first eigenfunction, such as a_1 and b_1 in Equation (6.11), and hence depends on the exact solution. While this information is not available a priori, it can be estimated once a finite element solution is available. In general, the more energy is associated with the leading singular term, the more layers of geometrically graded elements are needed. For practical engineering accuracy, one or two layers are sufficient in most problems.

Boundary layers

The existence of boundary layers is well known in connection with models of viscous fluid flow, but less well known, although similarly important, in connection with models of heat conduction and structural plates and shells. In this section the main ideas are outlined in a

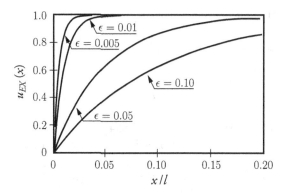

Figure 6.21 The solution $u_{EX}(x)$, given by Equation (6.71), in the neighborhood of $x = 0$ for various values of ϵ.

one-dimensional setting and illustrated by an example. Analogous considerations apply to problems in higher dimensions.

Consider the problem

$$-\kappa u'' + u = f(x), \quad u(0) = u'(\ell) = 0 \qquad (6.70)$$

on the interval $I = (0, \ell)$. Let $\kappa = \epsilon^2$ (constant) where $\epsilon \ll 1$. The exact solution of this problem has a boundary layer. Qualitatively speaking, we see that since κ is a small number, $u \approx f(x)$ on the inside of the interval I, but, given the boundary conditions, u has to be zero at $x = 0$. Therefore $u(x)$ changes rapidly over an interval $0 < x < d(\epsilon) \ll \ell$. This is seen in Figure 6.21 where the exact solution corresponding to $f(x) = 1$ is plotted for various ϵ values on the interval $0 < x/\ell < 0.20$. Although the problem is in Category A, the regularity of its exact solution given by

$$u_{EX}(x) = 1 - \cosh x/\epsilon + \tanh(\ell/\epsilon)\sinh x/\epsilon \approx 1 - e^{(-x/\epsilon)} \qquad (6.71)$$

deteriorates at the boundary as $\epsilon \to 0$. Therefore it might require unrealistically high polynomial degrees to obtain a close approximation to the solution near the boundary if only one finite element were used.

The optimal discretization strategy for problems with boundary layers is discussed in the context of the hp-version in [71]. The results of analysis indicate that the size of the element at the boundary is proportional to the product of the polynomial degree p and the parameter ϵ. Specifically, for the problem discussed here, the optimal mesh consists of two elements with the node points located at $x_1 = 0$, $x_2 = d$, $x_3 = \ell$, where $d = Cp\epsilon$ with $0 < C < 4/e$.

Example 6.4.4 The lateral surface of a cylindrical rod, made of a ceramic material, is exposed to flowing water, the temperature of which is $90\,°C$, while the ends of the rod are maintained at $10\,°C$. The diameter of the rod is $d = 10\,\mathrm{mm}$ and its length is $\ell = 200\,\mathrm{mm}$. The coefficient of thermal conduction of the ceramic material is $k = 1.0\,\mathrm{W/m\,K}$. The flow conditions are such that the coefficient of convective heat transfer is $h_c = 6.25 \times 10^4\,\mathrm{W/m^2\,K}$.

The objective is to determine the steady state temperature distribution in the rod and the flux at the ends.

Our mathematical model is formulated as a one-dimensional problem, using Equation (2.25):

$$Aku'' - c_b u = -c_b u_a$$

where the coefficients are $Ak = (d^2\pi/4)k = 7.854 \times 10^{-5}$ W m/K, $c_b = d\pi h_c = 1.963 \times 10^3$ W/m K and $u_a = 90\,°\mathrm{C}$. Dividing by c_b and letting $\epsilon^2 = Ak/c_b$ we get

$$-\epsilon^2 u'' + u = u_a \quad \text{where } \epsilon = 2.0 \times 10^{-4}\,\text{m}.$$

The boundary conditions are $u(0) = u(\ell) = 10\,°\mathrm{C}$. Alternatively, since the solution is symmetric with respect to $\ell/2$, it is sufficient to solve on $I = (0, \ell/2)$ with the boundary conditions $u(0) = 10\,°\mathrm{C}$, $u'(\ell/2) = 0$. Therefore the exact solution is

$$u_{EX} = 90 - 80\cosh x/\epsilon + 80\tanh \ell/2\epsilon \sinh x/\epsilon, \quad 0 \le x \le \ell/2 \quad (u \text{ in units of }°\mathrm{C}).$$

Since $\epsilon \ll \ell$, $\tanh \ell/2\epsilon \approx 1$,

$$u_{EX} \approx 90 - 80\mathrm{e}^{(-x/\epsilon)}, \quad 0 \le x \le \ell/2.$$

We solve this problem using a uniform p-distribution ranging from $p = 5$ to $p = 8$ and compute the relative error in u' in maximum norm on $\bar{I}$ as a function of d. The results are shown in Figure 6.22. The two-element mesh and the definition of d are shown in the inset in Figure 6.22.

It can be seen that, consistent with the results of analysis [71], the location of the minimum value of $\max |u'_{EX} - u'_{FE}|$ changes with p. It can be seen also that the minimum is more sharply

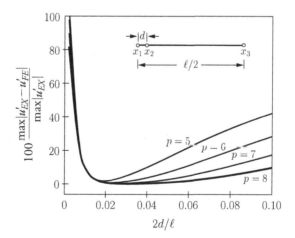

Figure 6.22 Relative error of the first derivative in maximum norm on $[0, \ell/2]$ vs. the distance d.

defined for lower p values than for higher p values. In other words, the sensitivity of the error to the position of node x_2 decreases with increasing p.

The exact value of the flux at $x = 0$ is

$$q_{EX}(0) = -ku'_{EX}(0) \approx -80/\epsilon = -4.0 \times 10^5 \text{ W/m}^2.$$

At $d = 1.982 \times 10^{-3}$ m and $p = 8$, $q_{FE}(0) = -4.0 \times 10^5$ W/m^2 to 5 significant figures, hence the relative error is less than $1.0 \times 10^{-3}\%$.

Remark 6.4.4 In higher dimensions the mesh has to be designed to resolve the boundary layer as well as singularities arising from corners, changes in boundary conditions and material properties.

6.4.3 A posteriori error estimation in the energy norm

Having computed a finite element solution, or a sequence of finite element solutions, it is necessary to estimate the errors of approximation. Whatever the goals of analysis are, to ensure that the errors in the data of interest are small, it is necessary to ensure that the error in the energy norm is small.

There are various approaches to estimating the error in the energy norm. One estimator, based on a hierarchic sequence of finite element spaces, was described in Section 2.7 and illustrated in Example 6.4.2; see Table 6.4 where the a priori estimate given by Equation (6.62) was used for estimating the exact value of the potential energy.

In the h-version the finite element solution corresponding to one finite element space S is available. Error estimators developed for the h-version share the common feature that the finite element solution is post-processed with the aim of obtaining a more accurate solution. The energy norm of the difference between the original and post-processed solutions is the estimated error. An error estimator is said to be robust if it works well for a wide range of problems in the sense that the effectivity indices are close to unity. In the following we outline two procedures for a posteriori error estimation in the energy norm. For a comprehensive treatment of this subject we refer to [2], [12], [16].

The Zienkiewicz–Zhu estimator

Assume that a finite element solution corresponding to $p = 1$ is available on a triangular mesh, such as the mesh shown in Figure 6.23. Therefore the finite element solution is known at each node point. The first derivatives are constant over each element and hence are discontinuous at inter-element boundaries. We know that, outside of the neighborhoods of singular points and interfaces, the exact solution has continuous derivatives. The goal is to estimate the value of the derivatives of the exact solution at each node point. This is done by evaluating the derivatives in the centroid of each triangle within the patch of elements associated with each node. A patch of elements, associated with node A, is shown in Figure 6.23. An estimate of the derivatives is computed for the interior node of the patch by means of least squares fitting. It can be shown that, under reasonable assumptions concerning the regularity of the exact solution, the error in the recovered gradients is proportional to h^2, whereas the error in the derivatives computed directly from the finite element solution is proportional to h.

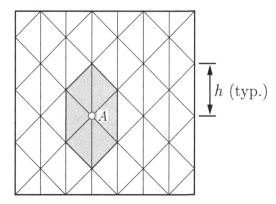

Figure 6.23 Example of a patch corresponding to node A.

The derivatives are computed at each node point and piecewise linear functions are constructed to represent the first derivatives on each triangle. The estimated error on each element is the energy norm of the difference between the piecewise constant derivatives computed from the finite element solution and the recovered derivatives computed by the patch recovery method.

This estimate, proposed in a series of papers by Zienkiewicz and Zhu [96] [97], [98] for $p = 1$ and $p = 2$, is called the Zienkiewicz–Zhu estimator or ZZ estimator or sometimes the Z^2 estimator. Experience has shown that this is a very robust estimator.

Estimators based on residuals

The finite element solution can be viewed as the exact solution of a problem that differs from the original problem in loading only. For example, if we wish to solve

$$-\Delta u = f \quad x \in \Omega \quad u = 0 \text{ on } \Gamma$$

and we have a finite element solution u_{FE} then we can compute $f_{FE} = -\Delta u_{FE}$ and have the following equation for the error:

$$-\Delta(u - u_{FE}) = f - f_{FE}. \tag{6.72}$$

In addition, there will be line loads on the boundaries of the elements because the stress or flux components at the common boundaries of elements, computed from u_{FE}, are generally discontinuous in the finite element solution. The discontinuity is equivalent to a line load acting on the element boundary. Each element must be equilibrated by assigning part of the discontinuities along its perimeter to its sides such that the line loads, together with the residuum $f - f_{FE}$, satisfy the equilibrium condition. The problem is then to find an approximation to the error $u^{(k)} - u_{FE}^{(k)}$ by solving Equation (6.72) element by element and computing the local estimate

$$\eta_k := \|u^{(k)} - u_{FE}^{(k)}\|_{E(\Omega)}, \quad k = 1, 2, \ldots, M(\Delta)$$

where η_k measures the contribution of the kth element to the total error measured in the energy norm. The estimator for the total error is then

$$\|u - u_{FE}\|_{E(\Omega)} \approx \left(\sum_{k=1}^{M(\Delta)} \eta_k^2 \right)^{1/2}.$$

For detailed analysis of residual error estimation we refer to [16].

Richardson extrapolation

Suppose that a finite element mesh is refined in such a way that the diameter of each element is halved. We denote the corresponding finite element solutions by u_h and $u_{h/2}$ and the corresponding errors by $e_h = u - u_h$ and $e_{h/2} = u - u_{h/2}$.

When u is smooth and $p = 1$ we have $e_h \approx \Psi_h h$ and $e_{h/2} \approx \Psi_{h/2} h/2$ where $\Psi_h \approx \Psi_{h/2}$ up to higher order terms in h. Denoting $\Psi \approx \Psi_h \approx \Psi_{h/2}$ we have

$$u_{h/2} - u_h \approx \Psi h/2. \tag{6.73}$$

Therefore $e_h \approx 2(u_{h/2} - u_h)$ and

$$\|u - u_h\|_{E(\Omega)} \approx 2\|u_h - u_{h/2}\|_{E(\Omega)}. \tag{6.74}$$

This is a classical error estimation based on extrapolation, known as Richardson extrapolation.[13]

6.4.4 Adaptive and feedback methods

The term adaptivity has various meanings. In the broad (engineering) sense adaptivity is understood to be a process devised for efficient control of the errors of idealization and the errors of discretization to guarantee that the data of interest are computed to within a specific tolerance. Adaptivity in a restricted sense is understood to be a process for efficient control of the errors of discretization in terms of the data of interest for linear problems only. Adaptivity in the sense of the first research papers published on the subject was understood to mean a process by which the errors of discretization are controlled in linear elliptic problems in the energy norm only. See, for example, [9].

The goal of adaptive methods is to create a sequence of finite element spaces S_i, $i = 2, 3, \ldots, n$, from information developed from the finite element solutions corresponding to the spaces S_j, $j = 1, 2, \ldots, i - 1$, such that the relative errors in the data of interest Φ_k are within prescribed tolerances τ_k, see Equation (1.3).

Adaptive control of error in the energy norm

The first goal of adaptivity is to ensure that the error measured in the energy norm is sufficiently small for the purposes of analysis. There are several adaptive methods in existence, greatly varying in performance and reliability. In general, adaptive processes involve the use of an error indicator, an error estimator and an extension operator.

---

[13] Lewis Fry Richardson (1881–1953).

Error indicators are estimates of the relative contribution of each element to the total error in the energy norm. If an indicator also measures the error contribution of each element to the total error then it is called a local error estimator. Local error estimators used in the h-version were described in Section 6.4.3. For a given number of degrees of freedom an ideal discretization has the property that each element contributes the same amount of error. The error estimator measures the error in the energy norm for the entire domain and provides information for the stopping criterion. An extension operator is an algorithm that produces a new finite element space based on the information provided by the estimator and the indicator. Ideally, an extension operator would yield the final discretization in a single step.

Extension operators are very different for the h-, p- and hp-extensions. In h-extensions the mesh is refined by one of several rules. For example, those elements are subdivided for which the error indicator is grater than (say) 75% of the maximal value of the error indicators computed for all elements. In p-extensions the polynomial degree is increased either selectively or uniformly. In hp-extensions the number of layers of geometrically graded elements is increased together with the polynomial degree of elements. Another kind of extension operator for the hp-method is described in [24a].

Adaptivity for problems in Category B is best based on the hp-extension for two reasons: the fast rate of convergence and the fixed pattern of mesh refinement, independent of the degree of singularity. Only the number of geometrically graded elements depends on the coefficient of the singular term, that is, the flux or stress intensity factor [65].

Control of errors in the data of interest

The errors in the data of interest depend on the error in the energy norm and the method used for computing the data from the finite element solution, called the method of extraction. Some data, such as displacements and strains, can be computed from the finite element solution directly, whereas other data, such as stress intensity factors, can be computed by indirect methods only.

Whatever the method of computation, it is necessary to show that the data of interest are substantially independent of the discretization. Dependence on the discretization (e.g., number of elements, polynomial degree) is a clear indication that the errors are large.

One of the major advantages of the p-version is that data computed from finite element solutions corresponding to a hierarchic sequence of finite element spaces, which is naturally produced when p is increased on a fixed mesh, provide convergence information from which it is possible to obtain reliable assessments of the accuracy of the solution.

A frequently occurring conceptual error in engineering practice is that the goals of computation are inconsistent with the exact solution. Singular points may be a property of the problem (e.g., models of cracks, laminated composites) or may have been introduced in order to simplify meshing: since it is much simpler to generate a finite element mesh for an object that has no fillets than one that does, analysts often omit fillets. In either case, when singular points exist in the region of interest then data that depend on unbounded derivatives should never be reported.

6.5 Chapter summary

In applications of the finite element method, a priori estimation of the regularity of the exact solution is important for the following reasons:

(a) The definition of a mathematical model must be consistent with the goals of computation. Therefore, when the first derivatives are not finite in one or more points then data that depend on the first derivatives, such as flux, strain and stress, should not be computed or reported at those points.

(b) The use of inadmissible data, such as point constraints, concentrated forces and concentrated fluxes, should be avoided or treated with care.

(c) The initial mesh layout should be defined with an understanding of what an optimal mesh would be, given information on the regularity of the exact solution that can be inferred from the input data.

The regularity of the solution is determined by the input data. Solution domains are typically bounded by piecewise analytic surfaces. The intersections of analytic surfaces are singular arcs, the endpoints of which are singular points. In addition, there may be abrupt changes in material properties and boundary conditions that further partition the analytic surfaces and therefore produce additional singular arcs and points.

Considering the problem $-\Delta u = f$ in two dimensions, the homogeneous part of the exact solution in the neighborhood of corner singularities and at the intersections of material interfaces with external boundaries or other material interfaces can be written in the form

$$u_{EX} = \sum_{i=1}^{\infty} A_i r^{\lambda_i} (\log r)^{\beta_i} \phi_i(\theta) \qquad r < \rho_0 \tag{6.75}$$

where r and θ are polar coordinates centered on the singular point, $\lambda_i > 0$, $\beta_i > 0$, $\phi_i(\theta)$ are analytic or piecewise analytic functions, A_i are coefficients that depend on f, called flux intensity factors, and ρ_0 is the radius of convergence. The regularity of the solution in the neighborhood of singular points is characterized by the smallest λ_i (corresponding to non-zero A_i) denoted by λ. In two- and three-dimensional elasticity similar, although more complicated, expressions can be written and λ_i may be complex. Here λ is understood to be the real part of the smallest fractional characteristic value.

Solutions where $\lambda < 1$ at one or more points are said to be strongly singular, otherwise weakly singular. When the solution is strongly singular the first (and higher) derivatives are infinite at the singular point. The finite element mesh should be laid out so that the elements in the vicinity of strongly singular points are the smallest. If the option of using p-extensions is provided by the finite element analysis software then mesh grading should be in geometric progression toward strongly singular points with a common factor of approximately 0.15.

7

Computation and verification of data

Following assembly and solution (outlined in Sections 2.5.6 and 2.5.8), the finite element solution is stored in the form of data sets that contain the coefficients of the shape functions, the mapping functions and indices that identify the polynomial space associated with each element.

Some of the data of interest, such as temperature, displacement, flux, strain, stress, can be computed from the finite element solution either by direct or indirect methods, while others, such as stress intensity factors, can be computed by indirect methods only. In this chapter the techniques used for the computation and verification of engineering data are described.

7.1 Computation of the solution and its first derivatives

If one is interested in the value of the solution at a point (x_0, y_0, z_0) then the domain has to be searched to identify the element in which that point lies. Suppose that the point lies in the kth element. The next step is to find the standard coordinates (ξ_0, η_0, ζ_0) from the mapping functions

$$x_0 = Q_x^{(k)}(\xi_0, \eta_0, \zeta_0), \quad y_0 = Q_y^{(k)}(\xi_0, \eta_0, \zeta_0), \quad z_0 = Q_z^{(k)}(\xi_0, \eta_0, \zeta_0). \tag{7.1}$$

Unless the mapping of the kth element happens to be linear, the inverse of the mapping function is not known explicitly. Therefore this step involves a root finding procedure, such as the Newton–Raphson method. If the point is a vertex or lies on an edge, or in three dimensions on a face, then it may be shared by more than one element.

The next step is to look up the parameters that identify the standard space associated with the element and the computed coefficients of the basis functions. With this information the solution and its derivatives can be computed. For example, let us assume that the solution is a

Introduction to Finite Element Analysis: Formulation, Verification and Validation, First Edition.
Barna Szabó and Ivo Babuška. © 2011 John Wiley & Sons, Ltd. Published 2011 by John Wiley & Sons, Ltd.

scalar function and the standard space $S^{p,q}(\Omega_{st}^{(q)})$ is associated with the kth element, denoted by Ω_k. Then the finite element solution at point $(x_0, y_0) \in \Omega_k$ is:

$$u_{FE}(x_0, y_0) = \sum_{i=1}^{n} a_i^{(k)} N_i(\xi_0, \eta_0) \qquad (7.2)$$

where $n = (p+1)(q+1)$ is the number of shape functions that span $S^{p,q}(\Omega_{st}^{(q)})$, $N_i(\xi, \eta)$ are the shape functions and $a_i^{(k)}$ are the corresponding coefficients. When the solution is a vector function then each component of $\vec{u}_{FE}$ is in the form of Equation (7.2).

Computation of the first derivatives of u_{FE} at the point (x_0, y_0) involves computation of the inverse of the Jacobian matrix at the corresponding point (ξ_0, η_0) and multiplying the derivatives of the finite element solution with respect to the standard coordinates. Referring to Equation (5.58),

$$\begin{Bmatrix} \dfrac{\partial u_{FE}}{\partial x} \\ \dfrac{\partial u_{FE}}{\partial y} \end{Bmatrix}_{(x_0, y_0)} = \begin{bmatrix} \dfrac{\partial x}{\partial \xi} & \dfrac{\partial y}{\partial \xi} \\ \dfrac{\partial x}{\partial \eta} & \dfrac{\partial y}{\partial \eta} \end{bmatrix}^{-1}_{(\xi_0, \eta_0)} \sum_{i=1}^{n} a_i^{(k)} \begin{Bmatrix} \dfrac{\partial N_i}{\partial \xi} \\ \dfrac{\partial N_i}{\partial \eta} \end{Bmatrix}_{(\xi_0, \eta_0)} \qquad (7.3)$$

where $x = Q_x^{(k)}(\xi, \eta)$, $y = Q_y^{(k)}(\xi, \eta)$. The flux vector (resp. stress tensor) is computed by multiplying the temperature gradient (resp. the strain tensor) by the thermal conductivity matrix (resp. material stiffness matrix). The transformation of vectors and tensors is described in Appendix C.

The derivatives of the finite element solution are discontinuous along inter-element boundaries. Therefore, if the point selected for the evaluation of fluxes, stresses, etc., is a node point, or a point on an inter-element boundary, then the computed value depends on the element selected for the computation. The degree of discontinuity in the normal and shearing stresses or the normal flux component at inter-element boundaries is an indicator of the quality of the approximation. In implementations of the h-version the derivatives are typically evaluated at the integration points and are interpolated over the elements. In graphical displays in the form of contour plots, discontinuities of the derivatives at element boundaries are often masked by means of smoothing the contour lines through averaging.

In the p-version the standard element is subdivided so as to produce a uniform grid, called a display grid, and the solution and its derivatives are evaluated at the grid points. Since the standard coordinates of the grid points are known, inverse mapping is not involved. The quality of contour plots depends on the quality of data being displayed and on the fineness of the display grid.

Searching for a maximum or minimum value also involves searching on a uniform grid defined on the standard elements. The fineness of the grid, and hence the number of points searched for the minimum or maximum, is controlled by a parameter. In conventional implementations of the h-version the search grid is typically defined by the integration points or the node points.

Exercise 7.1.1 Consider two plane elastic elements that have a common edge. Assume that different material properties were assigned to the elements. Show that the normal and shearing stresses corresponding to the exact solution have to be the same along the common edge. Hint: Consider equilibrium in the coordinate system defined in the normal and tangential directions.

7.2 Nodal forces

Recall the definition of nodal forces $\{f^{(k)}\}$ in Section 2.5.9

$$\{f^{(k)}\} = [K^{(k)}]\{a^{(k)}\} - \{\bar{r}^{(k)}\}, \qquad k = 1, 2, \ldots, M(\Delta) \tag{7.4}$$

where $[K^{(k)}]$ is the stiffness matrix, $\{a^{(k)}\}$ is the solution vector and $\{\bar{r}^{(k)}\}$ is the load vector corresponding to volume forces and thermal loads acting on element k.

7.2.1 Nodal forces in the h-version

When solving problems in elasticity using finite element analysis based on the h-version, nodal forces are treated in the same way as concentrated forces are treated in statics. Typical uses of nodal forces are: (a) isolating some region of interest from a larger structure and treating the isolated region as if it were a free body held in equilibrium by the nodal forces; and (b) computing stress resultants. The underlying assumption is that nodal forces reliably represent the load path, that is, the distribution of internal forces in a statically indeterminate structure. This assumption is usually justified by the argument that nodal forces satisfy the equations of static equilibrium for any element or group of elements.

The following discussion will show that satisfaction of equilibrium is related to the rank deficiency of unconstrained stiffness matrices. Consequently, equilibrium of nodal forces should not be interpreted as an indicator of the quality of the finite element solution and does not guarantee that the nodal forces are reliable approximations of the internal forces in a statically indeterminate structure. On the other hand, nodal forces are useful for the computation of stress resultants.

Let us assume, for example, that Ω_k is an eight-node isoparametric quadrilateral element. The number of degrees of freedom per field, denoted by n, is 8. The notation is shown in Figure 7.1.

In expanded notation the elements of the nodal force vector $\{f^{(k)}\}$ are

$$f_x^{(k,i)} = f_i^{(k)}, \quad f_y^{(k,i)} = f_{n+i}^{(k)}, \quad i = 1, 2, \ldots, n. \tag{7.5}$$

Similarly, the elements of the vector $\{\bar{r}\}$ are written as follows:

$$\bar{r}_x^{(k,i)} = \bar{r}_i^{(k)}, \quad \bar{r}_y^{(k,i)} = \bar{r}_{n+i}^{(k)}, \quad i = 1, 2, \ldots, n. \tag{7.6}$$

On adapting the notation used in Equation (5.62) to problems of planar elasticity we can write

$$f_x^{(k,i)} = \int_{\Omega_{st}^{(q)}} \left([M_1][J_k]^{-1}\{\mathcal{D}\}N_i\right)^T [E] \sum_{j=1}^{2n} \{\varepsilon_j^{(k)}\} a_j^{(k)} |J_k| d\xi d\eta - \bar{r}_x^{(k,i)} \tag{7.7}$$

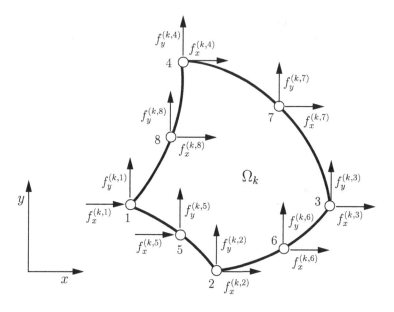

Figure 7.1 Nodal forces associated with the eight-node quadrilateral element: notation.

and

$$f_y^{(k,i)} = \int_{\Omega_{st}^{(q)}} \left([M_2][J_k]^{-1}\{\mathcal{D}\}N_i\right)^T [E] \sum_{j=1}^{2n} \{\varepsilon_j^{(k)}\} a_j^{(k)} |J_k| d\xi d\eta - \bar{r}_y^{(k,i)} \qquad (7.8)$$

where $[J_k]$ is the Jacobian matrix and

$$[M_1] = \begin{bmatrix} 1 & 0 \\ 0 & 0 \\ 0 & 1 \end{bmatrix}, \quad [M_2] = \begin{bmatrix} 0 & 0 \\ 0 & 1 \\ 1 & 0 \end{bmatrix}, \quad \{\mathcal{D}\} = \left\{ \begin{array}{c} \dfrac{\partial}{\partial \xi} \\ \dfrac{\partial}{\partial \eta} \end{array} \right\}$$

and

$$\{\varepsilon_j^{(k)}\} = \begin{cases} [M_1][J_k]^{-1}\{\mathcal{D}\}N_j & \text{for } j = 1, 2, \ldots, n \\ [M_2][J_k]^{-1}\{\mathcal{D}\}N_{j-n} & \text{for } j = n+1, n+2, \ldots, 2n. \end{cases}$$

The elements of the load vector corresponding to body forces and thermal loads are

$$\bar{r}_x^{(k,i)} = \int_{\Omega_{st}^{(q)}} N_i F_x |J_k| d\xi d\eta + \int_{\Omega_{st}^{(q)}} \left([M_1][J_k]^{-1}\{\mathcal{D}\}N_i\right)^T [E]\{\alpha\} T_\Delta |J_k| d\xi d\eta \qquad (7.9)$$

and

$$\bar{r}_y^{(k,i)} = \int_{\Omega_{st}^{(q)}} N_i F_y |J_k| d\xi d\eta + \int_{\Omega_{st}^{(q)}} \left([M_2][J_k]^{-1}\{\mathcal{D}\}N_i\right)^T [E]\{\alpha\}T_\Delta |J_k| d\xi d\eta. \qquad (7.10)$$

The equations of static equilibrium are

$$\sum_{i=1}^{n} f_x^{(k,i)} + \int_{\Omega_k} F_x \, dx dy = 0 \qquad (7.11)$$

$$\sum_{i=1}^{n} f_y^{(k,i)} + \int_{\Omega_k} F_y \, dx dy = 0 \qquad (7.12)$$

and

$$\sum_{i=1}^{n} \left(X_i f_y^{(k,i)} - Y_i f_x^{(k,i)}\right) + \int_{\Omega_k} (xF_y - yF_x) \, dx dy = 0 \qquad (7.13)$$

where X_i, Y_i are the coordinates of the ith node.

Satisfaction of Equation (7.11) and Equation (7.12) follows from the fact that

$$\sum_{i=1}^{n} N_i(\xi, \eta) = 1, \text{ therefore } \{\mathcal{D}\} \sum_{i=1}^{n} N_i(\xi, \eta) = 0.$$

Satisfaction of Equation (7.13) follows from the mapping functions (5.40) and (5.41) and the fact that infinitesimal rigid body rotations do not cause strain:

$$\left([M_1][J_k]^{-1}\{\mathcal{D}\}\sum_{i=1}^{n} Y_i N_i\right)^T \equiv \left\{\frac{\partial}{\partial x} \ 0 \ \frac{\partial}{\partial y}\right\} y = \{0 \ 0 \ 1\} \qquad (7.14)$$

$$\left([M_2][J_k]^{-1}\{\mathcal{D}\}\sum_{i=1}^{n} X_i N_i\right)^T \equiv \left\{0 \ \frac{\partial}{\partial y} \ \frac{\partial}{\partial x}\right\} x = \{0 \ 0 \ 1\}. \qquad (7.15)$$

On substituting Equations (7.14) and (7.15) into the expressions for $f_x^{(k,i)}$, $f_y^{(k,i)}$, $\bar{r}_x^{(k,i)}$ and $\bar{r}_y^{(k,i)}$, Equation (7.13) is obtained.

Note that the conditions for static equilibrium are satisfied independently of $\{a^{(k)}\}$. Therefore, equilibrium of nodal forces is unrelated to the finite element solution.

Exercise 7.2.1 Consider the problem of heat conduction in two dimensions. Assume that the nodal fluxes were computed analogously to Equation (7.4). Define the term which is analogous to $\{\bar{r}\}$ and show that the sum of nodal fluxes plus the integral of the source term over an element is zero. Use an eight-node quadrilateral element and a six-node triangular element to illustrate this point.

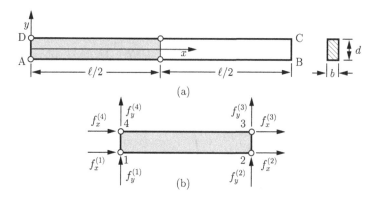

Figure 7.2 Example 7.2.1: notation.

7.2.2 Nodal forces in the *p*-version

We have seen in the case of the eight-node quadrilateral element that equilibrium of nodal forces was related to the facts that the sum of the shape functions is unity and the functions x and y could be expressed as linear combinations of the shape functions. When the hierarchic shape functions based on the integrals of Legendre polynomials are used, such as those illustrated in Figure 5.4, then the sum of the first four shape functions is unity independently of p. Therefore, in the equilibrium equations (7.11) through (7.13), $n = 4$. Here we consider only isoparametric and subparametric mappings. Other mappings involve additional considerations which are discussed in Section 5.4.4.

Example 7.2.1 A rectangular domain representing an elastic body of constant thickness is subjected to the boundary conditions $u_n = 0$, $u_t = \delta$ on boundary segments BC and DA shown in Figure 7.2(a). The subscripts n and t refer to the normal and tangent directions respectively. Boundary segments AB and CD are traction free.

We solve this as a plane stress problem using one element, represented by the shaded part of the domain, and product spaces. The antisymmetry condition is applied at $x = \ell/2$. The node numbering is shown in Figure 7.2(b). Since only one element is used, the superscript that identifies the element number is dropped. Letting $\ell = 1000$ mm, $d = 50$ mm, $b = 20$ mm, $E = 200$ GPA, $\nu = 0.295$, $\delta = 5$ mm, the computed values of the nodal forces are

Table 7.1 Example 7.2.1: nodal forces (kN). The notation is shown in Figure 7.2(b). Product spaces.

p	$f_x^{(1)}$	$f_x^{(2)}$	$f_x^{(3)}$	$f_x^{(4)}$	$f_y^{(1)}$	$f_y^{(2)}$	$f_y^{(3)}$	$f_y^{(4)}$
1	−1971.3	0.000	0.000	1971.3	−98.564	98.564	98.564	−98.564
2	−64.678	0.000	0.000	64.678	−3.234	3.234	3.234	−3.234
3	−50.546	0.000	0.000	50.546	−2.527	2.527	2.527	−2.527
4	−50.190	0.000	0.000	50.190	−2.509	2.509	2.509	−2.509
5	−50.010	0.000	0.000	50.010	−2.500	2.500	2.500	−2.500
6	−49.907	0.000	0.000	49.907	−2.495	2.495	2.495	−2.495
7	−49.843	0.000	0.000	49.843	−2.492	2.492	2.492	−2.492
8	−49.802	0.000	0.000	49.802	−2.490	2.490	2.490	−2.490

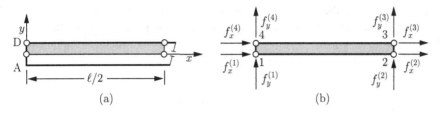

Figure 7.3 Example 7.2.1: the smallest solution domain.

as given in Table 7.1. In this example $\{\bar{r}\} = 0$. Therefore the nodal forces were computed from $\{f\} = [K]\{a\}$. The results shown in Table 7.1 indicate that equilibrium of nodal forces is satisfied at every p-level, independently of the accuracy of the finite element solution.

This problem could have been solved on the shaded domain shown in Figure 7.3, in which case the antisymmetry condition would have been prescribed on the x-axis. Using the notation shown in Figure 7.3(b), the computed nodal forces are given in Table 7.2 for $p = 8$. It can be seen that the nodal forces once again satisfy the condition of static equilibrium.

Exercise 7.2.2 The nodal forces given in Table 7.1 (resp. Table 7.2) were computed on the shaded domain shown in Figure 7.2(a) (resp. Figure 7.3(a)). Show that the stress resultants acting on boundary segments BC and DA are the same. Hint: Sketch and label the values of the nodal forces on the element shown in Figure 7.3(b) and its antisymmetric pair.

7.2.3 Nodal forces and stress resultants

The nodal forces are related to the extraction functions for stress resultants. We illustrate this on the basis of Example 7.2.1. We have

$$\int_{\Omega} ([D]\{v\})^T [E][D]\{u_{FE}\} b \, dx dy = \int_{\partial \Omega} (T_x^{(FE)} v_x + T_y^{(FE)} v_y) b \, ds \qquad (7.16)$$

where Ω is the domain of the element shown in Figure 7.2(b) and b is the thickness. If we are interested in the shear force acting on the side between nodes 2 and 3, denoted by $V_{2,3}$, then we select $v_x = 0$ on Ω and v_y a smooth function of Ω such that $v_y = 1$ on the side between nodes 2 and 3, and $v_y = 0$ on the side between nodes 4 and 1. Specifically, if we select $v_y = N_2(\xi, \eta) + N_3(\xi, \eta)$ then we have

$$V_{2,3} = \int_{\text{node 2}}^{\text{node 3}} T_y^{(FE)} b \, dy = \sum_{j=1}^{2n} (k_{n+2,j} + k_{n+3,j}) a_j = f_y^{(2)} + f_y^{(3)} \qquad (7.17)$$

where n is the number of degrees of freedom per field. In Example 7.2.1 product spaces were used, therefore $n = (p+1)^2$.

Table 7.2 Example 7.2.1: nodal forces (kN). Solution on the shaded domain shown in Figure 7.3(a). The notation is shown in Figure 7.3(b). Product space.

p	$f_x^{(1)}$	$f_x^{(2)}$	$f_x^{(3)}$	$f_x^{(4)}$	$f_y^{(1)}$	$f_y^{(2)}$	$f_y^{(3)}$	$f_y^{(4)}$
8	−12.454	−37.353	0.000	49.806	4.922	1.144	1.346	−7.412

Exercise 7.2.3 Solve the problem of Example 7.2.1 and compute the stress resultants $V_{4,1}$ and $M_{4,1}$ on the side between nodes 4 and 1 by direct integration. Keep refining the mesh until satisfactory convergence is observed. Compare the results with those computed from the nodal forces in Table 7.1. By definition

$$M_{4,1} = \int_{-d/2}^{+d/2} T_x y b \, dy.$$

This exercise shows that integration of stresses is much less efficient than extraction. This is because the presence of singularities at points A and D influences the accuracy of the numerical integration.

7.3 Verification of computed data

Having performed a finite element analysis, it is necessary to determine whether the data of interest computed from the finite element solution are sufficiently accurate for the purposes of the analysis. The accuracy of the data depends on the choice of the finite element space and the method by which the data are computed. We have seen a simple illustration of this in Example 2.5.8. Only direct methods of computation will be considered in this section.

In many practical problems the data of interest lie in some small subdomain of the solution domain. For example, we may be interested in the stresses in the vicinity of a fastener hole in a large plate. The neighborhood of the fastener hole is the region of primary interest, the rest of the plate is the region of secondary interest. Errors in the computed data may be caused by insufficient discretization in the region of primary or secondary interest, or both. Errors caused by insufficient discretization of the region of secondary interest are called pollution errors.

Since the exact solution is independent of the parameters that characterize the finite element space, verification involves steps to ascertain that (a) the correct input data were used in the analysis and (b) the data of interest are substantially independent of the parameters that characterize the finite element space. This involves obtaining two or more solutions corresponding to a sequence of finite element spaces and examination of feedback information, that is, information generated from finite element solutions. Finite element spaces generated by p-extension are hierarchic whereas finite element spaces generated in h-extensions, with the sequence of meshes created by mesh generators, are typically not hierarchic. The recommended steps are as follows:

1. Display the solution graphically and check whether the solution is reasonable. For example, plotting the deformed configuration with the scale factor set to 1 provides information on whether a large error occurred in specifying the loading and constraint conditions or material properties.

2. Estimate the relative error in the energy norm and its rate of convergence. The estimated relative error in energy norm is a useful indicator of the overall quality of the solution, roughly equivalent to estimating the root-mean-square error in stresses [80]. The estimated rate of convergence is an indicator of whether the rate of change of error is consistent with the asymptotic rate for the problem class. Substantial deviations from the values given in Table 6.2 and Table 6.3 typically indicate errors in the input data or in the finite element mesh. For example, some elements may be highly distorted.

3. Check for the presence of jump discontinuities in stresses and fluxes in regions where the stresses or fluxes are large. The normal flux and the normal and shearing stress components at inter-element boundaries must be continuous. In regions where the stresses or fluxes are small, some discontinuity is generally acceptable. Jump discontinuities in stresses and fluxes between the regions of primary and secondary interest are indications of pollution errors.

4. Show that the data of interest are substantially independent of the mesh and/or the polynomial degree of elements.

These steps are illustrated by the following example.

Example 7.3.1 Let us consider the problem of Example 4.3.5. Referring to the notation in Figure 4.11 and Figure 7.4, we let $a = 1$, $b/w = 1.5$ and $\sigma_\infty = 1$. The boundary conditions are the same as in Example 4.3.5. The maximum tensile stress occurs at point E and its exact value is $\sigma_x = 3$. The goal of the analysis is to approximate the maximum tensile stress by the finite element method and to verify that the error of approximation is small. Verification procedures and the use of feedback information are illustrated for three values of the length parameter b in the following.

The region of primary interest is the neighborhood of the circular hole. The initial mesh is laid out so that the size of elements in the neighborhood of the hole is approximately the size of the radius of the hole, as indicated in Figure 7.4. Since this problem belongs in Category A we expect exponential convergence in energy norm when p-extension is used. Using the method of Section 2.7, the estimated relative errors in energy norm were computed and plotted in Figure 7.5 for three values of the length parameter b, indicated in the figure. Also plotted are the true errors, indicated by the dashed lines. For $b = 10$ the estimated and true errors are so close that they cannot be distinguished on this diagram.

Note that the relative error in the energy norm is plotted on a logarithmic scale vs. $\sqrt{N}$. The reason for this is seen in Equation (6.63) and Table 6.2. Since exponential convergence is expected, the relative error curve should appear as a straight line in this diagram. As seen in the figure, this is indeed the case and the estimated relative errors are reassuringly small. The first step in the verification process indicates that the sequence of finite element spaces

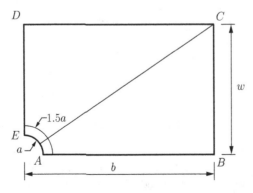

Figure 7.4 Four-element mesh, $b/w = 1.5$.

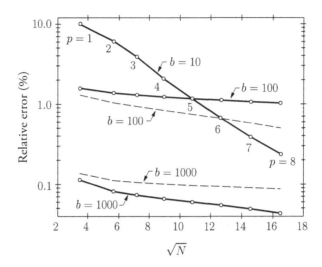

Figure 7.5 Four-element mesh, $b/w = 1.5$. Solid lines: estimated relative errors in energy norm, p-extension, trunk space. Dashed lines: true errors.

generated by uniform p-extension and the mesh shown in Figure 7.5 yields small relative errors in the energy norm at the expected rate of convergence for the three values of b.

The computed values of σ_x at point E are shown in Figure 7.6(a). It can be seen that for $b = 10$ the values of σ_x converge strongly to a limiting value of 3.0 which is the correct limit. This does not happen for $b = 100$ and $b = 1000$, however. The value of σ_x changes substantially with p in the range $p = 1, 2, \ldots, 8$ allowed by the implementation. This indicates that as b is increased, the four-element mesh becomes inadequate from the point of view of computing and verifying σ_x when the maximum value of p is limited to 8. The situation would be

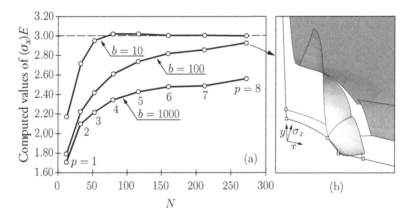

Figure 7.6 Four-element mesh, $b/w = 1.5$. (a) Computed values of σ_x in point E. (b) Computed values of σ_x in the neighborhood of the hole for $b = 100$, $p = 8$ trunk space.

different if, for example, the displacement of point C were to be computed. It is left to the reader to demonstrate this in Exercise 7.3.1.

This example demonstrates that it is not sufficient to have a small relative error in the energy norm if the goal is to compute the maximum stress or, more generally, any data computed from the first derivatives. In this example the size of the region of primary interest is fixed but the size of the region of secondary interest increases with b. Referring to Equations (4.34) to (4.36) we see that the stress varies like $1/r^2$ and $1/r^4$ in the neighborhood of the hole, but far from the hole the stress is very nearly constant. The center of the hole is a singular point that lies outside of, but close to, the solution domain and its influence on the stress distribution decays in proportion to $1/r^2$.

When $b = 10$ the estimated relative error in the energy norm and the error in $(\sigma_x)_E$ are both small, indicating that the finite element mesh is properly designed for approximating the exact solution and its first derivatives in the range $p = 1, 2, \ldots, 8$.

When $b = 100$ or $b = 1000$ the error in the energy norm is small but the error in $(\sigma_x)_E$ is large. The problem is that, given the restriction $p \leq 8$ imposed by the implementation, the elements in the region of secondary interest are too large. The pollution error caused by inadequate meshing is manifested by the large jump in σ_x at the boundary between the inner and outer layers of elements, as shown in Figure 7.6(b). On the other hand, the value of the potential energy is dominated by the nearly constant stress away from the hole and the large local error in the small neighborhood of the hole is insignificant in comparison.

In problems like this the finite element mesh should be graded in geometric progression with reference to the center of the hole, with a common factor of approximately 0.15. The exact solution of this problem is in Category A; however, when the size of the hole is small in relation to the rest of the domain then the mesh should be laid out as if it were in Category B in order to obtain a good approximation of the first derivatives over the entire domain.

Remark 7.3.1 It is clearly visible in Figure 7.6 that $(\sigma_x)_E$ has not converged to a limiting value for $b = 100$ and $b = 1000$. It is possible to construct meshes such that it would appear that the data has converged but the apparent limit is wrong. Therefore it is not sufficient to test for indications of convergence of the data of interest, it is also necessary to check for pollution errors.

Remark 7.3.2 There are some interesting features demonstrated in Fig 7.5:

(a) The distortion of the large elements increases with increasing b, therefore one would expect an increase in error. However, the relative error decreases. This is due to the fact that the energy norm of the solution over the entire domain increases faster with respect to b than the absolute error. Using the absolute error in the energy norm at $p = 8$, $b = 10$ for reference, the absolute error at $p = 8$ is 22.0 times greater at $b = 100$ and 38.5 times greater at $b = 1000$.

(b) The rate of convergence (the slope of the dashed line) is decreasing with increasing values of b.

(c) The error estimator overestimates the error for $b = 100$ and underestimates it for $b = 1000$.

Exercise 7.3.1 Using the four-element mesh shown in Figure 7.4, compute and verify the value of the displacement of point C for $b = 100$ and $b = 1000$ for the problem of Example 7.3.1. Let $\nu = 0.3$ and assume plane strain conditions. Report the value of $u_x G/(a\sigma_\infty)$ and its relative error. See Equation (4.37).

Exercise 7.3.2 Solve the problem of Example 7.3.1 using $b = 1000$ and a geometrically graded mesh. Compute and verify the value of σ_x at point E.

Exercise 7.3.3 Solve the problem of Example 7.3.1 using $b = 1000$ and a sequence of automatically generated meshes comprising six-node triangles. Compute and verify the value of σ_x at point E.

Example 7.3.2 Consider the L-shaped domain problem of Example 6.4.2. The exact solution has a strong singularity at the reentrant corner. Therefore it would not be meaningful to compute the maximum stress on the entire domain. However, if we exclude the elements that share the singular vertex then the stresses will converge on the rest of the domain.

Referring to the results of Exercise 6.3.6 and the definition of stress invariants in Appendix C, Section C.2, the first stress invariant corresponding to the exact solution for plane strain is

$$\sigma_x + \sigma_y + \sigma_z = 4(1 + \nu)A_1\lambda_1 r^{\lambda_1 - 1}$$

where A_1 is arbitrary and $\lambda_1 = 0.544\,483\,737$. The relative error in the first stress invariant corresponding to the 18-element geometrically graded mesh shown in the inset of Figure 6.16, p-extension and trunk space, is plotted in Figure 7.7. The results indicate strong convergence.

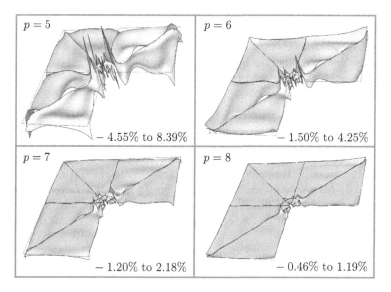

Figure 7.7 The L-shaped domain problem of Example 6.4.2. Relative errors in the first stress invariant outside of the innermost layer of elements.

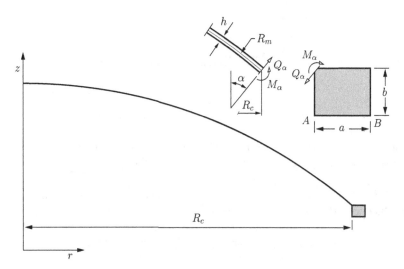

Figure 7.8 The Girkmann problem: notation. First published in [81]. Reproduced with kind permission from Springer Science+Business Media © 2010.

Exercise 7.3.4 Solve the problem of Example 7.3.2 using a sequence of automatically generated meshes comprising six-node triangles. Plot the relative error in the first stress invariant for $r > 0.0225a$ where r is the radial distance from the reentrant corner and a is the dimension shown in the inset of Figure 6.16.

Exercise 7.3.5 A spherical shell of thickness $h = 0.06$ m, crown radius $R_c = 15.00$ m, is connected to a stiffening ring at the meridional angle $\alpha = 2\pi/9$ (40°). The dimensions of the ring are $a = 0.60$ m, $b = 0.50$ m. The radius of the mid-surface of the spherical shell is $R_m = R_c/\sin\alpha$. The notation is shown in Figure 7.8. The z-axis is the axis of rotational symmetry.

The shell is made of reinforced concrete, assumed to be homogeneous, isotropic and linearly elastic, with Young's modulus $E = 20.59$ GPa and Poisson's ratio $\nu = 0$.

Consider gravity loading only. The equivalent (homogenized) unit weight of the material composed of the shell and the cladding is 32.69 kN/m^3. Assume that uniform normal pressure p is acting at the base AB of the stiffening ring. The resultant of p equals the weight of the structure.

1. Find the shearing force Q_α in kN/m units and the bending moment M_α in Nm/m units acting at the junction between the spherical shell and the stiffening ring.

2. Determine the location (meridional angle) and the magnitude of the maximum bending moment in the shell.

3. Verify that the results are accurate to within 5%.

This problem was first discussed by Girkmann[1] [30] and subsequently by Timoshenko and Woinowsky-Krieger [88]. Solutions by classical methods are presented in both references. In

---

[1] Karl Girkmann (1890–1959).

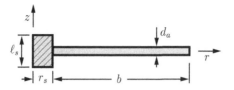

Figure 7.9 Notation for Exercise 7.3.6.

the classical solutions the stiffening ring was assumed to be weightless. Solutions by the h- and p-versions of the finite element method are discussed in [81].

Exercise 7.3.6 An annular aluminum plate with inner radius r_s and outer radius $r_s + b$, constant thickness d_a, was joined by shrink fitting to a stainless steel shaft. The configuration is shown in Figure 7.9. Denote the mechanical properties of aluminum as follows: modulus of elasticity, E_a; modulus of rigidity, G_a; mass density, ϱ_a; coefficient of thermal expansion, α_a; and the corresponding mechanical properties of stainless steel as E_s, G_s, ϱ_s, α_s. Let $\ell_s = 80$ mm, $r_s = 17.5$ mm, $d_a = 15$ mm, $b = 150$ mm, $E_a = 72.0 \times 10^3$ MPa, $G_a = 28.0 \times 10^3$ MPa, $\varrho_a = 2800$ kg/m^3, $\alpha_a = 23.6 \times 10^{-6}$/K, $E_s = 190 \times 10^3$ MPa, $G_s = 75.0 \times 10^3$ MPa, $\varrho_s = 7920$ kg/m^3, $\alpha_s = 17.3 \times 10^{-6}$/K.

Consider the following conditions: (a) The shaft and the aluminum plate were heated to 220 °C, the shaft was inserted and then the assembly was cooled to 20 °C. Assume that at 220 °C the clearance between the shaft and the plate was zero and $\alpha_a > \alpha_s$. (b) The assembly is spinning about the z-axis at an angular velocity ω.

Estimate the value of ω at which the membrane force in the aluminum plate, F_r, is approximately zero at $r = r_s$. By definition

$$F_r = \int_{-d_a/2}^{d_a/2} \sigma_r \, dz.$$

Specify ω in units of cycles per second (hertz). Note that in order to have consistent units, kg/m^3 must be converted to N s^2/mm^4.

7.4 Flux and stress intensity factors

In this section, procedures for the computation of the coefficients of the asymptotic expansions from finite element solutions are discussed. These coefficients are called flux or stress intensity factors depending on whether we speak of the Laplace or the Navier equation. The algorithm is based on two key results: (a) the existence of a path-independent integral and (b) the orthogonality of the characteristic functions. The procedure is illustrated on the basis of the Laplace equation.

7.4.1 The Laplace equation

Consider a two-dimensional domain Ω with boundary Γ. For any two functions in $E(\Omega)$ we have

$$\int_\Omega \Delta u \, v \, dx dy = \oint_\Gamma (\nabla u \cdot \vec{n}) v \, ds - \oint_\Gamma (\nabla v \cdot \vec{n}) u \, ds + \int_\Omega \Delta v \, u \, dx dy$$

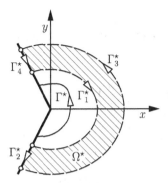

Figure 7.10 Definition of Ω^*.

where we have applied the divergence theorem twice. When u and v satisfy the Laplace equation, that is, $\Delta u = 0$ and $\Delta v = 0$, then this equations becomes:

$$\oint_\Gamma (\nabla u \cdot \vec{n}) v \, ds = \oint_\Gamma (\nabla v \cdot \vec{n}) u \, ds \tag{7.18}$$

which is applicable to Ω and any subdomain of Ω.

Path-independent integral

Now consider a subdomain Ω^* in the neighborhood of a corner point, shown as the shaded region in Figure 7.10. Assume that either $u = 0$ or $\nabla u \cdot \vec{n} = 0$ and either $v = 0$ or $\nabla v \cdot \vec{n} = 0$ on Γ_2^* and Γ_4^*. Then Equation (7.18) becomes

$$\int_{\Gamma_1^*} (\nabla u \cdot \vec{n}) v \, ds + \int_{\Gamma_3^*} (\nabla u \cdot \vec{n}) v \, ds = \int_{\Gamma_1^*} (\nabla v \cdot \vec{n}) u \, ds + \int_{\Gamma_3^*} (\nabla v \cdot \vec{n}) u \, ds$$

which is equivalent to

$$\int_{\Gamma_1^*} (\nabla u \cdot \vec{n}) v \, ds - \int_{\Gamma_1^*} (\nabla v \cdot \vec{n}) u \, ds = -\int_{\Gamma_3^*} (\nabla u \cdot \vec{n}) v \, ds + \int_{\Gamma_3^*} (\nabla v \cdot \vec{n}) u \, ds.$$

Observe that integration along Γ_1^* is clockwise about the corner point whereas integration along Γ_3^* is counterclockwise. Reversing the direction of integration along Γ_1^* so that both integrals are counterclockwise about the corner point, we find that the two integrals are equal, and since Ω^* is arbitrary, we may select an arbitrary counterclockwise path Γ^* and the integral expression

$$I_{\Gamma^*} := -\int_{\Gamma^*} (\nabla u \cdot \vec{n}) v \, ds + \int_{\Gamma^*} (\nabla v \cdot \vec{n}) u \, ds \tag{7.19}$$

will be path independent.

Orthogonality

Let $u = r^{\lambda_i}\phi_i(\theta)$, $v = r^{\lambda_j}\phi_j(\theta)$ and $\Gamma^* = \Gamma_\varrho$ where Γ_ϱ is a circular path of radius ϱ, centered on the corner point. Then on Γ_ϱ we have

$$\nabla u \cdot \vec{n} = \left(\frac{\partial u}{\partial r}\right)_{r=\varrho} = \lambda_i \varrho^{\lambda_i - 1}\phi_i(\theta)$$

$$\nabla v \cdot \vec{n} = \left(\frac{\partial v}{\partial r}\right)_{r=\varrho} = \lambda_j \varrho^{\lambda_j - 1}\phi_j(\theta).$$

Using $ds = \varrho\, d\theta$, Equation (7.19) can be written as

$$I_{\Gamma_\varrho} = (\lambda_j - \lambda_i)\varrho^{\lambda_i + \lambda_j} \int_{-\alpha/2}^{+\alpha/2} \phi_i(\theta)\phi_j(\theta)\, d\theta.$$

Since I_{Γ_ϱ} is path independent, the integral expression must be zero when $\lambda_j \neq \pm\lambda_i$. Note that since Ω^* does not include the corner point, solutions corresponding to the negative characteristic values are in the energy space. In the following we will denote

$$C_{ij} := \int_{-\alpha/2}^{+\alpha/2} \phi_i(\theta)\phi_j(\theta)\, d\theta \quad \text{and} \quad C_{ij}^- := \int_{-\alpha/2}^{+\alpha/2} \phi_i(\theta)\phi_j^-(\theta)\, d\theta \qquad (7.20)$$

where $\phi_j^-(\theta)$ is the characteristic function corresponding to $-\lambda_j$. The characteristic functions are orthogonal in the sense that $C_{ij} = 0$ when $\phi_j \neq \phi_i$ and $\phi_j \neq \phi_i^-$.

Remark 7.4.1 In the case of the Laplace operator all characteristic values are real and simple.

Exercise 7.4.1 Show that for the asymptotic expression given by Equation (6.11)

$$C_{ij} = \begin{cases} \alpha/2 & \text{if } i = j \\ 0 & \text{if } i \neq j. \end{cases}$$

Extraction of A_k

Using the orthogonality property of the characteristic functions, it is possible to extract the coefficients from the finite element solution. Let us consider the asymptotic expansion

$$u_{EX} = \sum_{i=1}^{\infty} A_i r^{\lambda_i} \phi_i(\theta). \qquad (7.21)$$

Suppose that we are interested in computing A_k. We then define the extraction function w_k as

$$w_k := r^{-\lambda_k}\phi_k^-(\theta)$$

FLUX AND STRESS INTENSITY FACTORS

which satisfies the Laplace equation and the boundary conditions and evaluate the path-independent integral on a circular path Γ_ϱ centered on the corner point:

$$I_{\Gamma_\varrho}(u_{EX}, w_k) = -\int_{\Gamma_\varrho} (\nabla u_{EX} \cdot \vec{n}) w_k \, ds + \int_{\Gamma_\varrho} (\nabla w_k \cdot \vec{n}) u_{EX} \, ds.$$

It is now left to the reader to show that, utilizing the orthogonality property of the characteristic functions, we have

$$A_k = -\frac{1}{2C_{kk}^- \lambda_k} I_{\Gamma_\varrho}(u_{EX}, w_k). \tag{7.22}$$

In the finite element method u_{EX} is replaced by u_{FE} to obtain an approximate value for A_k. This method of computing the coefficients, called the contour integral method, is very efficient, as illustrated by the following example.

Example 7.4.1 We refer to the results of Exercise 6.3.1 and construct a model problem on the domain shown in Figure 7.11(a) so that the exact solution is a linear combination of the first two terms of the asymptotic expansion:

$$u_{EX} = a_1 r^{\lambda_1} \underbrace{(\cos \lambda_1 \theta - \sin \lambda_1 \theta)}_{\phi_1(\theta)} + a_2 r^{\lambda_2} \underbrace{(\cos \lambda_2 \theta + \sin \lambda_2 \theta)}_{\phi_2(\theta)}$$

where $\lambda_1 = \pi/(2\alpha)$, $\lambda_2 = 3\pi/(2\alpha)$. Let $\alpha = 7\pi/4 = 315°$. The boundary conditions are: $\partial u/\partial n = 0$ on Γ_{AB}; $u = 0$ on Γ_{BC} and on Γ_{AC}; that is, the flux corresponding to u_{EX} is specified:

$$q_n = -a_1 \lambda_1 r^{\lambda_1 - 1}(\cos \lambda_1 \theta - \sin \lambda_1 \theta) - a_2 \lambda_2 r^{\lambda_2 - 1}(\cos \lambda_2 \theta + \sin \lambda_2 \theta).$$

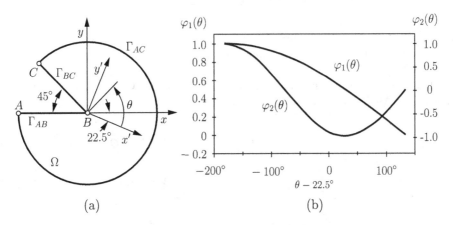

Figure 7.11 Domain for Example 7.4.1.

We normalize the characteristic functions as follows. Let θ_i be the angle where the absolute value of $\phi_i(\theta)$ is maximum:

$$|\phi_i(\theta_i)| = \max_{\theta \in I_\alpha} |\phi_i(\theta)| \quad \text{where } I_\alpha := \{\theta \mid -\alpha/2 \leq \theta \leq +\alpha/2\}.$$

The normalized characteristic functions are defined by:

$$\varphi_i(\theta) := \phi_i(\theta)/\phi_i(\theta_i). \tag{7.23}$$

Therefore the maximum value of $\varphi_i(\theta)$ on I_α is unity.

The functions $\varphi_1(\theta)$ and $\varphi_2(\theta)$ are shown in Figure 7.11(b). If we let $a_1 = 1.0$, $a_2 = 0$ then, using a 16-element mesh with one layer of geometrically graded elements around the corner point, trunk space, the estimated relative error in the energy norm is 17.21% at $p = 8$. Using the method of extraction described in this section, the computed value of the coefficient of the first normalized characteristic function is $A_1 = 1.348$. Its exact value is 1.414, that is, the relative error is 4.66%.

Exercise 7.4.2 For the problem described in Example 7.4.1, let $a_1 = 0$ and $a_2 = 1.0$. Determine the approximate value of the coefficient of the second normalized characteristic function A_2 and estimate the relative error. The exact value is -1.414.

7.4.2 Planar elasticity

The analysis of corner points in planar elasticity is analogous to that of the Laplace equation, but much more complicated. Complications arise from the fact that in planar elasticity the characteristic values may be complex, and not all of the real roots are simple. Furthermore, there are corner angles that require special treatment [58]. Full treatment of this subject is beyond the scope of this text. Only one important special case, crack-tip singularities, will be discussed. For cracks ($\alpha = 2\pi$) Equations (6.30) and (6.31) reduce to one equation:

$$\sin 2\lambda\pi = 0, \quad \text{therefore} \quad \lambda_n = \pm\frac{n}{2}, \; n = 1, 2, 3, \ldots$$

and hence all roots are real and simple. The goal is to compute the coefficients of the first terms of the symmetric (Mode I) and antisymmetric (Mode II) expansions, see Equations (7.24) through (7.26).

In linear elastic fracture mechanics it is customary write the Cartesian components of the corresponding Mode I stress tensor in the following form:

$$\sigma_x = \frac{K_I}{\sqrt{2\pi r}} \cos\frac{\theta}{2}\left(1 - \sin\frac{\theta}{2}\sin\frac{3\theta}{2}\right) + T + O(r^{3/2}) \tag{7.24}$$

$$\sigma_y = \frac{K_I}{\sqrt{2\pi r}} \cos\frac{\theta}{2}\left(1 + \sin\frac{\theta}{2}\sin\frac{3\theta}{2}\right) + O(r^{3/2}) \tag{7.25}$$

$$\tau_{xy} = \frac{K_I}{\sqrt{2\pi r}} \sin\frac{\theta}{2}\cos\frac{\theta}{2}\cos\frac{3\theta}{2} + O(r^{3/2}) \tag{7.26}$$

where $-\pi \leq \theta \leq \pi$, T is a constant, called the T-stress, see Exercise 6.3.5. The constant K_I is called the Mode I stress intensity factor.

The antisymmetric (Mode II) stress tensor components are usually written in the following form:

$$\sigma_x = -\frac{K_{II}}{\sqrt{2\pi r}} \sin\frac{\theta}{2}\left(2 + \cos\frac{\theta}{2}\cos\frac{3\theta}{2}\right) + O(r^{3/2}) \qquad (7.27)$$

$$\sigma_y = \frac{K_{II}}{\sqrt{2\pi r}} \sin\frac{\theta}{2}\cos\frac{\theta}{2}\cos\frac{3\theta}{2} + O(r^{3/2}) \qquad (7.28)$$

$$\tau_{xy} = \frac{K_{II}}{\sqrt{2\pi r}} \cos\frac{\theta}{2}\left(1 - \sin\frac{\theta}{2}\sin\frac{3\theta}{2}\right) + O(r^{3/2}) \qquad (7.29)$$

where K_{II} is called the Mode II stress intensity factor.

Computation of stress intensity factors

In this section the computation of stress intensity factors from finite element solutions by the contour integral method (CIM) is outlined. The CIM for the Navier equations, analogously to the CIM for the Laplace equation outlined in Section 7.4.1, is based on the existence of a path-independent integral and the orthogonality of eigenfunctions. Detailed derivation of the contour integral method for the Navier equations is available in [79], [80].

We will consider the Navier equation in two dimensions under the assumption that the volume forces are zero and the temperature and thickness are constants. The path-independent contour integral, analogous to Equation (7.19), is

$$I_\Gamma := \int_\Gamma (T_x^{(u)} w_x + T_y^{(u)} w_y)\,ds - \int_\Gamma (T_x^{(w)} u_x + T_y^{(w)} u_y)\,ds \qquad (7.30)$$

where the superscripts **u** (resp. **w**) represents the exact solution (resp. a test function that satisfies the Navier equation and the boundary conditions on the edges that intersect at the singular point), and Γ is an arbitrary contour that begins at one edge and runs in a counterclockwise direction to the other, as shown in Figure 7.10. We will assume that Γ is a circular contour, the radius of which is arbitrary, and we will be interested in the computation of K_I, K_{II} and T. The procedure is outlined in Appendix D.

Under special conditions the stress intensity factors can be determined from the energy release rate. The procedure for computing stress intensity factors from the energy release rate is described in Appendix D.

Example 7.4.2 Let us consider the problem described in Exercise 4.3.9 and assume that a crack 1.0 mm long developed at the location of the maximum tensile stress. The crack is oriented normal to the boundary. The goal of the computation is to determine the stress intensity factors K_I, K_{II} and the T-stress.

The location of the maximum tensile stress, found from the solution of Exercise 4.3.9, is indicated in Figure 7.12(b) where $\alpha = 12.8°$. A crack 1 mm long is introduced and a mesh, such as the mesh shown Figure 7.12(a), is generated. The mesh depends on user-specified settings for the mesh generator. Upon solving this problem and extracting K_I, K_{II} and T by

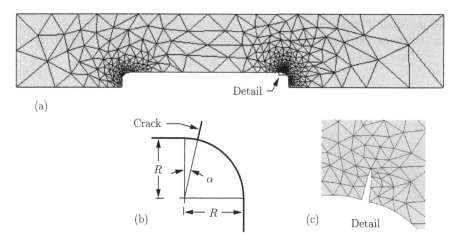

Figure 7.12 The finite element mesh for the notched plate with a crack. Number of elements: 627.

the contour integral method, the results shown in Table 7.3 are obtained. It can be seen that the computed data do not change significantly as p is increased from 2 to 8.

Exercise 7.4.3 Refer to Appendix D.2. Verify the accuracy of $K_I^2 + K_{II}^2$ in Table 7.3 by computing the energy release rate $\mathcal{G}$. Hint: Use the central difference formula to estimate $\mathcal{G}$:

$$\mathcal{G} \approx -\frac{\Pi(a + \Delta a) - \Pi(a - \Delta a)}{2\Delta a}.$$

Let $\Delta a = 0.01a$ and check the sensitivity of $\mathcal{G}$ to the choice of Δa. What happens if Δa is too small or too large?

Exercise 7.4.4 Estimate the critical crack length for the problem in Example 7.4.2. Assume that the location and orientation of the crack is the same as in Example 7.4.2 and the fracture

Table 7.3 Example 7.4.2: computed values of stress intensity factors (MPa$\sqrt{\text{mm}}$).

p	N	K_I	K_{II}	T
2	2 669	260.2	7.35	−40.78
3	5 885	259.7	7.32	−39.47
4	10 355	259.8	7.37	−39.25
5	16 079	259.9	7.35	−39.15
6	23 057	260.0	7.34	−39.10
7	31 289	260.1	7.33	−39.07
8	40 775	260.1	7.32	−39.04

toughness of the material is $900\,\text{MPa}\sqrt{\text{mm}}$. In this exercise fracture toughness is treated as a fixed number. In reality fracture toughness is a random variable that has a large dispersion. Therefore the critical crack length is a function of a random variable.

7.5 Chapter summary

At the end of the solution process the coefficients of the shape function, the mapping and the material properties are available at the element level. The data of interest are computed from this information either by direct or indirect methods.

In order to meet the requirements of verification it is necessary to show that the errors in the data of interest do not exceed stated tolerances. In practical problems the exact solution is typically unknown and it is not possible to determine the errors of approximation with precision. It is possible, however, to show that the necessary conditions are satisfied for the errors in the data of interest to be small.

Error estimation is based on a priori knowledge that the data of interest corresponding to the exact solution are finite and independent of the discretization parameter. Therefore a necessary condition for the error to be small is that the computed data should exhibit convergence to a limit value as the number of degrees of freedom is increased. An efficient and robust way to achieve this is to use properly designed meshes and increase the polynomial degree.

8

What should be computed and why?

Up to this point we have been concerned with the formulation of mathematical models, their numerical solution by the finite element method, computation of data from the numerical solution and their verification. In this chapter we consider the question of what should be computed and why. This question has to be addressed in the process of conceptualization. Because the subject is very large and diverse, we will consider the question with reference to a specific problem class only: mathematical models formulated for the purpose of estimating the service life of structural and mechanical components subjected to cyclic loading of variable amplitude. We will be concerned with components made of metals only. The ability to make reliable predictions of the service life of safety-critical mechanical systems is obviously of great importance in the formulation and justification of decisions concerning design, certification and maintenance.

An interesting aspect is that cyclic loading causes damage accumulation in the material through dislocations followed by the formation, coalescence and propagation of cracks. Mathematical models based on continuum mechanics, where it is assumed that the displacement field is a continuous function, do not account for these phenomena. Nevertheless, models of damage accumulation are typically based on the linear theory of elasticity or small-strain plasticity. Such models, coupled with experimentation, have been demonstrated to be useful for predicting the formation and propagation of cracks. The main points are outlined in this chapter.

Another interesting and highly relevant aspect of this problem class is that vast amounts of experimental data have been collected over many years, yet proper interpretation of the data remains an open question. Experimental measurements of metal fatigue typically have substantial statistical dispersion. This makes the separation of uncertainties associated with modeling assumptions (epistemic uncertainties) and statistical (aleatory) uncertainties a challenging problem. The conceptual framework of verification and validation outlined in

Chapter 1 and the methods available for estimating and controlling the errors of discretization outlined in Chapters 2 through 7 provide means for the re-evaluation of existing data and improvement of mathematical models formulated for the prediction of damage accumulation caused by cyclic loading.

8.1 Basic assumptions

Mathematical models constructed for the prediction of damage accumulation caused by cyclic loading are based on the following assumptions:

1. There exist one or more functionals, computable from the solution of mathematical models based on infinitesimal strain, small-deformation theory, that can be correlated with crack initiation events and crack propagation rates with sufficient accuracy to suit the purposes of engineering decision-making.

2. There exist one or more procedures suitable for the generalization of the results of fatigue experiments performed under a particular cyclic loading, characterized by a mean value and constant amplitude, to cyclic loading characterized by arbitrary mean value and constant amplitude.

3. There exist one or more procedures suitable for correlating damage accumulation with variable amplitude cyclic loading.

Each of these assumptions implies the existence of one or more mathematical models. In the following we will be concerned primarily with the formulation of models pertaining to Assumption 1. These models differ by the definition of functionals that are correlated with damage accumulation events. The functionals are called driving forces of damage accumulation or drivers of damage accumulation (DDA). Assumptions 2 and 3 will be discussed only briefly.

8.2 Conceptualization: drivers of damage accumulation

Consider the model problem of a solid body of constant thickness with a notch of radius $\varrho \geq 0$. A schematic illustration is shown in Figure 8.1. The solution domain is denoted by Ω. We assume that outside of the close neighborhood of the notch, denoted by $\Omega^\star$, the assumptions of small-strain, small-deformation models of continuum mechanics are justified. In high-cycle fatigue, that is, when the number of cycles to failure is greater than 10^4, the assumptions of the linear theory of elasticity are generally considered to be applicable on $\Omega - \Omega^\star$. In low cycle fatigue, inelastic deformation may occur on $\Omega - \Omega^\star$.

The domain $\Omega^\star$, bounded by Γ and Γ_{SS}, is divided into two parts. The immediate neighborhood of the notch, bounded by Γ and Γ_{PZ} and indicated as the hatched region in Figure 8.1, is called the process zone and denoted by Ω_{PZ}. Cracks nucleate and propagate in the process zone and therefore the assumptions of continuum mechanics are not applicable there. In the domain $\Omega^\star - \Omega_{PZ}$, bounded by Γ, Γ_{PZ} and Γ_{SS}, the assumptions of continuum mechanics are applicable, but the small-strain assumptions are not.

In high-cycle fatigue the size of the process zone is typically small, of the order of the size of a representative volume element (RVE) which is the size of the smallest body that can be modeled by the methods of continuum mechanics.

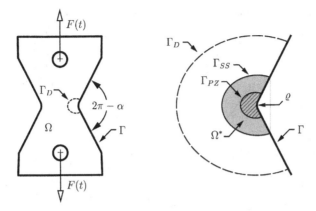

Figure 8.1 Notched plate: notation.

Let us assume that a mathematical model was formulated that accounts for the physical processes leading to the formation of discontinuities and cracks in the process zone and the solution of this model is available over the entire domain. We denote this solution by $\vec{u}_{PZ}(\vec{x}, t) = \{u_x(\vec{x}, t)\, u_y(\vec{x}, t)\}_{PZ}$ where $\vec{x}$ is the position vector and t is time. We denote the solution of the continuum mechanics problem based on small-strain theory by $\vec{u}_{SS}(\vec{x}, t) = \{u_x(\vec{x}, t)\, u_y(\vec{x}, t)\}_{SS}$ where the subscript is a reminder that this solution is based on small-strain theory.

The justification for using $\vec{u}_{SS}$ for the prediction of damage accumulation rests on the assumption that there is some domain Ω^*, the boundary of which is represented by Γ_{SS} in Figure 8.1, outside of which the difference between $\vec{u}_{PZ}$ and $\vec{u}_{SS}$ is negligible:

$$\|\vec{u}_{PZ} - \vec{u}_{SS}\|_{\max} \approx 0 \quad \text{on } \Omega - \Omega^*. \tag{8.1}$$

In other words, it would be possible to compute $\vec{u}_{PZ}$ on Ω^* from $\vec{u}_{SS}$ by prescribing $\vec{u}_{SS}$ on Γ_{SS}.

In principle it is possible to formulate a mathematical problem that would model the formation of voids and cracks. In practice this is not feasible because such a model would be very complicated and require information about the physical properties of the material on length scales smaller than the RVE. These properties are difficult to obtain and their statistical dispersion is large. Therefore substantial uncertainties would be introduced into mathematical models of physical phenomena on length scales smaller than the RVE.

In order to avoid this problem, we assume that damage accumulation is related to and can be predicted from functionals computable from $\vec{u}_{SS}$. In other words, failure initiation and crack propagation events which are related to the solution of the highly nonlinear problem inside of Γ_{PZ} can be determined from $\vec{u}_{SS}$ even though the assumptions of the small-strain continuum theory are not applicable inside of Γ_{SS}. This involves the correlation of functionals computed from $\vec{u}_{SS}$ with failure initiation events observed in physical experiments. This is the conceptual basis of classical models of fatigue as well as models of crack propagation discussed in the following sections.

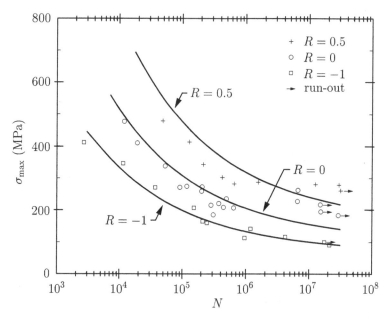

Figure 8.2 Typical S–N curves for 7050-T7451 aluminum plates, based on Reference [43].

8.3 Classical models of metal fatigue

The first investigations of fatigue in metals were motivated by occurrences of unexpected failure in railcar axles in the mid-1800s. Railcar axles are round beams subjected to pure bending between the bearings. As the axle rotates, the stresses caused by the bending moment vary sinusoidally, each rotation corresponding to one cycle.

On examining the failure surfaces it was found that cracks originated at some small imperfection at the surface and then propagated at an accelerating rate until the axle broke. A number of experiments were conducted in which similar patterns of failure were observed. The maximum normal stress, computed by the bending formula, was plotted against the number of cycles at which the specimens failed. There is substantial scatter in the data points in fatigue experiments. Curves fitted to the data points are called Wöhler curves[1] or more commonly S–N curves where S represents the maximum cyclic stress σ_{max} and N represents the number of cycles to failure.

Typical S–N curves for 7050-T7451 aluminum, based on [43], are shown in Figure 8.2 where R, called the stress ratio or cycle ratio, is the ratio of the minimum reference stress or load to the maximum reference stress or load. When $R = -1$ the minimum reference stress is compressive and has the same absolute value as the maximum reference stress. By definition, the fatigue strength is the curve corresponding to $R = -1$. The specimens were 7.62 mm (0.3 inch) diameter round bars. The surface condition of the bars is not specified.

---

[1] August Wöhler (1819–1914).

Axial loading was applied at 13.3 Hz. It can be seen that the variance of log life ($\log_{10} N$) tends to increase with decreasing stress level.

The empirical formula given in [43], converted to SI units, is

$$\log_{10} N = 9.73 - 3.24 \log_{10}(0.1450 \sigma_{max}(1-R)^{0.63} - 106.9) \tag{8.2}$$

where $\sigma_{max} > 106.9(1-R)^{-0.63}$ is in units of MPa. This formula, constructed by nonlinear regression, can be understood as an approximation of the mean of the random variable $\log_{10} N(\sigma_{max}, R)$ which will be denoted by μ. The standard deviation in log life is defined by

$$s = \sqrt{\frac{\sum_{i=1}^{m}(\log_{10} N_i(\sigma_{max}^{(i)}, R_i) - \mu_i)^2}{m-1}} \tag{8.3}$$

where m is the number of data points. Based on the data shown in Figure 8.2, $s = 0.471$. If it is assumed that $\log_{10} N_i(\sigma_{max}^{(i)}, R_i)$ has normal probability density so $\mathcal{N}_i \equiv \exp(\log_{10} N_i)$ has log-normal probability density.

When the maximum stress in polished specimens of steel and titanium subjected to sinusoidal loading ($R = -1$) is below a certain value, then it has been traditionally assumed that the specimens will not fail under any number of cycles. This value is called the endurance limit or fatigue limit. Some materials, such as aluminum, do not have a fatigue limit. For those materials the fatigue strength at 10^7 or 5×10^7 cycles is typically reported as the fatigue limit. A rough estimate of the fatigue limit is one-half of the ultimate tensile strength.

The assumption that a fatigue limit exists was made on the basis of practical considerations. It takes 28 hours to reach 10^7 cycles with a conventional testing machine operating at 100 Hz. Therefore much testing could not be performed beyond 10^7 cycles. It is now possible to reach 10^{10} cycles in less then a week by means of piezoelectric fatigue testing machines operating at 20 kHz. Based on fatigue data collected in the gigacycle range, the existence of a fatigue limit is in question [17]. The implication is that irreversible changes occur in the material even at low stress levels, very likely in the form of dislocations. These changes gradually accumulate and eventually result in the formation of cracks.

In the classical treatment of fatigue the maximum normal stress or maximum shearing stress was assumed to be the driving force for damage accumulation. This should be understood in the context of conceptualization: the investigators had in mind the objective of predicting the service life of mechanical components in which either the normal or shearing stress was dominant and, furthermore, the stress distribution in the vicinity of the stress maxima could be represented by smooth functions. They had at their disposal simple formulas that correlate known forces and moments with normal and shearing stresses. Therefore the goal of computation was to determine the maximum stress at notches, fillets, holes, etc., which was then correlated with calibration data in the form of S–N curves to predict the number of cycles at which fatigue failure would be expected to occur.

Example 8.3.1 We consider a shaft of diameter D with a U-notch. The shaft is subjected to bending moment M. The notation is shown in Figure 8.3.

242 WHAT SHOULD BE COMPUTED AND WHY?

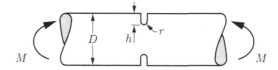

Figure 8.3 Shaft of diameter D with U-notch: notation.

The maximum stress is the nominal stress multiplied by the geometric stress concentration factor K_t

$$\sigma_{max} = K_t \sigma_{nom} \tag{8.4}$$

where in this example σ_{nom} is computed by the classical bending formula

$$\sigma_{nom} = \frac{32M}{\pi(D-2h)^3}. \tag{8.5}$$

K_t depends on the ratios h/r and h/D. Formulas for K_t can be found, for example, in [95], Table 17.1, entry 15. These formulas are based on data collected from the literature fitted with empirical curves. It is stated that "over the majority of the ranges specified by the variables, the curves fit the data points within much less than 5%." However, the accuracy of the data in the literature had not been recorded. For example, the formula for K_t in [95], Table 17.1, entry 15, for the case $h = r$ is

$$K_t = 3.04 - 7.236 \frac{2h}{D} + 9.375 \left(\frac{2h}{D}\right)^2 - 4.179 \left(\frac{2h}{D}\right)^3. \tag{8.6}$$

Exercise 8.3.1 For the problem described in Example 8.3.1 let $D = 30$ mm and $h = r = 5$ mm. Determine K_t by finite element analysis and verify that it is accurate to within 5%. Compare your result with K_t computed from Equation (8.6).

Definition 8.3.1 The geometric stress concentration factor denoted by K_t is the ratio of the maximum stress in the vicinity of a notch, hole, fillet, screw thread or other feature that causes locally increased stress to the nominal stress. As the name implies, K_t can be determined from the geometrical description of the part and the type of loading. It does not depend on material properties.

Definition 8.3.2 Nominal stress, defined for machine elements subjected to tension, bending and torsion, is understood in machine design to be the maximum normal or shearing stress at a notch, fillet, hole or other stress riser computed by formulas based on the assumption that the strain distribution over the cross section is a linear function. This was illustrated in Example 8.3.1. Because this definition cannot be generalized to arbitrary domains, we will understand nominal stress to mean the maximum stress that would exist at the location of a notch if the notch were not present, unless otherwise stated.

8.3.1 Models of damage accumulation

The calibration data presented in the form of S–N curves or equations such as Equation (8.2) are collected under laboratory conditions using constant amplitude loading at fixed R ratios. However, mechanical and structural components in service are subjected to variable amplitude loading represented by complex load–time functions, called load spectra. For example, the load spectra of rotorcraft components are characterized by blocks of high R-ratio, low-amplitude cycles interspersed with low-value minima that correspond to the start–stop cycle [41].

Several models devised to account for the accumulation of fatigue damage have been proposed for the prediction of fatigue life of mechanical and structural components subjected to variable amplitude loading. For example, Osgood [54] summarized 15 such models divided into three categories: models of linear cumulative damage, models of nonlinear cumulative damage and other models. Although often criticized for its shortcomings, the model most commonly used is known as Miner's rule, which falls into the category of linear cumulative damage models.

Miner's rule is based on the assumption that if n cycles of constant amplitude load characterized by ($\sigma_{\max}$, R) are imposed and the S–N data indicate that fatigue failure would occur at N cycles, then the fraction of fatigue life expended would be $n/N \leq 1$. Furthermore, when multiple load cycles n_i with corresponding limit values N_i are imposed and there are K such cycles, then the fraction of fatigue life expended (F) is

$$F = \sum_{i=1}^{K} \frac{n_i}{N_i} \leq 1. \tag{8.7}$$

Since $N_i = N_i(\sigma_{\max}^{(i)}, R_i)$ is a random variable, F is a function of random variables. Typical goals of computation are: (a) to estimate the probability Prob($F < x$) where $0 < x < 1$, given a load–time history; or (b) given a pattern of loading, estimate the number of times the pattern can be repeated before failure occurs ($F \geq 1$). An example of repeated load patterns is the ground–air–ground (GAG) cycle of aircraft. In either case the goal is to estimate the probability of an event. The procedure is illustrated in the following example.

Example 8.3.2 Consider the data given in Figure 8.2. Suppose that a structural component made of the same material is subjected to blocks of variable amplitude cyclic load. Each block is characterized by the data given in Table 8.1 where $\sigma_{\max}^{(i)}$ is in units of MPa.

Assume that the probability density function of $\log_{10} N_i(\sigma_{\max}^{(i)}, R_i)$ is normal (Gaussian) with the mean given by Equation (8.2) and the standard deviation computed using

Table 8.1 Data for Example 8.3.2.

i	$\sigma_{\max}^{(i)}$	R_i	n_i
1	200	−1	3000
2	400	0	2000
3	500	0.5	1000

Equation (8.3) is $s = 0.471$. Using Miner's rule, estimate the probability that fatigue failure will occur when M blocks are applied. Assume that M is an integer.

We first consider the case $M = 1$ and write Miner's rule in the form

$$F = \sum_{i=1}^{K} \frac{n_i}{M_i} \equiv \sum_{i=1}^{K} n_i N_i^{-1}(\sigma_{\max}^{(i)}, R_i). \tag{8.8}$$

In this example $K = 3$. Referring to Equation (8.2), the mean of $\log_{10} N_i^{-1}$, denoted by $\bar{\mu}_i$, is

$$\bar{\mu}_i = -9.73 + 3.24 \log_{10}(0.1450\sigma_{\max}^{(i)}(1 - R_i)^{0.63} - 15.5) \tag{8.9}$$

and the standard deviation is the same as that of $\log_{10} N_i$, that is, $s = 0.471$. Since F is a sum of random variables, correlation among the random variables N_i^{-1} ($i = 1, 2, 3$) has to be given. We will assume that N_i^{-1} and N_j^{-1} are independent random variables when $i \neq j$. The justification for this assumption is that each data point on which the S–N curve is based was obtained using a different sample.

We will use Monte Carlo simulation to generate a histogram and a cumulative distribution function (CDF) for F. The algorithm consists of the following steps:

1. Specify the number of Monte Carlo iterations J_{MC} and set the iteration counter j to 1. In this example we will use $J_{\text{MC}} = 1000$.

2. Using a random number generator for the normal distribution with mean $\mu = 0$ and standard deviation $s = 1$ generate a sequence of random numbers z_i ($i = 1, 2, \ldots, K$).

3. Compute $x_i := \log_{10} N_i^{-1} = sz_i + \bar{\mu}_i$ for $i = 1, 2, \ldots, K$. This is because, by assumption, x_i is normally distributed with mean $\bar{\mu}_i$ and standard deviation s. Therefore the random variable $z_i = (x_i - \bar{\mu}_i)/s$ is normally distributed with mean zero and standard deviation unity.

4. Compute $N_i^{-1} = 10^{x_i}$ for $i = 1, 2, \ldots, K$.

5. Compute

$$F_j = \sum_{i=1}^{K} \frac{n_i}{N_i}. \tag{8.10}$$

6. Increment j by 1. While $j < J_{\text{MC}}$ repeat steps 2 to 5.

7. Plot the histogram and the cumulative distribution function for F_j.

The histogram is shown in Figure 8.4. Values of $F_j \geq 1$ correspond to simulated failure events. For the data represented by the histogram, the estimated mean is 0.213, the estimated median is 0.152 and the estimated standard deviation is 0.261. Since these values depend on random numbers, small variations in these statistics can be expected. The computations were performed with MATLAB®.

The empirical cumulative distribution function for F is shown in Figure 8.5. The estimated probability that the part will not fail is 99%, that is, Prob($F < 1$) ≈ 0.99. This was computed from the cumulative distribution data by interpolation. Therefore the probability that the structural component will fail after one block of loading is approximately 0.01.

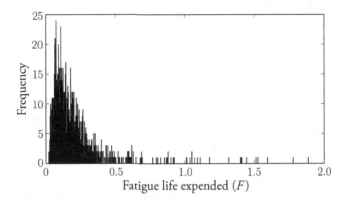

Figure 8.4 Example 8.3.2: histogram generated by the Monte Carlo method in 1000 iterations for $M = 1$.

Let us now consider the case $M > 1$. Since the sequence of loading is applied to one specimen or part, the random variables are fully correlated and therefore the foregoing procedure for computing the cumulative distribution function for F is the same as for $M = 1$ with the exception that Equation (8.10) is replaced by

$$F_j^{(M)} = \sum_{i=1}^{K} \frac{Mn_i}{N_i} = M \sum_{i=1}^{K} \frac{n_i}{N_i} = MF_j. \qquad (8.11)$$

Therefore the abscissas in Figure 8.4 and Figure 8.5 should be understood as fatigue life expended per block of loading. The probability that failure will not occur when M blocks of loading are applied can be read from Figure 8.5 as follows: letting $F = 1/M$, the ordinate of the cumulative distribution function at this point is the desired probability, that is, $\text{Prob}(F^{(M)} < 1)$.

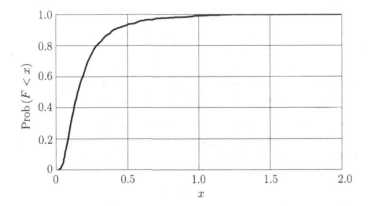

Figure 8.5 Example 8.3.2: empirical cumulative distribution function generated by the Monte Carlo method in 1000 iterations for $M = 1$.

For example, when $M = 3$ the ordinate at $F = 1/3$ read from Figure 8.5 is 0.81, therefore the probability that failure will not occur is estimated to be 81%.

Remark 8.3.1 The only source of uncertainty considered in Example 8.3.2 was the aleatory uncertainty of failure events of test coupons under cyclic loading. We did not address epistemic uncertainty associated with the interpretation of the outcome of fatigue experiments on test coupons. Rather we accepted the interpretation presented in [43] and we assumed that the probability density function of $\log_{10} N(\sigma_{\max}, R)$ is normal, the mean is represented by Equation (8.2) and the standard deviation is independent of the cycle ratio. Many other interpretations are possible. These interpretations are based on individual belief and judgment and are therefore subjective.

Another important assumption was that Miner's rule is a sufficiently accurate predictor of damage accumulation. Uncertainties associated with material properties, boundary conditions and errors in numerical approximation were neglected, implying the assumption that the dominant uncertainty is the aleatory uncertainty in the fatigue data. Of course, the effects of the various assumptions on the probability of failure can be examined through virtual experimentation.

Exercise 8.3.2 Solve the problem of Example 8.3.2 using $J_{\mathrm{MC}} = 10^2$ and $J_{\mathrm{MC}} = 10^4$ for one block of loading. Compare the predictions of probability of failure.

Exercise 8.3.3 In Example 8.3.2 the standard deviation $s = 0.471$ was assumed to be independent of R. Suppose that upon examining the data shown in Figure 8.2 an expert recommends using $s = s(R) = 0.5106 + 0.1814R$. Solve the problem of Example 8.3.2 for one block using the recommended $s(R)$ values and compare the predictions of probability of failure.

Exercise 8.3.4 In Example 8.3.2 the stresses shown in Table 8.1 were assumed to be exact. Suppose that the stresses were underestimated by 10%. Predict the probability of failure for one block when the underestimated values of $\sigma_{\max}^{(i)}$ are used.

8.3.2 Notch sensitivity

It was observed that when notches, fillets and holes of small radius are present then the S–N curves are not satisfactory predictors of crack initiation events. Referring to Figure 1.2, this should be understood to mean that models based on the assumption that the maximum stress is the driving force for damage accumulation could not adequately predict the fatigue life of mechanical components with notches of small radius and therefore had to be rejected. Several alternative models have been proposed of which we mention two in the following:

1. Neuber[2] [49] proposed that the driving force should be the maximum normal or shearing stress averaged over a characteristic material-dependent distance. He introduced the notion of effective stress concentration factor K_e. By definition

$$K_e = q(K_t - 1) + 1 \tag{8.12}$$

---

[2] Heinz Neuber (1906–1989).

where q is called the notch sensitivity index, defined by Neuber as follows:

$$q := \frac{1}{1+f\sqrt{\varrho'/\varrho}}, \quad f := \frac{\pi}{\pi - \omega}, \quad \varrho \geq \varrho_0 > 0 \tag{8.13}$$

where ϱ is the notch radius, ϱ' is an experimentally determined material constant (in length units) and $0 \leq \omega < \pi$ is called the flank angle. The lower limit of ϱ, denoted by ϱ_0, imposes a restriction on the range of validity of Equation (8.12). In the notation used in Figure 8.1, $\omega = 2\pi - \alpha$. The parameter ϱ' has been correlated with the ultimate tensile stress (UTS). For example, Kuhn and Hardrath [40] found that for steels with UTS ranging between 345 and 1725 MPa the estimated range of ϱ' is 430 to 0.9 μm respectively. They reported that scatter in ϱ' increases with decreasing ϱ.

2. Peterson[3] [60] retained the formalism of Equation (8.12) but proposed an alternative definition for the notch sensitivity index which we denote by $\bar{q}$:

$$\bar{q} := \frac{1}{1+\alpha/\varrho}, \quad \varrho \geq \varrho_0 > 0 \tag{8.14}$$

where α is an experimentally determined material constant and ϱ_0 is a lower bound on the notch radius. Peterson gave approximate values for α for steels as a function of their UTS: for UTS ranging between 345 and 1725 MPa the estimated range of α is 380 to 33 μm respectively. The experimentally determined values of $\bar{q}$ for aluminum and steel published in [60] indicate that ϱ_0 is greater than approximately $\alpha/4$.

The differences between K_e values based on the notch sensitivity indices proposed by Neuber and Peterson are not large. For low-strength steels (UTS about 400 MPa) K_e computed using Neuber's definition is about 9% lower than K_e computed from Peterson's definition. For high-strength steels the differences are negligibly small [69].

Neuber and Peterson introduced material-dependent parameters. It is of course possible to improve the fitting of calibration data with the use of additional parameters. However, calibration does not distinguish between aleatoric and epistemic uncertainties. Therefore, making successful predictions within or close to the range of parameters for which the model was calibrated is not validation.

Remark 8.3.2 The ASTM grain size number[4] of conventionally heat-treated AF 1410 steel is approximately 10.8 [90]. This corresponds to an average grain size of approximately 8.5 μm. The estimate of ϱ' given in [40] is smaller than this grain size.

Remark 8.3.3 Neuber and Peterson were concerned with the fatigue strength of notched machine elements. They assumed that there is a clearly defined notch of radius $\varrho > 0$ and the maximum stress depends on ϱ. There are many practical problems where the maximum stress is not related to a notch radius, however. Consider, for example, the T-junction of two pipes of identical cross-sections as shown in Figure 8.6. The intersection of the external surfaces is filleted. The fillet is idealized as a "rolling ball fillet," that is, the fillet surface would be

---

[3] Rudolph Earl Peterson (1901–1982).
[4] ASTM Standard E11296 (2004).

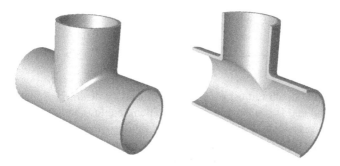

Figure 8.6 T-junction of two pipes of identical dimensions.

in contact with a ball of radius $r_f > 0$ rolled along the intersection curve while maintaining contact with the external surfaces. The intersection of the internal surfaces is chamfered. The chamfer radius is $r_c \geq 0$.

Suppose that the T-junction is subjected to internal pressure p. The maximum stress will occur along the intersection of the internal surfaces and depend only weakly on r_c.

Remark 8.3.4 Averaging stresses over a material-dependent distance for notched bars, shafts and beams can be understood also as averaging over an area or volume. Therefore Neuber's conceptualization admits alternative interpretations on general domains.

Exercise 8.3.5 Estimate the maximum principal stress in the T-junction shown in Figure 8.6 and verify that the error is not greater than 5%. Assume that the outside diameter of both pipes is $D_o = 105.0$ mm, the inside diameter is $D_i = 95.0$ mm and the centerlines intersect at right angles. The radius of the rolling ball fillet is 8.0 mm and Poisson's ratio is 0.3. Perform the computations for two chamfer radii: $r_c = 0$ and $r_c = 5.0$ mm. The T-junction is loaded by internal pressure $p_0 = 1.0$ MPa and corresponding normal tractions $T_n = p_0 D_i^2/(D_o^2 - D_i^2)$ acting on the cross-sections. Select length dimensions for the pipes such that the effects of those dimensions on the data of interest are negligible.

8.3.3 The theory of critical distances

The theory of critical distances (TCD) is based on the assumption that there is a material property called critical distance. According to Taylor [86], TCD covers a group of methods, based on the assumption that brittle failure and fatigue failure are predictable from two material parameters: a critical distance d and a critical stress σ_0. In particular, the line, area and volume methods described in [86] differ by whether the stress is averaged over a line, area or volume, respectively. From the perspective of validation, each of these methods can be understood as a model of driving force for damage accumulation.

Since these definitions of driving force do not depend on a fillet or hole characterized by a radius ϱ, they are applicable to problems such as the T-joint problem shown in Figure 8.6 and where the fillet radius is zero, for example, cracks.

Example 8.3.3 Let us consider a conical feature machined into a polished steel plate. The cross-section is shown in Figure 8.7(a). The thickness of the plate is 3.020 mm, and the

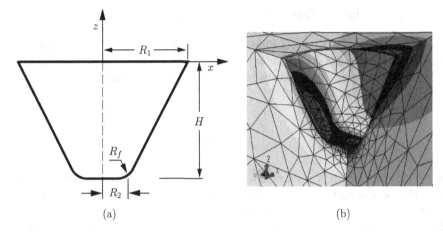

Figure 8.7 (a) Cross-section of a conical feature: notation. (b) Contours of the first principal stress, quarter symmetry.

dimensions of the conical feature are $R_1 = 1.016$ mm, $R_2 = 0.330$ mm, $r_f = 0.391$, $H = 0.889$ mm. Poisson's ratio is 0.280. The plate is subjected to tension in the x direction.

The contours of the first principal stress are shown in Figure 8.7(b). It can be seen that the maximum stress is not significantly influenced by the fillet, therefore Neuber's or Peterson's model is not applicable in this case. On the other hand TCD is applicable.

Exercise 8.3.6 Consider the problem of a circular hole in an infinite plate subjected to unidirectional tension. The notation is shown in Figure 4.11 and the classical solution for σ_x is given by Equation (4.34). The geometric stress concentration factor is $K_t = \sigma_{\max}/\sigma_\infty = 3$ where $\sigma_{\max} = \sigma_x(a, \pm\pi/2)$.

(a) Show that

$$\sigma_x^{(d)} := \frac{1}{d}\int_a^{a+d} \sigma_x(r, \pi/2)\,dr = \sigma_\infty\left(1 + 2\frac{1}{1+d/a} + O(d/a)\right) \approx K_e\sigma_\infty$$

where $K_e = \bar{q}(K_t - 1) + 1$ is the effective stress concentration factor and $\bar{q}$ is Peterson's notch sensitivity index given by Equation (8.14) with $\alpha = d$ and $\varrho = a$.

(b) Compute the relative error

$$e_r = 100\frac{\sigma_x^{(d)} - K_e\sigma_\infty}{\sigma_x^{(d)}}$$

for $d/a = 0.1$. The results of this exercise show that the effective stress concentration factor defined by Peterson can be viewed as an approximation to the line or area method of TCD.

8.4 Linear elastic fracture mechanics

Linear elastic fracture mechanics (LEFM) is based on the assumption that the driving force for crack propagation under cyclic loading is the amplitude of the stress intensity factor. Paris' law[5] [59] establishes a relationship between crack growth rate and the amplitude of the stress intensity factor at cycle ratio $R = 0$:

$$\frac{da}{dN} = C(\Delta K)^m \qquad (8.15)$$

where a is the crack length, N is the number of cycles, C and m are material constants and ΔK is the amplitude of the stress intensity factor. The constant C is in units of $(\text{MPa})^{-m}\, \text{m}^{1-m/2}$ or equivalent, the constant m is dimensionless.

This is a phenomenological representation of experimental observations schematically illustrated in Figure 8.8. Equation (8.15) is applicable in the range of ΔK where the growth rate curve plotted on a log–log scale can be well approximated by a straight line. This range is called Region *II* or the Paris ΔK region. Region *I* is called the threshold region and Region *III* is called the stable tearing crack growth region [69]. ΔK is usually understood to mean ΔK_I.

The threshold value of ΔK is indicated by $(\Delta K)_{\text{th}}$. At $(\Delta K)_{\text{th}}$ the crack length increment per cycle is of the order of magnitude of the interatomic spacing of the material. Whether there exists a threshold value $(\Delta K)_{\text{th}}$ below which crack growth will not occur is not a settled question. The available experimental data have a large scatter in this region.

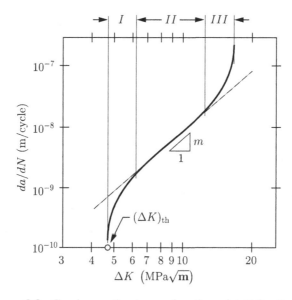

Figure 8.8 Crack growth rate as a function of ΔK for $R = 0$.

---

[5] Paul C. Paris (1930–).

There are several models formulated to account for the effects of the cycle ratio R. For example, the Walker correction (from the AFGROW Manual) is

$$\frac{da}{dN} = C \begin{cases} \left(\Delta K (1-R)^{n-1}\right)^m & \text{for } R \geq 0 \\ \left(|K_{\max}|(1-R)^{n-1}\right)^m & \text{for } R < 0 \end{cases} \qquad (8.16)$$

where $0 < n \leq 1$ is an experimentally determined parameter called the Walker exponent.

Remark 8.4.1 Typical values of the constant m are between 2.5 and 3.5. An error of $\mp 5\%$ in ΔK is magnified to -16.4% to 18.6% in the estimated crack growth rate in the Paris region when $m = 3.5$.

Calibration

The experimental procedures for the determination of fatigue crack growth rates as a function of ΔK are described in an ASTM standard.[6] One of the specimens used for the determination of crack growth rates as well as fracture toughness is the compact tension specimen shown in Figure 8.9.

While this is a widely used and generally accepted method, there are some conceptual difficulties associated with the interpretation of test data: the stress intensity factor is defined for the stress field in the neighborhood of crack tips in two-dimensional plane stress or plane strain domains. However, physical measurements can be performed in three dimensions only. The singularity at the intersection of the crack front with the free surface, represented by point A in Figure 8.9(b), has a very different character from the singularity of the two-dimensional crack-tip problem, given by Equations (7.24) through (7.26). This is because the stress-free

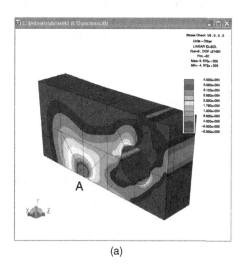
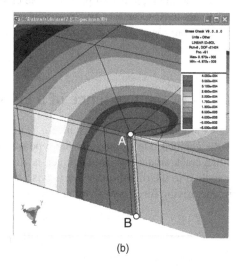

(a)   (b)

Figure 8.9 Compact tension specimen, quarter symmetry.

---

[6] ASTM E647-08e1 (2008) Standard Test Method for Measurement of Fatigue Crack Growth Rates. doi: 10.1520/E0647-08.

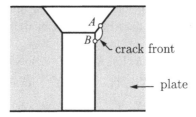

Figure 8.10 Example 8.4.1: small crack at a countersunk hole.

boundary conditions have to be satisfied not only at the crack faces but also at the surface of the specimen. When the specimen is thin these singularities are in close proximity and therefore tend to dominate the stress field in the neighborhood of the crack front.

Point B in Figure 8.9(b) lies on the plane of symmetry. Generally it is assumed that when the specimen in sufficiently thick then plane strain conditions exist. However, in the plane of symmetry of the compact tension specimen the solution is likely to be closer to the solution of the generalized plane strain problem described in Section 3.4.4. It can be asserted that plane strain conditions do not exist in the plane of symmetry. It cannot be asserted that generalized plane strain conditions exist because the planar surfaces of the specimen to which the crack front is normal are not sufficiently far from the plane of symmetry.

Example 8.4.1 One important area of practical concern is the lack of satisfactory predictability of crack growth rates from small cracks. A typical problem is illustrated in Figure 8.10 where a small crack has formed at a countersunk fastener hole. Note that the crack front is curved and the intersection points of the crack front with the boundary surfaces (labeled A and B) are in close proximity. The singularities associated with points A and B strongly influence the solution at the crack front and therefore the assumptions on which LEFM is based are not applicable.

8.5 On the existence of a critical distance

Let us assume that critical distance d is a material property and therefore it is independent of the geometric attributes and the type of loading. Considering a two-dimensional domain with a crack, assuming that the crack is along the x-axis, periodic loading of amplitude ΔT_y is applied in the direction of the y-axis and the origin of the coordinate system is the crack tip, the driving force is the average stress $\Delta \sigma_d$ induced by ΔT_y. By definition

$$\Delta \sigma_d = \frac{1}{d} \int_0^d \Delta \sigma_y(x, 0)\, dx. \tag{8.17}$$

The periodic loading ΔT_y induces variations in the stress intensity factor ranging between a minimum value $(K_I)_{\min} \geq 0$ and a maximum value $(K_I)_{\max} > (K_I)_{\min}$. Let $\Delta K_I \equiv (K_I)_{\max} - (K_I)_{\min}$ and denote by $(\Delta K_I)_{\text{th}}$ the threshold value of ΔK_I. In other words, a crack will not propagate when the amplitude of the average stress satisfies the following condition:

$$\Delta \sigma_d \leq \frac{1}{d} \int_0^d \frac{(\Delta K_I)_{\text{th}}}{\sqrt{2\pi x}} dx + O(d^{1/2}) \approx (\Delta K_I)_{\text{th}} \sqrt{\frac{2}{\pi d}}. \quad (8.18)$$

Equating $\Delta \sigma_d$ to the endurance limit $\Delta \sigma_0$ at a fixed R value, we have the following estimate for the critical distance:

$$d \approx \frac{2}{\pi} \left(\frac{(\Delta K_I)_{\text{th}}}{\Delta \sigma_0} \right)^2_R. \quad (8.19)$$

If $(\Delta K_I)_{\text{th}}$ and $\Delta \sigma_0$ are material properties then d is a material property also. One has to bear in mind, however, that the endurance limit varies with the size of specimens: when the size increases, the endurance limit decreases. This contradicts the assumption that $\Delta \sigma_0$ is a material property. Therefore d is a material property only if $(\Delta K_I)_{\text{th}}$ varies with size in the same way as $\Delta \sigma_0$. Due to large statistical dispersion in the data it is not possible to obtain accurate estimates for $(\Delta K_I)_{\text{th}}$ and $\Delta \sigma_0$. It is unlikely that the proposition that d is a material property can be validated with respect to reasonable criteria for rejection.

Example 8.5.1 The threshold stress intensity factor for conventionally processed AF 1410 steel[7] is approximately 16.0 MPa and its endurance limit is approximately 950 MPa. Therefore from Equation (8.19) we have $d = 180$ μm. This estimate of d is much larger than the estimates given in [40] and [60].

Exercise 8.5.1 Assume that the error in $(\Delta K_I)_{\text{th}}$ is $\pm 15\%$ and the error in $\Delta \sigma_0$ is $\pm 10\%$. Estimate the range of error for the critical distance d.

8.6 Driving forces for damage accumulation

In the foregoing sections we summarized widely used methods for the prediction of the formation and propagation of cracks caused by cyclic loading. There are many successful applications of these methods; however, those applications typically involve geometric configurations and loading conditions similar to those used in calibration. Making reliable predictions for geometric configurations and loading conditions that are very different from those used in calibration, such as corrosion defects and very small cracks, remains a challenging problem. This indicates that epistemic and aleatory uncertainties are mixed in the calibration data. Therefore the problem of identification of the driving force for damage accumulation remains an open question.

Many plausible conceptualizations are possible. We now have computational methods at our disposal which were not available at the time when the classical models of fatigue and crack propagation were formulated. We also have a well-developed conceptual framework for validation. Therefore new possibilities exist for the interpretation and generalization of

---

[7] This is a high-strength steel used in many safety-critical aerospace applications. Its ultimate tensile strength is approximately 1670 MPa.

existing new fatigue data. For example, the following family of models does not require the assumption that a material-dependent critical distance exists.

Let G define a family of possible driving forces in terms of some functional $\mathcal{F}(\vec{u}_{SS}) > 0$ and a condition C:

$$G(\mathcal{F}, \vec{u}_{SS}, C, T) = \int_{\Omega_C} \mathcal{F}(\vec{u}_{SS}(\vec{x}, t), T) \, dV, \quad \vec{x} \in \mathbb{R}^3 \qquad (8.20)$$

where $\vec{u}_{SS}$ is the solution of a continuum mechanics problem based on small-strain theory and T represents temperature. The domain of integration Ω_C depends on the condition by which Ω_C is defined. Formulation of this condition is part of the conceptualization process. For example, one may define $\mathcal{F}$ to be the von Mises stress and Ω_C as a function of the first principal stress σ_1:

$$\Omega_C = \{\vec{x} \mid \sigma_1 > \beta \sigma_{\text{yld}} \text{ when } F(t) = F_{\max}\} \qquad (8.21)$$

where σ_{yld} is the yield stress, $0 < \beta \leq 1$ is a dimensionless parameter and $F(t)$ is the load–time history. Of course, the definition of Ω_C may depend on the definition of $\mathcal{F}$. The choice of Ω_C as a function of stress or strain is related to the experimentally observed fact that the larger the volume subjected to elevated stress or strain, the lower the endurance limit and hence the likelihood of failure is greater.

In general, the solution $\vec{u}_{SS}(\vec{x}, t)$ is not known; only an approximation to $\vec{u}_{SS}$, which will be denoted by $\vec{u}_{FE}(\vec{x}, t)$, is known. Replacement of $\vec{u}_{SS}$ by $\vec{u}_{FE}$ is permissible only when it can be guaranteed that

$$|G(\mathcal{F}, \vec{u}_{SS}, C, T) - G(\mathcal{F}, \vec{u}_{FE}, C, T)| \leq \tau |G(\mathcal{F}, \vec{u}_{SS}, C, T)| \qquad (8.22)$$

where τ is a specified tolerance. Numerical accuracy is essential because, unless the accuracy of the computed data is known, it is not meaningful to compare experimental observations with predictions based on a mathematical model. This point was discussed in Section 1.2.2.

Example 8.6.1 Let us consider the conical feature described in Example 8.3.3 and assume that $\mathcal{F} = 1$ and let $Q = \sigma_1/\sigma_{\text{yld}}$ where σ_1 is the first principal stress and $\sigma_{\text{yld}} = 1517$ MPa is the yield stress. Let us assume further that a uniaxial stress of 1379 MPa is applied to the plate. In this case the driving force is the volume of the material where $\sigma_1 > \alpha \sigma_{\text{yld}}$. Note that this definition of driving force at least qualitatively accounts for the experimentally observed fact that the larger the volume of material exposed to elevated stress, the lower the endurance limit. The volume corresponding to $\alpha = 0.95$ is illustrated in Figure 8.11. The planes normal to the x- and y-axes are planes of symmetry.

8.7 Cycle counting

In order to estimate damage accumulation by means of Miner's rule, or some other model of cumulative damage, it is necessary to convert load–time histories into sets of stress reversals characterized by pairs of ($\sigma_{\max}$, R). Standard practices for cycle counting in fatigue analysis

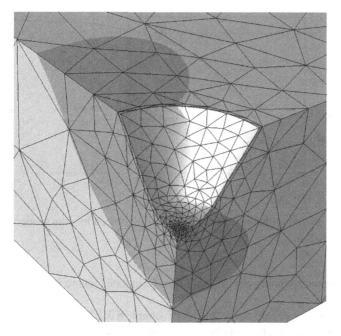

Figure 8.11 Example 8.6.1: over the dark grey region $\sigma_1 \geq 0.95\sigma_{yld}$, quarter symmetry.

are described in an ASTM standard.[8] Discussion of cycle counting methods is beyond the scope of this book, except to note that the definition of a cycle depends on the method used for cycle counting and therefore the choice of counting method will influence estimates of damage accumulation.

8.8 Validation

We have seen that the mathematical model used for predicting the probability of failure caused by cyclic loading is based on a number of assumptions in the following categories:

1. Definition of driving force. The most widely used classical definitions and new possibilities were outlined.

2. Statistical characterization of calibration data. This was discussed in Example 8.3.2.

3. Choice of the model of cumulative damage. We discussed Miner's rule only but noted that several other models exist.

4. The method of cycle counting.

The purpose of validation is to test whether a model meets necessary conditions for acceptance. The outcome of validation experiments is evaluated with reference to one or more

---

[8] ASTM E1049-85 (Reapproved 2005), Standard Practices for Cycle Counting in Fatigue Analysis.

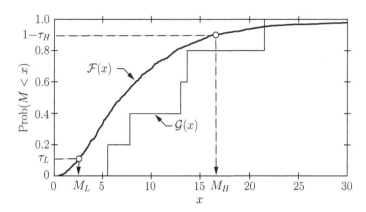

Figure 8.12 Predicted probability of the number of repeated loading sequences to failure M based on Example 8.3.2.

metrics and the corresponding criteria which are formulated taking into account the intended use of the model.

Let us consider once again the problem of Example 8.3.2. Suppose that validation experiments are planned in which the sequence of loading shown in Table 8.1 will be repeated until failure occurs. We are asked to make a prediction on the outcome of the experiment.

We denote the number of times the sequence is repeated by M and, using the procedure outlined in Example 8.3.2, estimate the cumulative distribution function $\mathcal{F}(x) = \mathrm{Prob}(M < x)$. The function $\mathcal{F}(x)$ is the prediction based on the set of assumptions incorporated in the model. For instance, the predicted probability of the number of repeated loading sequences to failure M based on Example 8.3.2 is shown in Figure 8.12. The model will be rejected if the predicted probability of the outcome of the experiments is lower than specified threshold values.

To formulate the criterion for rejection we select threshold values τ_L, τ_H and compute the low (resp. high) estimate for M, denoted by M_L (resp. M_H), such that

$$\mathrm{Prob}(M < M_L) = \tau_L \quad \text{and} \quad \mathrm{Prob}(M < M_H) = 1 - \tau_H.$$

This is illustrated schematically in Figure 8.12. Let us first assume that only one validation experiment was performed. Then the model is rejected if the outcome of the experiment falls outside the interval (M_L, M_H). The model would not be rejected if sufficiently small values were chosen for τ_L and/or τ_H.

Suppose now that $\tau_L = \tau_H = 0.1$ was selected, five experiments were performed and the outcome of the experiments is represented by the cumulative distribution function $\mathcal{G}(x)$ shown in Figure 8.12. We see that four of the five data points are within the interval (M_L, M_H). In this case there is no reason for rejecting the model. This is because the choice of τ is such that about 20% of the outcomes will fall outside the interval (M_L, M_H). Therefore the results are consistent with the prediction of the model.

While the choice of τ_L, τ_{HL} is arbitrary and therefore subjective, this method provides a reasonably objective basis for evaluating and ranking alternative models in the following sense. By changing one or more assumptions incorporated in the model, such as the definition of driving force, the statistical characterization of the calibration data, the model of cumulative damage and/or the method of cycle counting, we will have different predictions. In other words, each model will produce a cumulative distribution function (CDF) $\mathcal{F}(x)$. Various metrics can be devised for quantifying the difference between the predicted CDF and the experimental CDF. For example, denoting the inverse of $\mathcal{F}(x)$ (resp. $\mathcal{G}(x)$) by $\mathcal{F}^{-1}(y)$ (resp. $\mathcal{G}^{-1}(y)$) in [15], the following metric was used:

$$d_\epsilon = \max_{y \in I_\epsilon} |\mathcal{F}^{-1}(y) - \mathcal{G}^{-1}(y)| \quad \text{where } I_\epsilon = \{y \mid \epsilon < y < 1 - \epsilon\}. \tag{8.23}$$

This metric removes from consideration predictions of outcomes whose estimated probabilities are smaller than ϵ. This is justified by the observation that predictions in the range of very small probabilities tend to be unreliable. Of course, the metric depends on the choice of ϵ. In [28] the L_1 norm was used:

$$d_{L_1} = \int_0^1 |\mathcal{F}^{-1}(y) - \mathcal{G}^{-1}(y)| \, dy. \tag{8.24}$$

It is possible to define other metrics, such as the L_2 norm. Any metric d should satisfy the requirements that (a) $d \geq 0$ with $d = 0$ only if $\mathcal{F}(y)^{-1} - \mathcal{G}^{-1}(y) = 0$ and (b) $d < \infty$ for any $\mathcal{F}^{-1}(y)$.

Suppose that predictions based on more than one model were made. We would prefer the model whose predictions are closest to the outcome of experiments as measured by means of a suitably chosen metric.

Remark 8.8.1 More often than not, the data of interest cannot be observed in physical experiments. Therefore, in many cases, what is observed in a validation experiment and what the mathematical model is called upon to predict are not the same data. In other words, predictions of the data of interest can be tested by indirect means only.

Remark 8.8.2 There is a large and rapidly growing literature on validation. A special issue of *Computer Methods in Applied Mechanics and Engineering* is dedicated to this subject.[9] The papers in this issue and the references cited provide a good starting point for further reading on this subject.

8.9 Chapter summary

The key question of what should be computed and why has to be addressed in the process of conceptualization. This question was discussed in the context of the prediction of fatigue damage caused by cyclic loading. Specifically, the goal considered was to predict the statistical attributes of percent fatigue life expended (F) of mechanical components subjected to some specified variable amplitude cyclic loading. The information available consists of geometric description, material properties, load–time history, constraint definition and experimental

---

[9] Validation Challenge Workshop, eds. I. Babuška, K. Dowding and T. Paez, Vol. 197, issues 29–32, May 2008.

data of fatigue tests of pristine, notched or pre-cracked test specimens subjected to constant amplitude cyclic loading.

In order to generalize the laboratory data to practical situations characterized by variable amplitude random loading of parts that have notches, fillets, holes, scratches from incidental damage, etc., and make reliable predictions of F, it is necessary to formulate a mathematical model comprising the following elements:

1. Analysis of the object of interest based on small-strain theory of continuum mechanics. Aleatory and epistemic uncertainties are associated with geometric description, material properties, loading and constraints.

2. A definition of driving force for damage accumulation. In classical models of fatigue, the driving force is assumed to be the normal or shearing stress and the data are given in terms of S–N curves. In linear elastic fracture mechanics the assumed driving force is the stress intensity factor and the damage is measured by crack size. Many alternatives exist, as discussed in Section 8.6.

3. Calibration data obtained from laboratory experiments for fixed cycle ratios and load maxima. The load maxima and the cycle ratio are the independent variables, the number of cycles to failure (or increment in crack size) N, a random variable, is the dependent variable. It has substantial statistical dispersion even under carefully controlled laboratory conditions.

4. A rule that generalizes the calibration data for arbitrary cycle ratios. Equation (8.2) is an example.

5. A rule that correlates random load–time histories with counts of cycles characterized by stress maxima and cycle ratios. Several rules for counting cycles are in use.

6. A rule that predicts F, given the available calibration data and a load–time history. Several models have been proposed for this purpose. We have discussed only one, known as Miner's rule.

Phenomenological approaches cannot be avoided. Many rules, formulas and theories have been proposed. For predictions to be reliable, the reliability of each of these rules must be established through application of the methodology of validation.

A survey of experience with mathematical models of damage accumulation employed in current professional practice indicates that, generally speaking, successful predictions have been made only when the predictions were within the range of calibration data; that is, predictions were confined to parts that are very similar to those for which calibration data are available. In the terminology of validation, models that fail to make predictions within reasonable tolerances outside of the range of calibration data are rejected. These models will not be rejected for sufficiently large tolerances. The consequence of using large tolerances is that large factors of safety have to be specified. Not having reliable means to make predictions outside of the range of calibration data is evidence that significant epistemic uncertainties exist and the epistemic and aleatory uncertainties are mixed in the calibration data.

Ideally, all uncertainties should be aleatory and the factor of safety should be selected so as to account for aleatory uncertainties only. In reality the factor of safety must be sufficiently large to account for both epistemic and aleatory uncertainties. This imposes various penalties such as a weight penalty, increased frequency of inspection and premature retirement of components.

The goal is to reduce epistemic uncertainties through systematic examination of alternative models in validation experiments. To support such an effort, it will be necessary to design, implement and deploy a system of procedures for evaluating alternative models based on the concepts of validation using appropriate metrics and criteria. Some considerations relating to the design of such a system were presented in this chapter.

9

Beams, plates and shells

Dimensional reduction was discussed in Chapter 3 in connection with planar and axisymmetric models. Another very important class of dimensionally reduced models is discussed in this chapter. The formulation presented here differs from that in Chapter 4 where generalized formulations were derived from the differential equations. Here the generalized formulations for two- and three-dimensional elasticity are the starting points for the derivation.

9.1 Beams

In order to present the main points in a simple setting, mathematical models for structural beams are derived from the generalized formulation of the problem of two-dimensional elasticity. The formulation of models for beams in three dimensions is analogous but, of course, more complicated. Referring to Figure 9.1, the following assumptions are made:

1. The xy plane is a principal plane, that is, loads applied in the xy plane will not cause displacement in the direction of the z-axis.

2. The x-axis is coincident with the centroidal axis of the cross-section which will be denoted by ω.

3. The material is elastic and isotropic.

The displacement vector components are written in the following form:

$$u_x = u_{x|0}(x) + u_{x|1}(x)y + u_{x|2}(x)y^2 + \cdots + u_{x|m}(x)y^m \quad (9.1)$$

$$u_y = u_{y|0}(x) + u_{y|1}(x)y + u_{y|2}(x)y^2 + \cdots + u_{y|n}(x)y^n \quad (9.2)$$

where the functions $u_{x|k}(x)$, $u_{y|k}(x)$ are called field functions; their multipliers (powers of y) are called director functions. Writing the displacement components in this form allows us to consider a hierarchic family of models for beams characterized by the pair of indices (m, n).

Introduction to Finite Element Analysis: Formulation, Verification and Validation, First Edition.
Barna Szabó and Ivo Babuška. © 2011 John Wiley & Sons, Ltd. Published 2011 by John Wiley & Sons, Ltd.

BEAMS, PLATES AND SHELLS

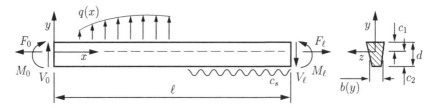

Figure 9.1 Notation.

The highest member of the hierarchy is the mathematical model based on two-dimensional elasticity, which corresponds to $m, n \to \infty$.

In the following we will denote the exact solution of a hierarchic beam model characterized by m and n by $\vec{u}_{EX}^{(m,n)}$ and the exact solution of the mathematical model based on two-dimensional elasticity by $\vec{u}_{EX}^{(2D)}$.

Any of the boundary conditions described in connection with two-dimensional elasticity may be specified; however, we will be concerned only with the loadings and constraints typically used in the analysis of beams. Specifically, the following types of load will be considered:

1. The traction component $T_n = T_y$ acting on the surface $y = c_1$ (see Figure 9.1):

$$T_n = \frac{q(x)}{b(c_1)} \qquad (9.3)$$

where q_y is the distributed load (in N/m units).

2. Normal tractions acting on cross-sections written in terms of the axial force F and the bending moment M. The sign conventions are indicated in Figure 9.1:

$$T_n = \frac{F}{A} - \frac{My}{I} \qquad (9.4)$$

where A is the area of the cross-section ω and I is the moment of inertia of ω with respect to the z-axis:

$$A := \int_\omega dy dz, \qquad I := \int_\omega y^2 dy dz. \qquad (9.5)$$

3. The shearing tractions $T_t(y, z)$ acting on cross-sections are either constant or defined by a function such that $T_t(c_1, z) = T_t(-c_2, z) = 0$ and the resultant is the shear force V:

$$T_t = -\frac{V}{A} \quad \text{or} \quad T_t = -\frac{VQ(y)}{Ib(y)} \qquad (9.6)$$

where $Q(y)$ is the static moment about the z-axis of that portion of the cross-section which extends from y to c_1:

$$Q(y) := \int_y^{c_1} sb(s)\,ds. \qquad (9.7)$$

The beam may be supported by an elastic foundation, with foundation modulus $c_s(x) \geq 0$ (N/m² units) and kinematic boundary conditions may be prescribed. We will assume that the elastic foundation reacts in the y direction only, that is, the foundation generates a distributed transverse load $q_s(x) = -c_s(x)u_y(x, -c_2)$ (N/m units).

We will write the generalized formulations for beam models as applications of the principle of minimum potential energy. The strain energy is

$$U := \frac{1}{2}\int_\Omega (\sigma_x\epsilon_x + \sigma_y\epsilon_y + \tau_{xy}\gamma_{xy})\,dV + \frac{1}{2}\int_0^\ell c_s u_y^2(x, -c_2)\,dx \qquad (9.8)$$

and the potential of external forces is

$$P := \int_0^\ell qu_y(x, c_1)\,dx + \sum_{i=1}^2 \int_{\omega_i} (T_x u_x + T_y u_y)\,dS \qquad (9.9)$$

where ω_i ($i = 1, 2$) are the cross-sections at $x = 0$ and $x = \ell$ respectively. The energy space is defined in the usual way. The exact solution of a particular model is the minimizer of the potential energy on the space of admissible functions:

$$\Pi\left(\vec{u}_{EX}^{(m,n)}\right) = \min_{\vec{u}^{(m,n)} \in \tilde{E}(\Omega)} \Pi\left(\vec{u}^{(m,n)}\right)$$

where $\Pi(\vec{u}) = U(\vec{u}) - P(\vec{u})$ and $\tilde{E}(\Omega)$ is the space of admissible functions.

Remark 9.1.1 Proper selection of a beam model characterized by the indices (m, n) is problem dependent. When the field functions are smooth and the thickness is small then m and n can be small numbers. The assumptions incorporated into the commonly used beam models imply that the field functions do not change significantly over distances comparable to the thickness.

Remark 9.1.2 The distinction between the notions of mathematical model and its discretization is blurred by conventions in terminology: It is customary to refer to the various beam-plate and shell formulations as theories or models. In fact, these models are semidiscretizations of the fully three-dimensional model. Therefore errors that can be attributed to the choice of indices, such as the indices (m, n) in Equations (9.1) and (9.2), and analogous indices defined for plate and shell models discussed in Sections 9.2 and 9.3, are related to discretization of the fully three-dimensional model rather than model definition.

9.1.1 The Timoshenko beam

Let us consider the simplest model, the model corresponding to the indices $m = 1, n = 0$. We introduce the notation

$$u_{x|0}(x) = u(x), \quad u_{x|1}(x) = -\beta(x), \quad u_{y|0}(x) = w(x).$$

The function β represents a positive (i.e., counterclockwise) angle of rotation (with respect to the z-axis). Since a positive angle of rotation multiplied by positive y would result in negative displacement in the x direction, $u_{x|1} = -\beta$. The strain components are as follows:

$$\epsilon_x = \frac{\partial u_x}{\partial x} = u' - \beta' y, \quad \epsilon_y = \frac{\partial u_y}{\partial y} = 0, \quad \gamma_{xy} = \frac{\partial u_x}{\partial y} + \frac{\partial u_y}{\partial x} = -\beta + w' \quad (9.10)$$

where the primes represent differentiation with respect to x. Assuming that the normal stresses σ_y and σ_z are negligibly small in relation to σ_x, the stress components are

$$\sigma_x = E\epsilon_x = E(u' - \beta' y), \quad \sigma_y = 0, \quad \tau_{xy} = G\gamma_{xy} = G(-\beta + w').$$

Therefore the strain energy is

$$U = \frac{1}{2} \int_0^\ell \left(\int_\omega [E(u' - \beta' y)^2 + G(-\beta + w')^2] \, dy \, dz \right) dx + \frac{1}{2} \int_0^\ell c_s w^2 \, dx$$

where the volume integral was decomposed into an area integral over the cross-section and a line integral over the length of the beam. Using the notation introduced in Equation (9.5) and noting that since the y- and z-axes are centroidal axes,

$$\int_\omega y \, dy \, dz = 0,$$

the strain energy can be written as

$$U := \frac{1}{2} \int_0^\ell [EA(u')^2 + EI(\beta')^2 + GA(-\beta + w')^2] \, dx + \frac{1}{2} \int_0^\ell c_s w^2 \, dx. \quad (9.11)$$

For the reasons discussed in the following section, the strain energy expression is usually modified by multiplying the shear term by a factor known as the shear correction factor, denoted by κ. The modified expression for the strain energy is

$$U_\kappa := \frac{1}{2} \int_0^\ell [EA(u')^2 + EI(\beta')^2 + \kappa GA(-\beta + w')^2] \, dx + \frac{1}{2} \int_0^\ell c_s w^2 \, dx. \quad (9.12)$$

The potential of external forces is

$$P := \int_0^\ell qw \, dx + F_\ell u(\ell) + M_\ell \beta(\ell) - V_\ell w(\ell) - F_0 u(0) - M_0 \beta(0) + V_0 w(0) \quad (9.13)$$

where F, M, V are respectively the axial force, bending moment and shear force, collectively called stress resultants, which are defined as follows:

$$F := \int_\omega \sigma_x \, dy dz, \quad M := -\int_\omega \sigma_x y \, dy dz, \quad V := -\int_\omega \tau_{xy} \, dy dz. \tag{9.14}$$

The subscripts 0 and ℓ refer to the location $x = 0$ and $x = \ell$, respectively.

The potential energy is defined by

$$\Pi_\kappa := U_\kappa - P. \tag{9.15}$$

This formulation is known as the Timoshenko beam model.[1] Particular applications of the principle of minimum potential energy depend on the specified loading and boundary conditions.

Shear correction

In this formulation the shear strain is constant on any cross-section (see Equation (9.10)). This is inconsistent with the assumption that no shear stresses are applied at the top and bottom surfaces of the beam, that is, at $y = c_1$ and $y = -c_2$, see Figure 9.1. From equilibrium considerations, it is known that the shear stress distribution is reasonably well approximated for a wide range of practical problems by

$$\tau_{xy} = -\frac{VQ}{Ib}$$

where V is the shear force, $Q = Q(y)$ is the function defined by Equation (9.7), I is the moment of inertia about the z-axis and $b(y)$ is the width as shown in Figure 9.1. The derivation of this formula can be found in standard texts on strength of materials.

We will consider a rectangular cross-section of depth d and width b in the following. In this case the shear stress distribution is

$$\tau_{xy} = \frac{3}{2} \frac{V}{db} \left(1 - 4\frac{y^2}{d^2}\right)$$

and the strain energy corresponding to this shear stress in a beam of length Δx is

$$\Delta U_\tau = \frac{1}{2} \frac{b\Delta x}{G} \int_{-d/2}^{+d/2} \tau_{xy}^2 \, dy = \frac{1}{2} \frac{\Delta x}{G} \frac{V^2}{db} \frac{6}{5}.$$

We will adjust the shear modulus G in the (1, 0) beam model so that the strain energy will be the same:

$$\Delta U_\tau^{(1,0)} = \frac{1}{2} \frac{b\Delta x}{\kappa G} \int_{-d/2}^{+d/2} \tau_{xy}^2 \, dy = \frac{1}{2} \frac{b\Delta x}{\kappa G} \int_{-d/2}^{+d/2} \left(\frac{V}{db}\right)^2 dy = \frac{1}{2} \frac{\Delta x}{\kappa G} \frac{V^2}{db}$$

---

[1] Stephen P. Timoshenko (1878–1972).

where κ is the shear correction factor. Letting $\Delta U_\tau = \Delta U_\tau^{(1,0)}$ we find that $\kappa = 5/6$. For this reason the strain energy expression given by Equation (9.11) is replaced by Equation (9.12). Most commonly $\kappa = 5/6$ is used, independently of the cross-section, although it is clear from the derivation that κ depends on the cross-section. For further discussion on shear correction factors we refer to [24].

In static analyses of slender beams the strain energy is typically dominated by the bending term, thus the solution is not influenced significantly by κ. As the length to depth ratio decreases, the influence of κ increases. In vibrating beams the influence of κ increases with the natural frequency.

Exercise 9.1.1 Based on the formulation outlined in Section 4.4.1, write down the generalized formulation of the undamped elastic vibration problem for the Timoshenko beam. Assume $u_{x|0} = 0$. Would you expect the natural frequencies computed for the Timoshenko model to be larger or smaller than the corresponding natural frequencies computed for the beam model (2, 3)? Why?

Numerical solution

The numerical solution of this problem by the finite element method is very similar to the solution of the one-dimensional model problem described in Section 2.5. The main difference is that here we have three field functions $u(x)$, $\beta(x)$ and $w(x)$, which are approximated by the finite element method. Let us write

$$u = \sum_{i=1}^{M_u} a_i \Phi_i(x), \quad \beta = \sum_{i=1}^{M_\beta} b_i \Phi_i(x), \quad w = \sum_{i=1}^{M_w} c_i \Phi_i(x)$$

and denote

$$a := \{a_1 \, a_2 \, \ldots \, a_{M_u}\}^T, \quad b := \{b_1 \, b_2 \, \ldots \, b_{M_\beta}\}^T, \quad c := \{c_1 \, c_2 \, \ldots \, c_{M_w}\}^T.$$

The structure of the unconstrained stiffness matrix becomes visible if we write the strain energy in the following form:

$$U_\kappa = \frac{1}{2}\{a^T \, b^T \, c^T\} \begin{bmatrix} K_u & 0 & 0 \\ 0 & K_\beta & K_{\beta w} \\ 0 & K_{\beta w}^T & K_w \end{bmatrix} \begin{Bmatrix} a \\ b \\ c \end{Bmatrix}.$$

Given the assumption that the shearing tractions on the top and bottom surfaces of the beam are zero and the x-axis is coincident with the centroidal axis (see Figure 9.1), the function u, representing axial deformation, is not coupled with the rotation β and the transverse displacement w and therefore can be solved independently of β and w.

At the element level, assuming constant sectional and material properties, K_β can be constructed directly from the matrices given by Equations (2.73) and (2.76) and K_w can be constructed from (2.73). For example, for $p_k = 2$ matrix $[K_\beta^{(k)}]$ is

$$[K_\beta^{(k)}] = \begin{bmatrix} \dfrac{EI}{\ell_k} + \dfrac{\kappa GA\ell_k}{3} & -\dfrac{EI}{\ell_k} + \dfrac{\kappa GA\ell_k}{6} & -\dfrac{\kappa GA\ell_k}{2\sqrt{6}} \\ & \dfrac{EI}{\ell_k} + \dfrac{\kappa GA\ell_k}{3} & -\dfrac{\kappa GA\ell_k}{2\sqrt{6}} \\ (\text{sym.}) & & \dfrac{2EI}{\ell_k} + \dfrac{\kappa GA\ell_k}{5} \end{bmatrix}. \quad (9.16)$$

The terms of the coupling matrix $K_{\beta w}^{(k)}$ are

$$k_{ij}^{(\beta w)} = GA \int_{-1}^{+1} \frac{dN_i}{d\xi} N_j \, d\xi.$$

Remark 9.1.3 A different polynomial degree may be assigned to each field function on each element. In the following we will assume that all fields have the same polynomial degree p_k on an element but p_k may vary from element to element.

Exercise 9.1.2 Write down the terms of the element-level matrices K_w and $K_{\beta w}^{(k)}$ for $p_k = 2$.

Example 9.1.1 Consider a beam of constant cross-section with built-in (fixed) support (i.e., $u_x = u_y = 0$) at $x = 0$ and simple support ($u_y = 0$) at $x = \ell$. There is an intermediate support ($u_y = 0$) at $x = \ell/2$, as shown in Figure 9.2(a). The beam is loaded by a constant distributed load $q = -q_0$. The goal is to determine the location and magnitude of the maximum bending moment. In this case the axial displacement u is zero and the potential energy is defined as follows:

$$\Pi = \frac{1}{2} \int_0^\ell [EI(\beta')^2 + \kappa GA(-\beta + w')^2] \, dx + \int_0^\ell q_0 w \, dx. \quad (9.17)$$

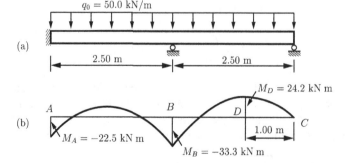

Figure 9.2 (a) Problem definition. (b) Bending moment diagram.

The spaces of admissible functions are defined as follows:

$$\tilde{E}_\beta := \left\{ \beta \,\bigg|\, \int_0^\ell [(\beta')^2 + \beta^2] \, dx \leq C < \infty, \; \beta(0) = 0 \right\}$$

$$\tilde{E}_w := \left\{ w \,\bigg|\, \int_0^\ell (w')^2 \, dx \leq C < \infty, \; w(0) = 0, \; w(\ell/2) = 0, \; w(\ell) = 0 \right\}.$$

The problem is to find β_{EX} and w_{EX} by minimization of the potential energy:

$$\Pi(\beta_{EX}, w_{EX}) = \min_{\substack{\beta \in \tilde{E}_\beta \\ w \in \tilde{E}_w}} \Pi(\beta, w).$$

Assume that the beam is an S200 × 27 American Standard steel beam.[2] The section properties are: $A = 3490$ mm, $I = 24.0 \times 10^6$ mm^4. Let $\ell = 5.00$ m, $E = 200$ GPa, $\nu = 0.3$ and $q_0 = 50.0$ kN/m. The weight of the beam ($27 \times 9.81 \times 10^{-3} = 0.265$ kN/m) is negligible in relation to the applied load.

The solution was obtained using two finite elements and the shear correction factor $\kappa = 5/6$. At $p \geq 4$ the exact solution is obtained (up to round-off errors). The results are shown in Figure 9.2(b).

Shear locking

Let us consider a beam of rectangular cross-section of dimension $b \times d$ subject to a distributed load $q = d^3 f(x)$ and assume that the axial force is zero. In this case the potential energy is

$$\Pi = \frac{1}{2} \int_0^\ell \left[\frac{Ebd^3}{12} (\beta')^2 + \kappa Gbd(-\beta + w')^2 \right] dx - d^3 \int_0^\ell f(x) w \, dx. \tag{9.18}$$

On factoring d^3 we have

$$\Pi = \frac{d^3}{2} \int_0^\ell \left[\frac{Eb}{12} (\beta')^2 + \frac{\kappa Gb}{d^2} (-\beta + w')^2 \right] dx - d^3 \int_0^\ell f(x) w \, dx. \tag{9.19}$$

For sufficiently small d values the term $\kappa Gb/d^2$ is much larger than $Eb/12$. Since the minimum of Π is sought, this forces $\beta \to w'$ as $d \to 0$:

$$\lim_{d \to 0} \int_0^\ell (-\beta + w')^2 \, dx = 0. \tag{9.20}$$

The effect of the constraint $\beta \approx w'$ is that the number of degrees of freedom is reduced. Convergence can be very slow when low polynomial degrees are used. This is called shear locking. As $d \to 0$, the solution of the Timoshenko model converges in the energy norm to the solution of the Bernoulli–Euler model which is described in the next section.

---

[2] In this designation S indicates the cross-section; 200 is the nominal depth in mm; and 27 is the mass per unit length (kg/m).

9.1.2 The Bernoulli–Euler beam

The Bernoulli–Euler beam model[3] is the limiting case of the Timoshenko model, and all higher order models, with respect to $d \to 0$. Assuming that $u = 0$ and $c_s = 0$ and letting $\beta = w'$, for homogeneous boundary conditions the potential energy expression (9.15) becomes

$$\Pi = \frac{1}{2}\int_0^\ell EI(w'')^2\,dx - \int_0^\ell qw\,dx. \tag{9.21}$$

This is the generalized formulation corresponding to the familiar fourth-order ordinary differential equation found in introductory texts on strength of materials:

$$(EIw'')'' = q(x) \tag{9.22}$$

which is the Bernoulli–Euler beam model. To show this, let $w \in \tilde{E}(I)$ be the minimizer of Π and let $v \in E^0(I)$ be an arbitrary perturbation of w. For the sake of simplicity let us assume that the boundary conditions are homogeneous, that is, the prescribed displacements, rotations, moments and shear forces are zero. Then

$$\Pi(w + \varepsilon v) = \frac{1}{2}\int_0^\ell EI(w'' + \varepsilon v'')^2\,dx - \int_0^\ell q(w + \varepsilon v)\,dx$$

will be minimum at $\varepsilon = 0$:

$$\left(\frac{\partial \Pi}{\partial \varepsilon}\right)_{\varepsilon=0} = 0.$$

Therefore

$$\int_0^\ell EIw''v''\,dx - \int_0^\ell qv\,dx = 0.$$

Integrating by parts twice, we get

$$(\underbrace{EIw''}_{M} v')\big|_0^\ell - (\underbrace{(EIw'')'}_{V} v)\big|_0^\ell + \int_0^\ell [(EIw'')'' - q]v\,dx = 0.$$

This must hold for any choice of v that does not perturb the prescribed essential boundary conditions. Since $M = EIw''$ and $V = (EIw'')'$, the boundary terms vanish and we have the strong form of the Bernoulli–Euler beam model given by Equation (9.22).

Exercise 9.1.3 Show that Equation (9.22) can be obtained without the assumption that moments or shear forces prescribed on the boundaries are zero. For the definition of moments and shear forces refer to Equation (9.14). Hint: If a non-zero moment and/or shear force is prescribed on a boundary then the expression for the potential energy given by Equation (9.21) must be modified to account for this term.

---

[3] Leonhard Euler (1707–1783), James Bernoulli (1654–1705).

Exercise 9.1.4 Derive the strong form of the Bernoulli–Euler beam model for the case $c_s \neq 0$.

Exercise 9.1.5 Assume that the boundary conditions are homogeneous and $u = 0$ and $c_s = 0$. Show that the strong form of the Timoshenko model is

$$(EI\beta')' + \kappa GA(-\beta + w') = 0 \tag{9.23}$$

$$\left[\kappa GA(-\beta + w')\right]' = -q \tag{9.24}$$

and hence show that in Example 9.1.1 the exact solution is obtained when $p \geq 4$. Hint: The procedure is analogous to that described in Section 9.1.2.

Numerical solution

For the Bernoulli–Euler beam model the energy space is

$$E(I) = \left\{ w \mid \int_0^\ell (w'')^2\, dx \leq C < \infty \right\}, \qquad I := \{x \mid 0 < x < \ell\}. \tag{9.25}$$

This implies that $w \in E(I)$ is continuous and its first derivative is also continuous. Functions that are continuous up to and including their nth derivatives are said to belong in the space $C^n(\Omega)$. Until now we have considered functions in $C^0(\Omega)$ only. Functions that lie in $E(I)$, defined by Equation (9.25), must also lie in $C^1(I)$.

The first four shape functions defined on the standard beam element $(-1 < \xi < +1)$ are shown in Figure 9.3. Note that N_2 and N_4 are scaled by $\ell_k/2$. This is because

$$\theta := \frac{dw}{dx} = \frac{2}{\ell_k}\frac{dw}{d\xi}.$$

To define $N_i(\xi)$, $i \geq 5$, it is convenient to introduce the function $\psi_j(\xi)$ which is analogous to $\phi_j(\xi)$ given by Equation (5.25):

$$\psi_j(\xi) := \sqrt{\frac{2j-3}{2}} \int_{-1}^{\xi}\int_{-1}^{s} P_{j-2}(t)\, dt\, ds, \quad j = 4, 5, \ldots \tag{9.26}$$

where $P_{j-2}(t)$ is the Legendre polynomial of degree $j - 2$. For example,

$$\psi_4(\xi) = \frac{1}{8}\sqrt{\frac{5}{2}}(\xi^2 - 1)^2.$$

Then, for $i \geq 5$,

$$N_i(\xi) = \psi_{i-1}(\xi).$$

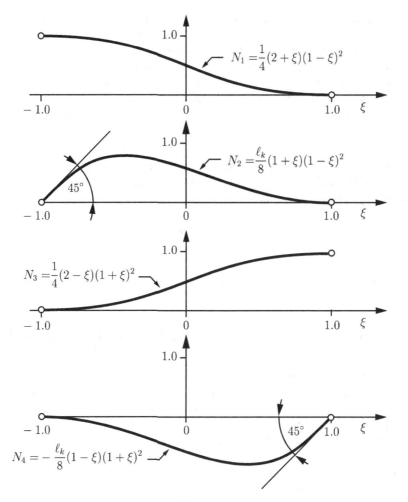

Figure 9.3 The first four C^1 shape functions in one dimension.

The stiffness matrix for a Bernoulli–Euler beam element of length ℓ_k, constant EI and $p = 5$ is

$$[K] = \frac{EI}{\ell_k^3} \begin{bmatrix} 12 & 6\ell_k & -12 & 6\ell_k & 0 & 0 \\ & 4\ell_k^2 & -6\ell_k & 2\ell_k^2 & 0 & 0 \\ & & 12 & -6\ell_k & 0 & 0 \\ & \text{(sym.)} & & 4\ell_k^2 & 0 & 0 \\ & & & & 8 & 0 \\ & & & & & 8 \end{bmatrix}. \quad (9.27)$$

Remark 9.1.4 The simple model given by Equation (9.22) has been used successfully for the solution of practical problems for more than 200 years. It is remarkably accurate for the computation of deflections, rotations, moments and shear forces when the solution does not change substantially over distances of size d. This is not the case, for example, for vibrating beams when the mode shapes have wavelengths close to d.

Timoshenko proposed the model that bears his name for modeling high-frequency vibrations. Of course, the Timoshenko model has limitations as well. Furthermore, the accuracy of a particular model cannot be ascertained unless it can be shown that the data of interest are substantially independent of the model characterized by the indices $(m\ n)$, the boundary conditions and other modeling decisions. For these reasons there is a need for implementation of a model hierarchy.

Remark 9.1.5 The Bernoulli–Euler and Timoshenko beam models are used with the objective to determine reactions, shear force and bending moment diagrams. Having determined the bending moment M, the normal stress is computed from the formula

$$\sigma = -\frac{My}{I} \qquad (9.28)$$

where y is a centroidal axis, see Figure 9.1. For the Bernoulli–Euler model this formula is derived from

$$\sigma \equiv \sigma_x = -Ew''y \quad \text{and} \quad M = -\int_A \sigma_x y\, dA = Ew''\int_A y^2\, dA = EIw''.$$

For the Timoshenko model the derivation is analogous. Stresses computed in this way can be very accurate away from the supports and points where concentrated forces are applied but very inaccurate in the neighborhoods of those points where the assumptions incorporated into these models do not hold. Nevertheless, engineering design is based on the maximum stress computed from the formula (9.28) subject to the requirement that $\|\sigma\|_{\max}$ must be less than an allowable value which is approximately two-thirds of the yield stress. This makes sense only if we understand that the actual goal is to design beams such that $\|M\|_{\max}$ is much less than the moment that would cause extensive plastic deformation.

Exercise 9.1.6 Determine $\psi_5(\xi)$ and verify k_{66} in Equation (9.27).

Exercise 9.1.7 Verify the value of k_{34} in Equation (9.27).

Exercise 9.1.8 The beam shown Figure 9.4 is simply supported on the left and fixed on the right. A positive rotation θ_A is imposed on the simply supported end.

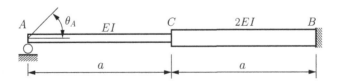

Figure 9.4 Problem definition and bending moment diagram.

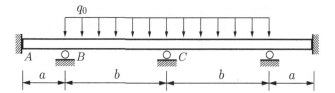

Figure 9.5 Problem definition for Exercise 9.1.9.

1. Using the Bernoulli–Euler beam model, determine the displacement and rotation of the centroidal axis at point C in terms of θ_A and a.

2. Given the exact value of the strain energy

$$U_{EX} = \frac{12}{11}\frac{EI}{a}\theta_A^2$$

what is the moment M_A that has to be imposed to cause the θ_A rotation of the centroidal axis? Is the moment positive or negative? (Hint: The work done by the moment equals the strain energy.)

3. If the Timoshenko beam model were used, would the exact value of M_A be larger or smaller? Explain.

4. If the number of degrees of freedom is increased by uniform mesh refinement, does the strain energy increase, decrease, or remain the same? Explain.

Exercise 9.1.9 The multi-span beam shown in Figure 9.5 is fixed at the ends and simply supported at three points. The bending stiffness EI is constant.

1. Taking advantage of the symmetry, state the principle of minimum potential energy and specify the space of admissible functions for the Bernoulli–Euler beam model.

2. Find an expression for the rotation at support B in terms of the parameters q_0, EI, a and b for the Bernoulli–Euler beam model.

3. Is it possible to assign polynomial degrees to the elements such that the finite element solution is the exact solution (up to round-off errors)? Explain.

Exercise 9.1.10 Based on the formulation outlined in Section 4.4.1, write down the generalized formulation of the undamped elastic vibration problem for the Bernoulli–Euler beam model.

Exercise 9.1.11 Model the problem shown in Figure 7.2 and described in Example 7.2.1 as a beam.

(a) Using the Bernoulli–Euler beam model, with one element on the interval $0 < x < \ell/2$, find the rotation at $x = \ell/2$ in terms of δ/ℓ. Hint: Write down the constrained stiffness matrix and load vector for $p = 3$. Impose the antisymmetry condition at $x = \ell/2$.

(b) Using the result obtained for part (a), compute the bending moment M_0 acting on the boundary at $x = 0$ by extraction. Hint: Select $v = -N_2(\xi)$ for the extraction function where $N_2(\xi)$ is defined in Figure 9.3.

9.2 Plates

The formulation of plate models is analogous to the formulation of beam models. The middle surface of the plate is assumed to lie in the xy plane. The two-dimensional domain occupied by the middle surface is denoted by Ω and the boundary of Ω is denoted by Γ. The thickness of the plate is denoted by d and the side surface of the plate is denoted by S, that is, $S = \Gamma \times (-d/2, d/2)$. The displacement vector components are written in the following form:

$$u_x = u_{x|0}(x, y) + u_{x|1}(x, y)z + \cdots + u_{x|m_x}(x, y)z^{m_x}$$
$$u_y = u_{y|0}(x, y) + u_{y|1}(x, y)z + \cdots + u_{y|m_y}(x, y)z^{m_y} \quad (9.29)$$
$$u_z = u_{z|0}(x, y) + u_{z|1}(x, y)z + \cdots + u_{z|n}(x, y)z^n$$

where $u_{x|0}$, $u_{x|1}$, $u_{y|0}$, etc., are independent field functions. In the following we will refer to a particular plate model by the indices (m_x, m_y, n) and denote the corresponding exact solution by $\vec{u}_{EX}^{(m_x,m_y,n)}$. The exact solution of the corresponding problem of three-dimensional elasticity will be denoted by $\vec{u}_{EX}^{(3D)}$.

In analyses of plates the stress resultants rather than the stresses are of interest. The stress resultants are the membrane forces:

$$F_x := \int_{-d/2}^{+d/2} \sigma_x \, dz, \quad F_y := \int_{-d/2}^{+d/2} \sigma_y \, dz, \quad F_{xy} = F_{yx} := \int_{-d/2}^{+d/2} \tau_{xy} \, dz; \quad (9.30)$$

the transverse shear forces:

$$Q_x := -\int_{-d/2}^{+d/2} \tau_{xz} \, dz, \quad Q_y := -\int_{-d/2}^{+d/2} \tau_{yz} \, dz; \quad (9.31)$$

and the bending and twisting moments:

$$M_x := -\int_{-d/2}^{+d/2} \sigma_x z \, dz, \quad M_y := -\int_{-d/2}^{+d/2} \sigma_y z \, dz \quad (9.32)$$

$$M_{xy} = -M_{yx} := -\int_{-d/2}^{+d/2} \tau_{xy} z \, dz. \quad (9.33)$$

M_x, M_y are called bending moments; M_{xy} is called the twisting moment. The negative sign in the expressions for the shear force and bending moment components is made necessary by the conventions adopted for the stress resultants shown in Figure 9.6 and the convention that tensile stresses are positive. Since $\tau_{xy} = \tau_{yx}$, the convention adopted for the twisting moments results in $M_{yx} = -M_{xy}$.

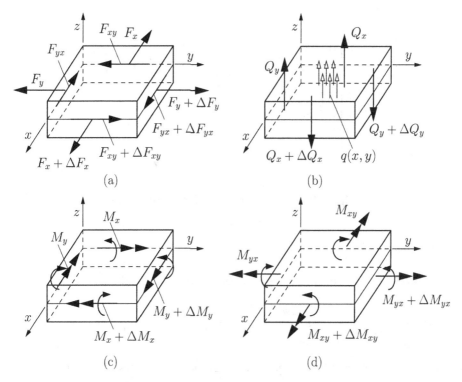

Figure 9.6 Sign convention for stress resultants.

The starting point for the formulation of plate models is the principle of virtual work or, equivalently, the principle of minimum potential energy written in terms of the field functions with the integration performed in the z direction. We will consider a restricted form of the three-dimensional elasticity problems, using constraints and loads typically used in connection with the analysis of plates.

The boundary conditions are usually given in terms of the normal–tangent (n, t) system. It is left to the reader in the following exercises to derive the transformations from the $x\,y$ to the $n\,t$ systems.

Exercise 9.2.1 Refer to Figure 9.7(a) and show that

$$Q_n = -\int_{-d/2}^{+d/2} \tau_{nz}\, dz = Q_x \cos\alpha + Q_y \sin\alpha. \qquad (9.34)$$

Exercise 9.2.2 For the infinitesimal plate element shown in Figure 9.7(b), subjected to bending and twisting moments only, show that

$$M_n = M_x \cos^2\alpha + M_y \sin^2\alpha + 2M_{xy} \sin\alpha \cos\alpha \qquad (9.35)$$

$$M_{nt} = -(M_x - M_y)\sin\alpha\cos\alpha + M_{xy}(\cos^2\alpha - \sin^2\alpha) \qquad (9.36)$$

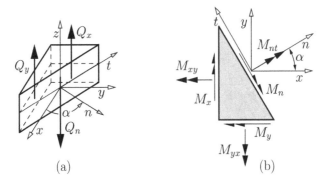

Figure 9.7 Transformation of stress resultants.

Hint: Use $M_{yx} = -M_{xy}$.

Exercise 9.2.3 Derive Equations (9.35) and (9.36) from the definitions

$$M_n := -\int_{-d/2}^{+d/2} \sigma_n\, z\, dz, \quad M_{nt} := -\int_{-d/2}^{+d/2} \tau_{nt}\, z\, dz$$

using the transformation given by Equation (C.12) and the formulas (9.32), (9.33).

Remark 9.2.1 The transformation of moments can be represented by a Mohr circle.[4] The maximum and minimum bending moments, denoted respectively by M_1 and M_2, called principal bending moments, occur at those values of α at which $M_{xy} = 0$. This can be seen by setting the first derivative of M_n with respect to α equal to zero. The principal moments are

$$M_1 = \frac{M_x + M_y}{2} + R\cos 2\alpha, \quad M_2 = \frac{M_x + M_y}{2} - R\cos 2\alpha \qquad (9.37)$$

where R is the radius of the Mohr circle:

$$R = \sqrt{\left(\frac{M_x - M_y}{2}\right)^2 + M_{xy}^2} \geq 0.$$

9.2.1 The Reissner–Mindlin plate

The plate model (1, 1, 0), known as the Reissner–Mindlin plate,[5] is widely used in finite element analysis. Its formulation is analogous to that of the Timoshenko beam. As in the Timoshenko beam model, the in-plane displacements are decoupled from the bending and shearing deformation. For simplicity we will be concerned with the bending and shearing

---

[4] Christian Otto Mohr (1835–1918).
[5] Eric Reissner (1913–1996), Raymond David Mindlin 1906–1987.

deformation only, that is, we let $u_{x|0} = u_{y|0} = 0$. We will use the notation

$$u_{x|1} = -\beta_x(x, y), \quad u_{y|1} = -\beta_y(x, y), \quad u_{z|0} = w(x, y)$$

hence the displacement vector components are of the form

$$u_x = -\beta_x(x, y)z, \quad u_y = -\beta_y(x, y)z, \quad u_z = w(x, y) \quad (9.38)$$

and the strain terms are

$$\epsilon_x = -\frac{\partial \beta_x}{\partial x}z, \quad \epsilon_y = -\frac{\partial \beta_y}{\partial y}z, \quad \epsilon_z = \frac{\partial w}{\partial z} = 0$$

$$\gamma_{xy} = -\left(\frac{\partial \beta_x}{\partial y} + \frac{\partial \beta_y}{\partial x}\right)z, \quad \gamma_{yz} = -\beta_y + \frac{\partial w}{\partial y}, \quad \gamma_{zx} = -\beta_x + \frac{\partial w}{\partial x}.$$

For the stress–strain law the plane stress relationships are used in the xy plane (3.59). In the yz and zx planes the shear modulus modified by the shear correction factor κ is

$$\sigma_x = \frac{E}{1-\nu^2}(\epsilon_x + \nu\epsilon_y), \quad \sigma_y = \frac{E}{1-\nu^2}(\nu\epsilon_x + \epsilon_y), \quad \sigma_z = 0$$

$$\tau_{xy} = G\gamma_{xy}, \quad \tau_{yz} = \kappa G\gamma_{yz}, \quad \tau_{zx} = \kappa G\gamma_{zx}.$$

Since the strain component ϵ_z and the stress component σ_z are both zero, this choice of stress–strain law may appear to be contradictory. The justification is that with this material stiffness matrix the exact solution of the Reissner–Mindlin model will approach the exact solution of the fully three-dimensional model as the thickness approaches zero:

$$\lim_{d \to 0} \frac{\|\vec{u}_{EX}^{(3D)} - \vec{u}_{EX}^{(110)}\|_E}{\|\vec{u}_{EX}^{(3D)}\|_E} = 0. \quad (9.39)$$

The model would not have this property, called asymptotic consistency, if the stress–strain law of three-dimensional elasticity were used.

The strain energy is

$$U_\kappa = \frac{1}{2}\int_V (\sigma_x\epsilon_x + \sigma_y\epsilon_y + \tau_{xy}\gamma_{xy} + \tau_{yz}\gamma_{yz} + \tau_{zx}\gamma_{zx})\,dx\,dy\,dz + \frac{1}{2}\int_\Omega c_s w\,dx\,dy$$

where $c_s(x, y) \geq 0$ is a spring coefficient (in N/m³ units). On substituting the expressions for stress and strain and integrating with respect to z, we get

$$U_\kappa = \frac{1}{2} \int_\Omega D\left[\left(\frac{\partial \beta_x}{\partial x}\right)^2 + 2\nu \frac{\partial \beta_x}{\partial x} \frac{\partial \beta_y}{\partial y} + \left(\frac{\partial \beta_y}{\partial y}\right)^2 + \frac{1-\nu}{2} \left(\frac{\partial \beta_x}{\partial y} + \frac{\partial \beta_y}{\partial x}\right)^2 \right.$$
$$\left. + \frac{6\kappa(1-\nu)}{d^2}\left(-\beta_y + \frac{\partial w}{\partial y}\right)^2 + \frac{6\kappa(1-\nu)}{d^2}\left(-\beta_x + \frac{\partial w}{\partial x}\right)^2 \right] dx\,dy$$
$$+ \frac{1}{2} \int_\Omega c_s w\, dx\,dy \qquad (9.40)$$

where D is the plate constant:

$$D := \frac{Ed^3}{12(1-\nu^2)}. \qquad (9.41)$$

The potential of external forces is:

$$P := \int_\Omega qw\, dx\,dy - \oint_\Gamma Q_n w\, ds + \oint_\Gamma M_n \beta_n\, ds + \oint_\Gamma M_{nt} \beta_t\, ds \qquad (9.42)$$

where the subscripts refer to the $n\,t\,z$ coordinate system shown in Figure 9.7(a). Noting that by definition $u_n = -\beta_n z$, $u_t = -\beta_t z$ and applying the rules of vector transformation we get

$$\beta_n = \beta_x \cos\alpha + \beta_y \sin\alpha \qquad (9.43)$$
$$\beta_t = -\beta_x \sin\alpha + \beta_y \cos\alpha. \qquad (9.44)$$

Particular applications of the principle of minimum potential energy depend on the boundary conditions. The commonly used boundary conditions are:

(a) Fixed: $\beta_n = \beta_t = w = 0$.

(b) Free: $M_n = M_{nt} = Q_n = 0$.

(c) Simple support can be defined in two different ways for the Reissner–Mindlin plate model:

  (i) Soft simple support: $w = 0$, $M_n = M_{nt} = 0$.

  (ii) Hard simple support: $w = 0$, $\beta_t = 0$, $M_n = 0$.

(d) Symmetry: $\beta_n = 0$, $M_{nt} = 0$, $Q_n = 0$.

(e) Antisymmetry: same as hard simple support.

Shear correction for plate models

For the Reissner–Mindlin plate model either the energy or the average mid-surface deflection can be optimized with respect to the fully three-dimensional model by the choice of the shear correction factor:

$$\kappa = \begin{cases} \dfrac{5}{6(1-\nu)} & \text{for optimal energy} \\ \dfrac{20}{3(8-3\nu)} & \text{for optimal displacement}. \end{cases} \quad (9.45)$$

For the model (1, 1, 1) there is one shear correction factor:

$$\kappa = \begin{cases} \dfrac{5}{6} & \text{for } \nu = 0 \\ \dfrac{12-2\nu}{\nu^2}\left(-1+\sqrt{1+\dfrac{20\nu^2}{(12-2\nu)^2}}\right) & \text{for } \nu \neq 0. \end{cases} \quad (9.46)$$

For all other models (m_x, m_y, n), where $m_x, m_y \geq 1$, $n \geq 2$, there is no shear correction factor (i.e., $\kappa = 1$) [4].

Exercise 9.2.4 Derive the expression (9.42) from Equation (4.16) assuming that only tractions are applied on the side surface $S = \Gamma \times (-d/2, d/2)$ of the plate. Hint: First show that

$$\int_{\partial \Omega_T} T_i u_i \, dS = \int_S (T_n u_n + T_t u_t + T_z u_z) \, dS.$$

Exercise 9.2.5 The mid-surface of a plate is an equilateral parallelogram (rhombus) characterized by the dimension ℓ and the angle β, as shown in Figure 9.8(a). The plate is of uniform thickness d. The elastic properties are: $E = 2.0 \times 10^5$ MPa, $\nu = 0.3$. The plate is uniformly loaded, that is, $q = q_0$ (constant), and simple support conditions are specified on all sides. Taking advantage of the two lines of symmetry, define the solution domain to be the triangle ABE. Let $\beta = \pi/6$ and $\ell/d = 100$.

This is one of the benchmark problems that have been used for illustrating the performance of various plate models and discretization schemes. The challenging aspect of the problem is the strong singularity at the obtuse corners B and D when a simple support is prescribed on the boundaries [10], [44].

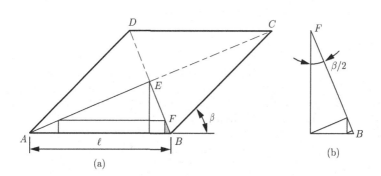

Figure 9.8 The rhombic plate problem, Exercise 9.2.5.

Using the Reissner–Mindlin plate model, estimate the displacement w and the principal bending moments M_1 and M_2 at point E, assuming (a) hard and (b) soft simple support prescribed on all sides. Report the displacement and moments in dimensionless form: $wD/(q_0\ell^4)$ and $M_i/(q_0\ell^2)$, $i = 1, 2$. Use the shear correction factor for optimal energy.

9.2.2 The Kirchhoff plate

The Kirchhoff plate model[6] is analogous to the Bernoulli–Euler beam model. The formulation is obtained by letting

$$\beta_x = \frac{\partial w}{\partial x}, \quad \beta_y = \frac{\partial w}{\partial y}$$

in Equation (9.40). Therefore the strain energy of the Kirchhoff plate model is

$$U_K = \frac{1}{2}\int_\Omega D\left[\left(\frac{\partial^2 w}{\partial x^2}\right)^2 + 2\nu\frac{\partial^2 w}{\partial x^2}\frac{\partial^2 w}{\partial y^2} + \left(\frac{\partial^2 w}{\partial y^2}\right)^2 + 2(1-\nu)\left(\frac{\partial^2 w}{\partial x \partial y}\right)^2\right]dxdy$$
$$+ \frac{1}{2}\int_\Omega c_s w\, dxdy. \tag{9.47}$$

The potential of external forces is obtained from (9.42) with the following modification:

$$\oint_\Gamma M_{nt}\beta_t\, ds \to \oint_\Gamma M_{nt}\frac{\partial w}{\partial s}\, ds = -\oint_\Gamma \frac{\partial M_{nt}}{\partial s}w\, ds + \underbrace{\oint_\Gamma \frac{\partial (M_{nt}w)}{\partial s}ds}_{0}$$

where we used $dt \equiv ds$ and the assumption that the product $M_{nt}w$ is continuous and differentiable. Therefore we have

$$P_K := \int_\Omega qw\, dxdy + \oint_\Gamma M_n\frac{\partial w}{\partial n}\, ds + \oint_\Gamma \left(Q_n - \frac{\partial M_{nt}}{\partial s}\right)w\, ds. \tag{9.48}$$

In the Kirchhoff model, simple support has only one interpretation: that of hard simple support. By definition

$$\Pi(w) = U_K(w) - P_K(w) \tag{9.49}$$

and

$$E(\Omega) = \{w \mid U_K(w) \leq C < \infty\}.$$

Define $\tilde{E}(\Omega) \subset E(\Omega)$ to be the space of functions that satisfy the prescribed kinematic boundary conditions. The problem is to find:

$$\Pi(w_{EX}) = \min_{w\in\tilde{E}(\Omega)} \Pi(w). \tag{9.50}$$

---

[6] Gustav Robert Kirchhoff (1824–1887).

PLATES   281

This formulation has great theoretical and historical significance (see, for example, [88]); however, it is not well suited for computer implementation because the basis functions have to be C^1 continuous. The difficulties associated with enforcement of C^1 continuity are discussed in the following section.

Exercise 9.2.6 Consider the Kirchhoff plate model and assume homogeneous boundary conditions. Following the procedure of Section 9.1.2, that is, letting

$$\frac{\partial \Pi(w + \epsilon v)}{\partial \epsilon}\bigg|_{\epsilon=0} = 0$$

where $\Pi(w)$ is defined by Equation (9.49) and v is an arbitrary test function in $E^0(\Omega)$, show that the function w that minimizes Π satisfies the biharmonic equation

$$\frac{\partial^4 w}{\partial x^4} + 2\frac{\partial^4 w}{\partial x^2 \partial y^2} + \frac{\partial^4 w}{\partial y^4} = \frac{q}{D}. \tag{9.51}$$

Exercise 9.2.7 We have seen in Exercise 9.2.6 that the strong form of the Kirchhoff plate model is the biharmonic equation given by Equation (9.51). Assume that all sides of the equilateral triangular plate and the rhombic plate shown in Figure 9.9 are simply supported. Considering the symmetric eigenfunctions only, characterize the smoothness of the homogeneous solution of the Kirchhoff plate model at each of the singular points. For the rhombic plate let $\beta = \pi/6$.

Hints: (a) The homogeneous biharmonic equation in polar coordinates is given by Equation (6.14). (b) Letting $w = r^{\lambda+1} F(\theta)$, the symmetric characteristic functions are of the form

$$F_i(\theta) = a_i \cos(\lambda_i - 1)\theta + b_i \cos(\lambda_i + 1)\theta,$$

see Equation (6.25). (c) The zero-moment condition on the simply supported edges ($M_n = 0$) is equivalent to

$$\left(\frac{1}{r^2}\frac{\partial^2 w}{\partial \theta^2}\right)_{\theta=\pm\alpha/2} = 0$$

where $\theta = 0$ is the internal bisector of the vertex angle α as defined in Figure 6.4.

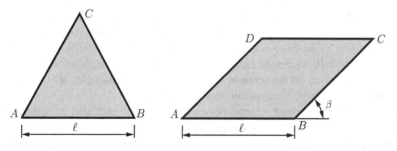

Figure 9.9 Exercise 9.2.7: equilateral triangle and rhombic plate, notation.

Exercise 9.2.8 The exact solution of the Kirchhoff model for a simply supported and uniformly loaded equilateral triangular plate is a polynomial of degree 5 (see, for example, [88]). In the center of the plate, the exact values of the displacement w and the bending moments are

$$\frac{wD}{q_0\ell^4} = \frac{1}{1728}, \quad \frac{M_x}{q_0\ell^2} = \frac{1+\nu}{72}, \quad M_y = M_x, \quad M_{xy} = 0$$

where D is the plate constant, q_0 is the value of the constant distributed load and ℓ is the dimension shown in Figure 9.9. Compare these results with their counterparts obtained with the Reissner–Mindlin model for (a) $\ell/d = 100$, $\nu = 0.3$ and (b) $\ell/d = 10$, $\nu = 0.3$ for hard and soft simple supports. Investigate the effects of shear correction factors for optimal energy and optimal displacement (see Section 9.2.1).

Exercise 9.2.9 Consider a uniformly loaded square plate with one side fixed, the opposite side simply supported, the other two sides free. State the principle of minimum potential energy for this problem for the Kirchhoff and Reissner–Mindlin models.

Exercise 9.2.10 Refer to Exercise 9.2.9. Noting that the Reissner–Mindlin model allows distinction between hard and soft simple supports but the Kirchhoff model does not, estimate the error of idealization of the Kirchhoff model in relation to the Reissner–Mindlin model through numerical experiments. Fix the size of the plate and vary the thickness. Compare the estimated limit values of the strain energy for soft and hard simple support.

9.2.3 Enforcement of C^1 continuity: the HCT element

The enforcement of C^1 continuity requires special consideration. This is because it is not possible to enforce exact and minimal C^1 continuity on piecewise polynomial basis functions on a finite element mesh. Enforcement of exact C^1 continuity implies enforcement of continuity of the second derivatives at the vertices also (see, for example, [20]). This causes problems at singular points where the second derivatives are discontinuous. To overcome this difficulty, composite elements and elements with singular functions have been developed. A composite triangular element, known as the Hsieh–Clough–Tocher (HCT) triangle [21], is described in the following.

The HCT triangle is a composite element, comprising three sub-triangles shown in Figure 9.10. On each sub-triangle an incomplete cubic polynomial approximation, composed of nine terms, is used. The polynomials are chosen so that along the external edges of the composite triangle the normal derivative varies linearly. Therefore there are nine coefficients per sub-triangle. C^0 continuity is enforced for the sub-elements by constructing basis functions corresponding to the three nodal displacements and 6 rotations for each sub-triangle, as indicated in Figure 9.10 where the circles represent transverse displacements and the arrows represent first derivatives (in the sense of the arrows). The sub-triangles are assembled, which is equivalent to satisfying C^0 continuity over the triangle.

At this point there are three internal degrees of freedom, indicated in Figure 9.10 by the closed circles and arrows, and nine external degrees of freedom, indicated by the open circles and arrows. In order to satisfy exact and minimal C^1 continuity, the continuity of the normal derivatives along the internal edges is enforced leading to three additional constraint equations.

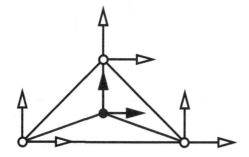

Figure 9.10 The Hsieh–Clough–Tocher (HCT) element.

Using these constraint equations, the internal degrees of freedom are eliminated. Thus the HCT triangle has nine degrees of freedom: three displacements and three first derivatives in each coordinate direction.

Remark 9.2.2 The definition of the HCT triangle given in [20] is different from the one described here in that all sub-triangles are complete polynomials of degree 3, in which case there are 12 degrees of freedom: the nine degrees of freedom shown in Figure 9.10 plus the first derivative in the direction of the normal at the mid-point of each side.

9.3 Shells

The formulation of mathematical models for structural shells is a very large and rather complicated subject that cannot be discussed in sufficient detail here. Only a brief overview of some of the salient points is presented. A structural shell is characterized by a surface, called mid-surface x_i, and the thickness d. Both are given in terms of two parameters α_1, α_2:

$$x_i = x_i(\alpha_1, \alpha_2), \quad d = d(\alpha_1, \alpha_2).$$

Note that the indices of the parameters α_i take on the values $i = 1, 2$ whereas the indices of the spatial coordinates x_i range from 1 to 3. Associated with each point of the mid-surface are three basis vectors. Two of the basis vectors lie in the tangent plane at the point (α_1, α_2):

$$b_i^{(1)} := \frac{\partial x_i}{\partial \alpha_1}, \quad b_i^{(2)} := \frac{\partial x_i}{\partial \alpha_2}.$$

Note that $b_i^{(1)}$ and $b_i^{(2)}$ are not necessarily orthogonal. However, in classical treatments of shells, the parameters α_1, α_2 are usually chosen so that the basis vectors are orthogonal. See, for example, [50]. The third basis vector $b_i^{(3)}$ is the cross-product of $b_i^{(1)}$ and $b_i^{(2)}$, therefore it is normal to the tangent plane. These are called curvilinear basis vectors. The normalized curvilinear basis vectors will be denoted by $\mathbf{e}_\alpha, \mathbf{e}_\beta, \mathbf{e}_n$. The Cartesian unit basis vectors will be denoted by $\mathbf{e}_x, \mathbf{e}_y, \mathbf{e}_z$. A vector $\mathbf{u}$ given in terms of the curvilinear basis vectors will be denoted by $\mathbf{u}_{(\alpha)}$, in Cartesian coordinates by $\mathbf{u}_{(x)}$. The transformation is

$$\mathbf{u}_{(x)} = [R]\mathbf{u}_{(\alpha)} \qquad (9.52)$$

where the columns of the transformation matrix $[R]$ are the unit vectors $\mathbf{e}_\alpha$, $\mathbf{e}_\beta$, $\mathbf{e}_n$.

The classical development of shell models was strongly influenced by the limitations of the methods available for solving the resulting systems of equations. The use of curvilinear coordinates allowed the treatment of shells with simple geometric description, such as cylindrical, spherical and conical shells, by classical methods subject to the condition that the thickness of the shell is small in relation to its other dimensions.

We will view shells as fully three-dimensional solids that may allow a priori restrictions on the transverse variation of displacements in certain regions. In virtually all practical applications there are regions where the assumptions incorporated into shell models do not hold. Shells have nozzles, support attachments, stiffeners, cut-outs, and regions where the curvature abruptly changes. In those regions the assumptions incorporated into shell models are not justified. However, from the point of view of stress analysis, those are typical regions of primary interest. Also, the thickness of shells that are of engineering interest is rarely very small in relation to the other dimensions. For example, in an experimental investigation of the intersection regions of thin-walled cylindrical shells four specimens were tested. The diameter to thickness ratios ranged from 7.68 to 100 [34].

The hierarchic shell models are generalizations of the hierarchic plate models represented by Equation (9.29). The displacement vector components are given in the following form:

$$
\begin{aligned}
u_\alpha &:= \sum_{i=0}^{m_\alpha} u_{\alpha|i}(\alpha,\beta)\phi_i(\nu) \\
u_\beta &:= \sum_{i=0}^{m_\beta} u_{\beta|i}(\alpha,\beta)\phi_i(\nu) \\
u_n &:= \sum_{i=0}^{m_n} u_{n|i}(\alpha,\beta)\phi_i(\nu)
\end{aligned}
\qquad (9.53)
$$

where $\phi_i(\nu)$ are called director functions. When the material is isotropic, $\phi_i(\nu)$ are polynomials; when the shell is laminated, $\phi_i(\nu)$ are piecewise polynomials (see, for example, [1]). Equation (9.53) represents a semi-discretization of the problem of fully three-dimensional elasticity, in the sense that $\phi_i(\nu)$ are fixed and thus the problem is reduced from a three dimensional problem to a two-dimensional one. A particular shell model is characterized by the set of numbers (m_α, m_β, m_n).

In the classical treatment of shells the curvilinear basis vectors are retained throughout the analysis. In the finite element method a generalized formulation is used, most commonly the principle of virtual work. The algorithmic structure becomes simpler if the displacement components, given with reference to the curvilinear basis vectors, are transformed to the (global) Cartesian reference frame by Equation (9.52).

The generic form of the virtual work of internal stresses is given by

$$B(\mathbf{u}_{(\alpha)}, \mathbf{v}_{(\alpha)}) := \int_\omega \int_{-d/2}^{+d/2} \left([\tilde{D}][R]\mathbf{v}_{(\alpha)}\right)^T [E][\tilde{D}][R]\mathbf{u}_{(\alpha)} \, dv d\omega \qquad (9.54)$$

where $[\tilde{D}]$ is the differential operator that transforms the displacement vector components given in terms of the curvilinear coordinates (α, β, v) to the Cartesian strain tensor components: $\{\epsilon\} = [\tilde{D}]\mathbf{u}_{(\alpha)}$. It is assumed that the material is isotropic. If the material is not isotropic and its reference frame differs from the global Cartesian coordinate system, then $[E]$ must be transformed into the global Cartesian coordinate system.

The generic form of the virtual work of external forces is

$$F(\mathbf{v}_{(\alpha)}) := \int_\omega \int_{-d/2}^{+d/2} \left([R]\mathbf{F}_{(\alpha)}\right)^T [R]\mathbf{v}_{(\alpha)} \, dv d\omega$$

$$+ \int_{\partial \omega} \int_{-d/2}^{+d/2} \left([R]\mathbf{T}_{(\alpha)}\right)^T [R]\mathbf{v}_{(\alpha)} \, dv ds$$

$$+ \int_\omega \int_{-d/2}^{+d/2} \left([\tilde{D}][R]\mathbf{v}_{(\alpha)}\right)^T [E]\{c\} T_\Delta \, dv d\omega \qquad (9.55)$$

where $\mathbf{F}_{(\alpha)}$ (resp. $\mathbf{T}_{(\alpha)}$) is the volume force (resp. surface traction vector) given in terms of the curvilinear basis, $\{c\}$ is the vector of coefficients of thermal expansion and T_Δ is the temperature change.

The Naghdi shell model

The Naghdi shell model[7] [46] is as an extension of the Timoshenko beam model and the Reissner–Mindlin plate model to shells: it is assumed that normals to the mid-surface prior to deformation remain straight lines but not necessarily normals after deformation. In other words, the kinematic assumptions account for some transverse shearing deformation. Specifically, the kinematic assumptions are the same as the kinematic assumptions for the hierarchic shell model (1,1,0):

$$u_\alpha = u_{\alpha|0}(\alpha, \beta) + u_{\alpha|1}(\alpha, \beta)v$$
$$u_\beta = u_{\beta|0}(\alpha, \beta) + u_{\beta|1}(\alpha, \beta)v \qquad (9.56)$$
$$u_n = u_{n|0}(\alpha, \beta).$$

The difference between the hierarchic shell model (1,1,0) and the Naghdi shell model is in the material stiffness matrix $[E]$. Whereas in Equation (9.54) the definition of $[E]$ is that of the three-dimensional stress–strain relationship, in the Naghdi model $[E]$ is replaced by the stress–strain relationship derived using the assumption that plane stress conditions exist. This is necessary for asymptotic consistency, see for example [70], [80].

---

[7] Paul M. Naghdi (1924–1994).

The Novozhilov–Koiter shell model

The Novozhilov–Koiter shell model[8] is an extension of the Euler–Bernoulli beam model and the Kirchhoff plate model to shells: it is assumed that normals to the mid-surface prior to deformation remain normals after deformation [50]. The formulation involves replacement of the field functions u_α and u_β in Equation (9.56) by a linear combination of the first derivatives of u_n. Consequently, the number of field functions is reduced to three and the second derivatives appear in Equation (9.54), which implies that the space of admissible functions has to be in class C^1. From the point of view of computer implementation, the advantages of using only three field functions are far outweighed by disadvantages of the requirement of C^1 continuity and restrictions on the kinematic boundary conditions. This model is mainly of theoretical and historical interest today.

9.3.1 Hierarchic "thin-solid" models

An alternative approach to hierarchic semi-discretization is to write the Cartesian components of the displacement vector in the form

$$u_x := \sum_{i=0}^{q} u_{x|i}(\alpha, \beta) \phi_i(\nu)$$

$$u_y := \sum_{i=0}^{q} u_{y|i}(\alpha, \beta) \phi_i(\nu) \qquad (9.57)$$

$$u_z := \sum_{i=0}^{q} u_{z|i}(\alpha, \beta) \phi_i(\nu)$$

where q is typically a small number. The generic form of the virtual work of internal stresses is given by the expression

$$B(\mathbf{u}_{(x)}, \mathbf{v}_{(x)}) := \int_{\omega} \int_{-d/2}^{+d/2} ([D]\mathbf{v}_{(x)})^T [E][D]\mathbf{u}_{(x)} \, d\nu d\omega. \qquad (9.58)$$

The generic form of the virtual work of external forces is:

$$F(\mathbf{v}_{(\alpha)}) := \int_{\omega} \int_{-d/2}^{+d/2} \mathbf{F}_{(x)} \cdot \mathbf{v}_{(x)} \, d\nu d\omega + \int_{\partial \Omega} \mathbf{T}_{(x)} \cdot \mathbf{v}_{(\alpha)} \, dS$$

$$+ \int_{\omega} \int_{-d/2}^{+d/2} ([D]\mathbf{v}_{(x)})^T [E] \mathbf{c}\tau \, d\nu d\omega \qquad (9.59)$$

where $\mathbf{F}_{(x)}$ (resp. $\mathbf{T}_{(x)}$) is the Cartesian volume force vector (resp. surface traction vector) given in terms of the curvilinear variables (α, β, ν) and $\mathbf{c}\tau$ is the thermal strain. Such models are called "thin-solid" models.

---

[8] Valentin Valentinovich Novozhilov (1910–1987), Warner Tjardus Koiter (1914–1997).

In computer applications, anisotropic trunk or anisotropic product spaces are used. The anisotropic trunk space on the standard hexahedral element $\Omega_{st}^{(h)}$ is defined by

$$S_{tr}^{ppq}(\Omega_{st}^{(h)}) := \text{span}(\xi^k \eta^\ell \zeta^m, \xi^p \eta \zeta^m, \xi \eta^p \zeta^m, (\xi, \eta, \zeta) \in \Omega_{st}^{(h)},$$
$$k, \ell = 0, 1, 2, \ldots, k + \ell \leq p, \ m = 0, 1, 2 \ldots, q), \quad p > 1. \quad (9.60)$$

The parameter q is fixed, and $p \geq q$ is increased until convergence is realized. The anisotropic product space on $\Omega_{st}^{(h)}$ is defined by

$$S_{pr}^{ppq}(\Omega_{st}^{(h)}) := \text{span}(\xi^k \eta^\ell \zeta^m, (\xi, \eta, \zeta) \in \Omega_{st}^{(h)},$$
$$k, \ell = 0, 1, 2, \ldots, p, \ m = 0, 1, 2 \ldots, q). \quad (9.61)$$

The definition of anisotropic spaces on the standard pentahedral element is analogous.

Advantages and disadvantages of hierarchic shell and thin-solid models

The advantages of thin-solid formulations over shell formulations are that they are easier to implement and continuity with other bodies, such as stiffeners, is easier to enforce. The disadvantages are that thin-solid formulations cannot be applied to laminated shells unless each lamina is explicitly modeled; the number of field functions must be the same for each displacement component; and the downward extension of the family of hierarchic models to satisfy the requirement of asymptotic consistency cannot be applied. The anisotropic spaces with $q = 1$ are similar but not equivalent to the Reissner–Mindlin plate model or the Naghdi shell model.

Exercise 9.3.1 The mid-surface of a hyperboloidal shell is given by

$$\frac{x^2}{R_t^2} + \frac{y^2}{R_t^2} - \frac{z^2}{(\alpha L)^2} = 1, \quad -L \leq z \leq L, \quad \alpha^2 := \frac{R_t^2}{R_c^2 - R_t^2}$$

where R_t is the throat radius and R_c is the crown radius. Let $R_t = 1.0$ m, $L = 1.0$ m and $\alpha = 1$. Denote the thickness of the shell by d. Assume also that the material is elastic with $E = 2.0 \times 10^5$ mPa, $\nu = 0.3$. Assume also that a normal pressure $p = p_0 \cos 2\theta$ is acting on the inside surface of the shell where θ is the angle measured from the positive x-axis as shown in Figure 9.11. Let $p_0 = 20.0$ kPa. The edge at $y = -L$ is fixed, that is, all displacement components are zero, and the edge at $y = L$ is free.

1. Construct a solid model of the shell with d defined as a parameter and construct a mesh similar to that shown in Figure 9.11.

2. Using the anisotropic product spaces $S_{pr}^{ppq}(\Omega_{st}^{(h)})$ and the anisotropic trunk spaces $S_{tr}^{ppq}(\Omega_{st}^{(h)})$, investigate the rates of convergence in the energy norm for $d = 0.01$ m and $d = 0.001$ m for $q = 1, 2, 3$.

This example shows the effects of locking: the relative error is substantially larger for $d = 0.001$ m than for $d = 0.01$ m, however, the estimated asymptotic rates of convergence are

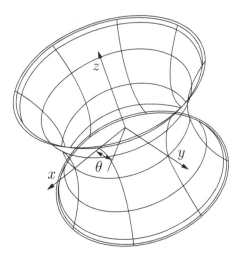

Figure 9.11 Hyperboloidal shell.

close to 1.0. The strong boundary layer at the fixed edge is clearly visible when the deformed shape is plotted.

9.4 The Oak Ridge experiments

In an investigation, performed at Oak Ridge National Laboratory (ORNL) in the 1970s, the results of finite element analysis were compared with the results of physical experiments [22], [34]. The motivation was that proven elastic stress analysis methods were not yet developed for various commonly used shell configurations in nuclear power plants and consequently reliable design information was not yet available. A project was undertaken in which four instrumented test articles, each subjected to 13 load cases, were tested and analyzed with the aim of assessing the utility of numerical simulation methods available at that time. In the terminology of Chapter 1, a set of validation experiments was performed. In the following the processes of conceptualization, verification and validation are discussed with reference to one test article and one load case only.

9.4.1 Description

Four test articles were manufactured and instrumented with great care. Only the first test article will be discussed in the following. This test article was made by welding two carbon steel pipes and then carefully machining it to the intended test dimensions. In order to reduce residual stresses, the test article was annealed several times in the machining process.

The experimental arrangement is shown in Figure 9.12. The horizontal part is called the cylinder, the vertical part is called the nozzle. The length of the cylinder was 39.0 in (991 mm). The length of the nozzle, measured from the point of intersection of the centerline of the nozzle with the centerline of the cylinder, was 19.5 in (495 mm). The outside diameter of the cylinder (resp. nozzle) was 10 in (254 mm) (resp. 5.0 in (127 mm)). The intended wall thickness of the cylinder (resp. nozzle) was 0.1 in (2.54 mm) (resp. 0.05 in (1.27 mm)).

Figure 9.12 Experimental arrangement. Previously published in [83]. Reproduced with permission from the *Journal of Applied Mechanics*, an ASME publication © 2005.

The test article was instrumented using 322 three-gauge foil rosettes[9] bonded on the inside and outside surfaces by epoxy adhesive and cured. The gauges in the rosettes were arranged in a - pattern (i.e., the directions of strain measurements were 120° apart). For details we refer to [22], [34].

As seen in Figure 9.12, the right end of the cylinder was rigidly clamped to a heavy flat plate bolted to a frame. Small flanges were machined into the ends of the cylinder and nozzle to support the seal and the clamping forces. Heavy loading fixtures were attached on the opposite end of the cylinder and on the end of the nozzle to provide seal and seating for the application of forces.

A total of 13 load cases that included pressure loading and axial forces, shear forces, bending moments and twisting moments was investigated. The forces and moments were applied to the cylinder and the nozzle by hydraulic rams acting through load cells. The pressure loading was applied by means of a hydraulic fluid. In order to compensate for the weight of the hydraulic fluid and the end fixtures, a balancing force was applied to the fixture at the free end of the cylinder through a cable that is visible in Figure 9.12.

For all 13 load cases the load was applied in increments of 20% of the full load, then decreased to zero again in 20% decrements. Only one of the loading cases, the pressure load, is discussed in the following. The maximum value of the pressure was 50.0 psi (344.8 kPa).

---

[9] Micro-Measurements type EA-06-030YB-120, option SE.

9.4.2 Conceptualization

For structural details, such as the test article discussed here and loaded in accordance with the ORNL test plan, mathematical models based on the theory of elasticity are typically assumed. The possibility of inelastic deformation at the junction of the shells cannot be excluded, however. Elsewhere, simplifications such as the Naghdi or Novozhilov–Koiter assumptions may be possible.

Keeping in mind that uncertainties exist in geometric description, material properties, loading and constraints, the development of a plan for the assessment of the effects of uncertainties on the data of interest is part of the conceptualization process.

Virtual experimentation

An important tool for the assessment of the influence of various modeling assumptions on the data of interest is virtual experimentation. In this project the data of interest were the von Mises stresses in the strain gauge locations. The following questions can be answered by virtual experimentation:

1. Are the equations of the linear theory of elasticity adequate for representing the response of the test article to the given set of loading conditions?

2. What is the size of the intersection region where the fully three-dimensional representation should be retained?

3. What is the simplest shell model that is capable of approximating the strain readings in the given locations?

4. What are the effects of variations in input data?

Utilizing symmetry, one-half of the solution domain is considered. The finite element mesh was constructed such that the size of the intersection region is controlled by parameters. The nozzle and the shell were partitioned into hexahedral elements and the intersection region was partitioned into hexahedral and pentahedral elements. The finite element mesh is shown in Figure 9.13.

The parameters that characterize the intersection region are denoted by d_s and d_n in Figure 9.13(b). The mesh layout isolates the line of intersection of the outer surfaces of the shell for the reasons discussed in connection with the L-shaped domain in Chapter 6. The solution is strongly singular along this line. The mesh layout also anticipates the existence of boundary layers at the ends of the shell and the nozzle.

Uniform pressure loading was applied on the inside surface. One end of the shell was fixed, the opposite end of the shell and the end of the nozzle were closed by end plates (not shown in Figure 9.13). The strain gauge locations on the inside and outside surfaces nearest to the intersection are labeled C, D, E, F in Figure 9.13(b).

The results of virtual experiments have shown that (a) the influence of a small plastic zone along the singular line at the intersection has negligibly small effect on the data of interest, (b) geometric nonlinearities can be neglected, (c) the data of interest are not sensitive to the size of the intersection region and (d) anisotropic trunk spaces defined by Equation (9.60) yield consistent results for any q value.

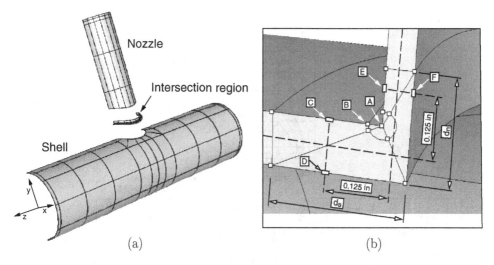

Figure 9.13 (a) Finite element mesh. (b) Detail at the intersection. Previously published in [83]. Reproduced with permission from the *Journal of Applied Mechanics*, an ASME publication © 2005.

9.4.3 Verification

Verification is a process comprising the steps described in Section 7.3. One of the steps is to show that the data of interest are substantially independent of the choice of discretization. The discretization and verification used in the original ORNL investigation and in the investigation reported in [83] are summarized in this section. It was concluded in both cases that the errors of discretization were negligibly small in comparison with other errors in both cases.

Verification in the ORNL investigation

The treatment of shell models by the finite element method was in its very early stages of development at the time of the ORNL investigation. Although the investigators were aware of some contemporary work on curved shell elements, shells were commonly approximated by assemblies of flat plate and membrane elements and most of the available experience was with those elements. For this reason the investigators decided to use a spatial assembly flat plate elements for the purpose of analyzing the shell intersection problem [22]. The HCT triangular element, described in Section 9.2.3, was chosen for the approximation of the displacement component normal to the mid-surface of the shell. Four HCT triangles were assembled to produce one four-sided element with five nodes.

The tangential (membrane) components of the displacement vector were approximated by a similar assembly of triangles. These vector components were approximated over each component triangle by quadratic polynomials constrained so that the basis functions were linear over the external edges. This is known as the constrained linear strain triangle (CLST). A quadrilateral membrane element has two degrees of freedom at each of its five nodes [22].

The five nodes of each four-sided element were located on the mid-surface of the shell. Therefore the assembled triangles were not co-planar and hence there is a third rotation

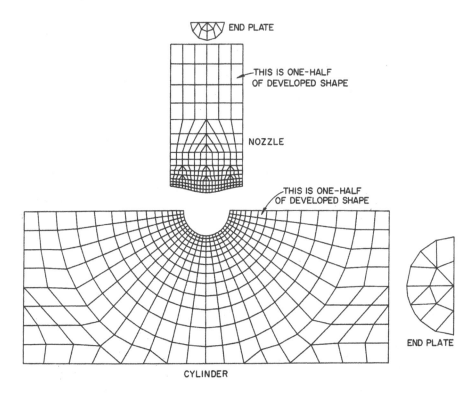

Figure 9.14 The finite element mesh used in the ORNL investigation. Previously published in [83]. Reproduced with permission from the *Journal of Applied Mechanics*, an ASME publication © 2005.

component, not present in the constituent triangles. The usual treatment is that at each node the rotation components are transformed into a Cartesian coordinate system, the origin of which lies on the shell surface at the node and one axis is coincident with the normal to the surface. The rotation component in the direction of the normal is then set to zero. This can cause various problems, however. For a discussion on this point we refer to [22], [34].

The finite element space used in the ORNL investigation is the span of the assembled basis functions on the mesh of four-sided elements shown in Figure 9.14. Verification was based on comparison of results using two meshes. It was concluded that the results were not sensitive to coarsening the mesh shown in Figure 9.14.

Verification based on *p*-extension

The mesh used in the investigation reported in [83] is shown in Figure 9.13. *p*-Extension was used to demonstrate that the computed data are substantially independent of the discretization. This is illustrated in Figure 9.15 where the von Mises stress is plotted vs. the number of degrees of freedom N on a semi-log scale for point B shown in Figure 9.13(b). The convergence of strain data computed at the other gauge locations considered here is similar.

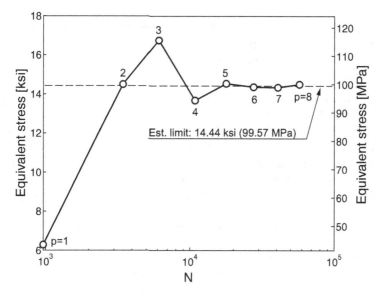

Figure 9.15 Convergence of the von Mises stress at point B shown in Figure 9.13(b). Previously published in [83]. Reproduced with permission from the *Journal of Applied Mechanics*, an ASME publication © 2005.

9.4.4 Validation: comparison of predicted and observed data

In comparing data computed from mathematical models with physical measurements, it is necessary to recognize the differences between the mathematical problem being solved and the physical system being modeled. These differences are enumerated in the following in relation to the ORNL test article discussed here.

Differences between the mathematical model and the physical reality

1. Geometric variations. The ORNL investigators were careful to minimize errors in manufacturing the test article. However, some variations in wall thickness and other dimensions occurred. The following quotation is from reference [22]:

    "A careful dimensional inspection of the machined model indicated that, despite the care taken in machining, there were wall thickness variations in both nozzle and cylinder with the nozzle thickness being as much as 15% greater (0.007 to 0.008 inches compared with the nominal 0.050 inches) in the fourth quadrant than in the second."

   In the mathematical model it is assumed that the shell and the nozzle are defined by perfect cylindrical surfaces and constant wall thickness.
   It was intended to manufacture the intersection with zero fillet radius. In reality, the milling tool leaves some fillet. In the mathematical model the fillet radius is zero. Variations in thickness d affect the distribution of membrane and bending forces within

the shell and the relationships between stresses and the membrane forces and bending moments. The stresses corresponding to membrane forces are proportional to $1/d$ whereas the stresses corresponding to bending are proportional to $1/d^2$.

The test article had small flanges at the ends. The mathematical model does not account for the flanges.

2. Variations in material properties. The investigators assumed that material properties are the nominal elastic constants of carbon steel. Modulus of elasticity $E = 30 \times 10^6$ psi (207 GPa); Poisson's ratio $\nu = 0.3$. In reality the modulus of elasticity and Poisson's ratio are stochastic functions of the spatial variables and their average values may differ from the nominal values by a few percent. There are no data on the variation of the modulus of elasticity and Poisson's ratio within the test article, but it is reasonable to expect that the mean value of E (resp. ν) is within about 2% (resp. 5%) of the nominal value. Therefore, systematic as well as random errors are present in making comparisons between measured strains and strains computed from the mathematical model.

The relationship between stress and strain in the test article will become nonlinear in locations where the strain corresponding to the proportional limit is exceeded.

3. Differences in constraint conditions. Rigid end fixtures were attached to the free end of the cylinder and the nozzle. Details on the end fixtures are not given in the ORNL report; however, the investigators assumed that the end fixtures were sufficiently rigid to constrain the ends of the cylinder and nozzle so as to maintain the ends as plane circles [22]. In the mathematical model constructed by the ORNL investigators, the end fixtures were represented by end plates. In the investigation described in [83] the fixture attached to the nozzle was also represented by an end plate; however, at the free end of the cylinder the radial and tangential displacement components were set to zero.

4. Differences in loading conditions. The test article was loaded through hydraulic rams acting on the end fixtures. The accuracy of the applied load and hence the accuracy of the stress resultants depends on the accuracy of the load cells. The transfer of the load through the end fixtures was through mechanical contact. The precise distribution of the tractions acting on the ends is not known. In the mathematical model used in the ORNL investigation, the applied loads were represented by nodal forces. In the investigation described in [83] the pressure loading was represented by constant normal traction.

It can be seen that even under very carefully controlled experimental conditions some degree of uncertainty concerning the physical system is present. Some of these uncertainties can be reduced, others either cannot be reduced or may not be feasible to reduce. For example, the mean value of the elastic constants can be determined by coupon tests. The dimensions of the test article can be measured with high accuracy. On the other hand, it would be very difficult to determine the distribution of the tractions or constraint conditions imposed by the end fixtures. In addition, some degree of uncertainty is associated with the instruments employed in making the observations and the effects of the environment on the instruments.

In view of these uncertainties one cannot expect very close correlation between computed and experimental data. The largest uncertainties in the ORNL experiments are related to the difficulties associated with manufacturing thin-walled objects to tight tolerances and, possibly, uncertainties in the mathematical representation of the constraint conditions.

Table 9.1 The von Mises stress (psi) in the gauge locations C, D, E, F shown in Figure 9.13(b). Fully three-dimensional model. Trunk space. The points are located in the plane of symmetry on the fixed-end side of the cylinder (1 psi = 6.895 kPa).

p	N	Pt. C	Pt. D	Pt. E	Pt. F
3	6 095	11 225	11 959	13 729	16 589
4	10 828	12 461	11 680	16 729	19 687
5	17 774	12 455	12 158	17 644	19 441
6	27 497	12 535	12 089	17 237	19 005
7	40 561	12 542	12 021	17 108	19 104
8	57 530	12 544	11 982	17 094	18 903
Expt.		15 569	14 554	16 981	13 100
Diff. (%)		24.1	21.5	−0.7	−30.7

Predictions and observations

The computed von Mises stress in the gauge locations C, D, E, F shown in Figure 9.13(b) and the von Mises stress computed from the experimental data are given in Table 9.1. The computed data were determined using the fully three-dimensional model, trunk space, with p ranging from 3 to 8. It can be seen that the stress data converge strongly, but the errors between the experimental measurements and predictions are large.

Assuming that the criterion for rejection was set at 20% error in the predicted von Mises stress, then the mathematical model has to be rejected. One instance of exceeding the tolerance set in the design of validation experiments is sufficient for rejection.

When a model fails to meet the criteria set for validation then the next problem is to identify possible reasons for rejecting the model. Virtual experimentation provides a very effective tool for such analyses.

Significant sensitivity to variations in wall thickness was found to exist, which leads to the conclusion that the likely reasons for rejection are the uncertainties in the input data. Therefore a model that accounts for the effects of uncertainties in the geometric description of the shells on the data of interest has to be considered. Owing to lack of sufficient statistical information, uncertainty quantification (UQ) often involves some reliance on expert opinion.

9.4.5 Discussion

The ORNL investigation highlights some of the difficulties and limitations of experimental validation of mathematical models for thin-shell problems. The experimental data are dominated by uncertainties caused by difficulties associated with the fabrication of thin-walled objects to exacting tolerances. Increasing the wall thickness would reduce errors caused by manufacturing tolerances but then the thin-shell model might not be applicable.

In correlation with experimental observations the predictions of a mathematical model are compared with data that are either observable directly or can be computed from observable data. In the model problem discussed here, the strain components in surface points located in the vicinity of the shell intersection were measured and the von Mises stresses were reported in the gauge locations.

In many cases the data of interest are not observable. For example, one may be interested in the maximum value of the integral of the normal stress over some small area, a functional that cannot be observed. Therefore, even if a mathematical model were shown to be successful in predicting certain measured data, it might not be suitable for computing other data of interest. Virtual experimentation is a very useful tool for the evaluation of the effects of uncertainties in the input data on the data of interest.

The validity of a mathematical model cannot be established by experimental correlation. The purpose of validation experiments is to determine whether certain necessary conditions are met.

Exercise 9.4.1 Discuss the relative importance of computational and physical experimentation in relation to the ORNL investigation of the shell intersection problem.

9.5 Chapter summary

In many cases it is advantageous to make certain a priori assumptions concerning the mode of deformation of an elastic body and use dimensionally reduced models instead of fully three-dimensional models. Whether a dimensionally reduced model should be used in a particular case depends on the goals of computation and the required accuracy. Generally speaking, dimensionally reduced models are well suited for structural analysis where the goals of computation are to determine structural stiffness, displacements and natural frequencies but are not well suited for strength analysis of structural connections and other details.

The accuracy of the data of interest depends not only on the discretization used, but also on how well the exact solution of a dimensionally reduced model approximates the exact solution of the corresponding fully three-dimensional model. The differences between the data of interest determined from the exact solution of a dimensionally reduced model and the corresponding fully three-dimensional model are errors of idealization, which were discussed in Section 1.1 and illustrated in Figure 1.1. The hierarchic view of models provides a conceptual framework for the control of modeling errors.

As the thickness is reduced, the exact solutions of the hierarchic beam, plate and shell models converge, respectively, to the exact solution of the Bernoulli–Euler beam model, the Kirchhoff plate model and the Novozhilov–Koiter shell model. These models, unlike the hierarchic models, require both the displacement functions and their first derivatives to be continuous. Therefore the exact solutions of the models in the hierarchic family that lie in $C^0(\Omega)$ converge to a solution that lies in $C^1(\Omega)$. Unless the polynomial degree is sufficiently high to satisfy or well approximate $C^1(\Omega)$ continuity, h-convergence will be slow or possibly non-existent. On the other hand, p-convergence will occur, but entry into the asymptotic range will occur only when $p \geq 4$.

10

Nonlinear models

It was noted in Chapter 1 and in Section 3.7 that the formulation of a mathematical model involves making various restrictive assumptions and therefore any mathematical model should always be viewed as a special case of a more comprehensive model. In order to test whether those assumptions are reasonable in the context of a particular application, it is necessary to perform virtual experimentation. The computational tools used for virtual experimentation must have the capability to solve nonlinear problems.

The subject of formulation and numerical treatment of nonlinear models is very large. A brief introduction to this important subject is presented in this chapter with emphasis on the algorithmic aspects. For additional discussion and details we refer to other books such as [66], [72].

10.1 Heat conduction

Mathematical models of heat conduction often involve radiation heat transfer and the coefficients of heat conduction are typically functions of the temperature. The formulation of mathematical models that account for these phenomena is outlined in the following.

10.1.1 Radiation

When two bodies exchange heat by radiation then the flux is proportional to the difference of the fourth power of their absolute temperatures:

$$q_n = \kappa f_s f_\epsilon (u^4 - u_R^4) \tag{10.1}$$

where u, u_R are the absolute temperatures of the radiating bodies, $\kappa = 5.699 \times 10^{-8}$ W/(m^2 K^4) is the Stefan–Boltzmann constant,[1] $0 \le f_s \le 1$ is the radiation shape factor and $0 < f_\epsilon \le 1$ is the surface emissivity which is defined as the relative emissive power of a body compared to that of an ideal blackbody. The surface emissivity is also equal to the absorption coefficient, defined as that fraction of thermal energy incident on a body which is absorbed. In general, surface emissivity is a function of the temperature.[2]

Equation (10.1) can be viewed as a convective boundary condition where the coefficient of convective heat transfer depends on the temperature of the radiating bodies. Writing

$$q_n = \kappa f_s f_\epsilon (u^4 - u_R^4) = \underbrace{\kappa f_s f_\epsilon (u^2 + u_R^2)(u + u_R)}_{h_r(u)} (u - u_R),$$

we have the form of Equation (3.14) with h_c replaced by $h_r(u)$.

10.1.2 Nonlinear material properties

When the conductivity or other material coefficients are functions of the temperature then solutions are obtained by iteration. The bilinear form is written as the sum of a linear form and a nonlinear form:

$$B_L(u, v) + B_{NL}(u, v) = F_L(v) + F_{NL}(u, v).$$

To obtain the initial solution $u^{(1)}$ the terms $B_{NL}(u, v)$ and $F_{NL}(u, v)$ are neglected and a linear solution is obtained. For subsequent solutions the following problem is solved:

$$B_L(u^{(k)}, v) = F_L(v) + F_{NL}(u^{(k-1)}, v) - B_{NL}(u^{(k-1)}, v) \qquad k = 2, 3, \ldots.$$

The process is terminated when $\|u^{(k)} - u^{(k-1)}\|_E \le \tau$ where τ is a prescribed tolerance. This process is not guaranteed to converge. If it fails to converge then the forcing function should be applied in small steps. In most practical problem convergence occurs when the steps, called increments, are sufficiently small.

10.2 Solid mechanics

In many practical problems the assumptions that (a) the strains are small, (b) the deformation of the body is so small that the equilibrium equations written for the deformed configuration would not be significantly different from the equilibrium equations written for the undeformed configuration, (c) the stress–strain laws are linear and (d) mechanical contact can be approximated by linear boundary conditions, such as linear springs, do not hold. In this section the formulation of mathematical models that account for these phenomena is outlined.

---

[1] Josef Stefan (1835–1893), Ludwig Boltzmann (1844–1906).
[2] For example, the surface emissivity of polished stainless steel is 0.22 at 373.15 K and 0.45 at 698.15 K.

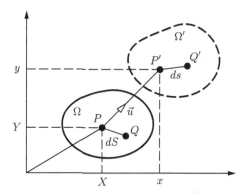

Figure 10.1 Notation.

10.2.1 Large strain and rotation

Deformation is characterized by the strain–displacement relationships. Consider an elastic body Ω in a reference state, as shown in Figure 10.1. The material points in the reference state are identified by the position vectors X_i, called Lagrangian coordinates, and in the deformed state by the position vectors x_i, called Eulerian coordinates. The displacement vector may be written as a function of X_i, in which case the uppercase letter U_i will be used. Alternatively, the displacement vector may be written as a function of x_i in which case the lowercase letter u_i will be used.

An infinitesimal "fiber" in the undeformed configuration has the length dS and in the deformed configuration it has the length ds. Let us assume first that the displacement is a function of the Eulerian coordinates x_i. In this case

$$dX_i = dx_i - du_i = dx_i - u_{i,j}dx_j = (\delta_{ij} - u_{i,j})dx_j.$$

Therefore

$$dS^2 = dX_k dX_k = (\delta_{ki} - u_{k,i})dx_i(\delta_{kj} - u_{k,j})dx_j$$
$$= (\delta_{ij} - u_{j,i} - u_{i,j} + u_{k,i}u_{k,j})dx_i dx_j$$

and

$$ds^2 - dS^2 = dx_k dx_k - dX_k dX_k$$
$$= \delta_{ij}dx_i dx_j - (\delta_{ij} - u_{j,i} - u_{i,j} + u_{k,i}u_{k,j})dx_i dx_j$$
$$= (u_{i,j} + u_{j,i} - u_{k,i}u_{k,j})dx_i dx_j.$$

The Almansi strain tensor,[3] denoted by e_{ij}, is defined by the relationship

$$ds^2 - dS^2 = 2e_{ij}dx_i dx_j.$$

---

[3] Emilio Almansi (1869–1948).

Therefore we have the definition of the Almansi strain tensor:

$$e_{ij} := \frac{1}{2}(u_{i,j} + u_{j,i} - u_{k,i}u_{k,j}). \tag{10.2}$$

In unabridged notation this can be written in two dimensions as

$$\begin{aligned} e_{xx} &= \frac{\partial u_x}{\partial x} - \frac{1}{2}\left[\left(\frac{\partial u_x}{\partial x}\right)^2 + \left(\frac{\partial u_y}{\partial x}\right)^2\right] \\ e_{xy} &= \frac{1}{2}\left(\frac{\partial u_x}{\partial y} + \frac{\partial u_y}{\partial x}\right) - \frac{1}{2}\left[\frac{\partial u_x}{\partial x}\frac{\partial u_x}{\partial y} + \frac{\partial u_y}{\partial x}\frac{\partial u_y}{\partial y}\right] \\ e_{yy} &= \frac{\partial u_y}{\partial y} - \frac{1}{2}\left[\left(\frac{\partial u_x}{\partial y}\right)^2 + \left(\frac{\partial u_y}{\partial y}\right)^2\right]. \end{aligned} \tag{10.3}$$

When the displacement vector components are written in terms of the Lagrangian coordinates X_i then $x_i = X_i + U_i$. Therefore

$$dx_i = dX_i + dU_i = dX_i + U_{i,j}dX_j = (\delta_{ij} + U_{i,j})dX_j,$$

thus

$$\begin{aligned} ds^2 = dx_k dx_k &= (\delta_{ki} + U_{k,i})dX_i(\delta_{kj} + U_{k,j})dX_j \\ &= (\delta_{ij} + U_{j,i} + U_{i,j} + U_{k,i}U_{k,j})dX_i dX_j \end{aligned}$$

and

$$\begin{aligned} ds^2 - dS^2 &= dx_k dx_k - dX_k dX_k \\ &= (\delta_{ij} + U_{j,i} + U_{i,j} + U_{k,i}U_{k,j})dX_i dX_j - \delta_{ij}dX_i dX_j \\ &= (U_{i,j} + U_{j,i} + U_{k,i}U_{k,j})dX_i dX_j. \end{aligned}$$

The Green strain tensor, denoted by E_{ij}, is defined by the relationship

$$ds^2 - dS^2 = 2E_{ij}dX_i dX_j.$$

Therefore we have the definition of the Green strain tensor:

$$E_{ij} := \frac{1}{2}(U_{i,j} + U_{j,i} + U_{k,i}U_{k,j}). \tag{10.4}$$

In unabridged notation this can be written in two dimensions as

$$E_{XX} = \frac{\partial U_X}{\partial X} + \frac{1}{2}\left[\left(\frac{\partial U_X}{\partial X}\right)^2 + \left(\frac{\partial U_Y}{\partial X}\right)^2\right]$$

$$E_{XY} = \frac{1}{2}\left(\frac{\partial U_X}{\partial Y} + \frac{\partial U_Y}{\partial X}\right) + \frac{1}{2}\left[\frac{\partial U_X}{\partial X}\frac{\partial U_X}{\partial Y} + \frac{\partial U_Y}{\partial X}\frac{\partial U_Y}{\partial Y}\right] \quad (10.5)$$

$$E_{YY} = \frac{\partial U_Y}{\partial Y} + \frac{1}{2}\left[\left(\frac{\partial U_X}{\partial Y}\right)^2 + \left(\frac{\partial U_Y}{\partial Y}\right)^2\right].$$

When the first derivatives $u_{i,j}$ (resp. $U_{i,j}$) are much smaller than unity then the product terms $u_{k,i}u_{k,j}$ (resp. $U_{k,i}U_{k,j}$) are negligible and the strain is called infinitesimal, small or linear strain. In such cases the Almansi and the Green strain tensors reduce to the definition of linear strain:

$$\epsilon_{ij} := \frac{1}{2}(u_{i,j} + u_{j,i}). \quad (10.6)$$

Exercise 10.2.1 Demonstrate in two dimensions that under arbitrary rigid body rotation all components of the Almansi strain tensor are zero. Hint: Let the reference configuration of a two-dimensional body be

$$X = R\cos\theta, \quad Y = R\sin\theta.$$

If this body is rotated by angle α, its new position will be

$$x = R\cos(\theta + \alpha), \quad y = R\sin(\theta + \alpha).$$

Therefore

$$u_x = x - X = R\cos(\theta + \alpha) - R\cos\theta$$
$$u_y = y - Y = R\sin(\theta + \alpha) - R\sin\theta.$$

Use

$$\frac{\partial u_x}{\partial x} = \frac{\partial u_x}{\partial R}\frac{\partial R}{\partial x} + \frac{\partial u_x}{\partial \theta}\frac{\partial \theta}{\partial x}$$
$$= [\cos(\theta + \alpha) - \cos\theta]\cos(\theta + \alpha) + [\sin(\theta + \alpha) - \sin\theta]\sin(\theta + \alpha)$$
$$= 1 - \cos\alpha$$

etc. to complete the exercise.

Exercise 10.2.2 Following the procedure indicated in Exercise 10.2.1, demonstrate that the Green strain tensor also vanishes under rigid body rotation.

Exercise 10.2.3 Following the procedure indicated in Exercise 10.2.1, compute the linear strain tensor defined by Equation (10.6).

Exercise 10.2.4 Consider a thin wire of length ℓ oriented along the x-axis. Write down expressions for the Almansi strain e_x and the Green strain E_x in terms of an imposed displacement Δ. (a) Show that $e_x \to 1/2$ as $\Delta \to \infty$. (b) Plot the Almansi strain and the Green strain in the range $-\ell < \Delta \leq 10\ell$. Hint: Let

$$X = \frac{1+\xi}{2}\ell, \qquad x = \frac{1+\xi}{2}(\ell + \Delta), \qquad -1 \leq \xi \leq +1.$$

10.2.2 Structural stability and stress stiffening

Investigation of buckling and stress stiffening is generally performed for structures which are beam-like and slender, or shell-like and thin. Such structures are usually stiffened and typically there are topological details, loads or boundary conditions for which the assumptions of beam, plate or shell theories do not hold. For this reason we consider the elastic stability of a fully three-dimensional body and construct a mathematical model which is not encumbered by the various restrictions implicit in beam, plate or shell models. From this model various dimensionally reduced models can be deduced as special cases.

We assume that a three-dimensional elastic body is subjected to an initial stress field σ_{ij}^0 which satisfies the equations of equilibrium of linear elasticity

$$\sigma_{ij,j}^0 + F_i^0 = 0 \tag{10.7}$$

where F_i^0 is the body force, and the traction boundary condition

$$\sigma_{ij}^0 n_j = T_i^0 \quad \text{on } \partial\Omega_T \cup \partial\Omega_s \tag{10.8}$$

where T_i^0 represents tractions imposed either directly on $\partial\Omega_T$ or through a displacement δ_j^0 imposed on a distributed elastic spring on $\partial\Omega_s$.

The initial stress field may be caused by body forces, surface tractions, temperature, or may be residual stress caused by the manufacturing process. The generalized form of Equation (10.7) is, for all $v_i \in E^0(\Omega)$,

$$\frac{1}{2}\int_\Omega \sigma_{ij}^0(v_{i,j} + v_{j,i})\,dV = \int_\Omega F_i^0 v_i\,dV + \int_{\partial\Omega_T \cup \partial\Omega_s} T_i^0 v_i\,dS \tag{10.9}$$

where dS represents the differential surface.

Let us assume that the reference configuration is perturbed by a small change in the body force $(\bar{F}_i)$; the temperature $(\bar{T}_\Delta)$; the surface tractions $(\bar{T}_i)$ on $\partial\Omega_T$; the displacement imposed on the distributed spring $\bar{\delta}_i$ on $\partial\Omega_s$; on boundary segment $\partial\Omega_u$. The corresponding kinematically admissible displacement field is denoted by $\bar{u}_i$. It is assumed in the following that $\bar{F}_i$, $\bar{T}_\Delta$, $\bar{T}_i$ and $\bar{\delta}_i$ are independent of $\bar{u}_i$. It is assumed further that the stress $\bar{\sigma}_{ij}$ caused by the perturbation is negligible in relation to the initial stress σ_{ij}^0.

When the reference configuration is not stress-free then the work done by σ_{ij}^0 due to the product terms of the Green strain tensor may not be negligible. Therefore the strain energy is

$$U(\bar{u}_i) := \frac{1}{2}\int_\Omega C_{ijkl}\bar{\epsilon}_{ij}\bar{\epsilon}_{kl}\,dV + \frac{1}{2}\int_{\partial\Omega_s} k_{ij}\bar{u}_i\bar{u}_j\,dS$$

$$+ \frac{1}{2}\int_\Omega \sigma_{ij}^0 \bar{u}_{\alpha,i}\bar{u}_{\alpha,j}\,dV. \tag{10.10}$$

The third integral in Equation (10.10) represents the work done by σ_{ij}^0 due to the product terms of the Green strain tensor. The work done by σ_{ij}^0 due to the linear strain terms is exactly canceled by the work done by F_i^0, T_i^0 and δ_i^0 in the sense of Equation (10.9). The potential energy is then

$$\Pi(\bar{u}_i) := U(\bar{u}_i) - \int_\Omega \bar{F}_i \bar{u}_i\,dV - \int_{\partial\Omega_T} \bar{T}_i \bar{u}_i\,dS - \int_{\partial\Omega_s} k_{ij}\bar{\delta}_j \bar{u}_i\,dS$$

$$- \int_\Omega \tilde{T}_\Delta C_{ijkl}\alpha_{kl}\bar{u}_{i,j}\,dV \tag{10.11}$$

where k_{ij} is a positive-semidefinite spring rate matrix, and α_{kl} represents the coefficients of thermal expansion. For isotropic materials $\alpha_{kl} = \alpha\delta_{kl}$.

We seek $\bar{u}_i \in \tilde{E}(\Omega)$ such that $\Pi(\bar{u}_i)$ is stationary; that is,

$$\delta\Pi(\bar{u}_i) := \left(\frac{\partial \Pi(\bar{u}_i + \varepsilon v_i)}{\partial \varepsilon}\right)_{\varepsilon=0} = 0 \quad \text{for all } v_i \in E^0(\Omega). \tag{10.12}$$

From this it follows that the principle of virtual work in the presence of an initial stress field is: find $\bar{u}_i \in \tilde{E}(\Omega)$ such that

$$\int_\Omega C_{ijkl}\bar{u}_{k,l} v_{i,j}\,dV + \int_{\partial\Omega_s} k_{ij}\bar{u}_j v_i\,dS + \int_\Omega \sigma_{ij}^0 \bar{u}_{\alpha,j} v_{\alpha,i}\,dV =$$

$$\int_\Omega \bar{F}_i v_i\,dV + \int_{\partial\Omega_T} \bar{T}_i v_i\,dS + \int_{\partial\Omega_s} k_{ij}\bar{\delta}_j v_i\,dS + \int_\Omega \tilde{T}_\Delta C_{ijkl}\alpha_{kl} v_{i,j}\,dV \tag{10.13}$$

for all $v_i \in E^0(\Omega)$.

The effect of the initial stress σ_{ij}^0 depends on its sense and magnitude: if σ_{ij}^0 is predominantly positive (i.e., tensile) then the stiffness increases. This is called stress stiffening. If, on the other hand, σ_{ij}^0 is predominantly negative, then the stiffness decreases. Of great practical interest is the critical value of the initial stress at which the stiffness is zero. Define

$$\sigma_{ij}^0 := \lambda \sigma_{ij}^*. \tag{10.14}$$

In stability problems σ_{ij}^* is the pre-buckling stress state. In stress stiffening problems σ_{ij}^* is some reference stress state and λ is some fixed number.

Define the bilinear form $B_\lambda(\bar{u}_i, v_i)$ by

$$B_\lambda(\bar{u}_i, v_i) := B(\bar{u}_i, v_i) - \lambda G(\bar{u}_i, v_i) \qquad (10.15)$$

where

$$B(\bar{u}_i, v_i) := \int_\Omega C_{ijkl} \bar{u}_{i,j} v_{k,l}\, dV + \int_{\partial\Omega_s} k_{ij} \bar{u}_j v_i\, dS$$

$$G(\bar{u}_i, v_i) := -\int_\Omega \sigma^\star_{ij} \bar{u}_{\alpha,i} v_{\alpha,j}\, dV$$

$$F(v_i) := \int_\Omega \bar{F}_i v_i\, dV + \int_{\partial\Omega_T} \bar{T}_i v_i\, dS + \int_{\partial\Omega_s} k_{ij} \bar{\delta}_j v_i\, dS + \int_\Omega \bar{T}_\Delta C_{ijkl} \alpha_{kl} v_{i,j}\, dV.$$

The problem is then to find $\bar{u}_i \in \tilde{E}(\Omega)$ such that

$$B_\lambda(\bar{u}_i, v_i) = F(v_i) \quad \text{for all } v_i \in E^0(\Omega). \qquad (10.16)$$

The set of λ for which Equation (10.16) is uniquely solvable for all F is called the resolvent set. The complement of the resolvent set is the spectrum. In addition to the point spectrum, which consists of the eigenvalues λ_i ($i = 1, 2, \ldots$), the spectrum may also include values that lie in a continuous spectrum. Fortunately, in problems of engineering interest that require consideration of elastic stability (the analysis of thin-walled structures) the lowest values of λ lie in a point spectrum.

The effect of stress stiffening is illustrated by considering the free vibration of elastic structures subjected to initial stress. The mathematical model of free vibration is: find ω and $\bar{u}_i \in E^0(\Omega)$, $\bar{u}_i \neq 0$, such that

$$B_\lambda(\bar{u}_i, v_i) - \omega^2 \int_\Omega \rho \bar{u}_i v_i\, dV = 0 \qquad (10.17)$$

where ρ is the specific density of the material and ω is the natural frequency. The natural frequency is now a function of λ. If $\sigma^\star_{ij}$ is predominantly compressive (negative) then the structural stiffness is decreased as $\lambda > 0$ is increased. If λ is in the point spectrum then there are functions $\bar{u}_i$ which satisfy Equation (10.17) for $\omega = 0$. That is, the natural frequency is zero. If $\sigma^\star_{ij}$ is tensile (positive) then the structural stiffness is increased as λ is increased. See Exercise 10.2.7.

Remark 10.2.1 Using the procedures of variational calculus, the strong form of the equations of equilibrium is found to be

$$\bar{\sigma}_{ij,j} + \bar{F}_i + (\sigma^0_{kj} \bar{u}_{i,k})_{,j} = 0 \qquad (10.18)$$

where $\bar{\sigma}_{ij} := C_{ijkl}(\bar{\epsilon}_{kl} - \bar{T}_\Delta \alpha_{kl})$. The corresponding natural boundary conditions are:

$$(\bar{\sigma}_{ij} + \sigma^0_{kj} \bar{u}_{i,k}) n_j = \bar{T}_i \quad \text{on } \partial\Omega_T \qquad (10.19)$$

and

$$(\bar{\sigma}_{ij} + \sigma^0_{kj}\bar{u}_{i,k})n_j = k_{ij}(\bar{\delta}_j - \bar{u}_j) \quad \text{on } \partial\Omega_s. \tag{10.20}$$

Example 10.2.1 Consider an elastic column of uniform cross-section, area A, moment of inertia I, length ℓ and modulus of elasticity E. The centroidal axis of the column coincides with the x_1-axis. A compressive axial force P is applied, hence $\sigma^0_{11} = -P/A$. The displacement field is assumed to be of the following form:

$$\bar{u}_1 = -\frac{dw}{dx_1}x_2, \quad \bar{u}_2 = w(x_1), \quad \bar{u}_3 = 0. \tag{10.21}$$

Therefore the only non-zero strain component is

$$\epsilon_{11} = \frac{d^2w}{dx_1^2}x_2 \tag{10.22}$$

and the term $\sigma^0_{ij}u_{\alpha,i}u_{\alpha,j}$ can be written as

$$\sigma^0_{ij}\bar{u}_{\alpha,i}\bar{u}_{\alpha,j} = -\frac{P}{A}\left[\left(\frac{d^2w}{dx_1^2}\right)^2 x_2^2 + \left(\frac{dw}{dx_1}\right)^2\right]. \tag{10.23}$$

In this case the strain energy can be written in the following form:

$$U = \frac{1}{2}\int_0^\ell \left(E - \frac{P}{A}\right) I \left(\frac{d^2w}{dx_1^2}\right)^2 dx_1 - \frac{P}{2}\int_0^\ell \left(\frac{dw}{dx_1}\right)^2 dx_1 \tag{10.24}$$

where I is the moment of inertia of the cross-section:

$$I := \int_A x_2^2 \, dx_2 dx_3. \tag{10.25}$$

The term P/A can be neglected in relation to the modulus of elasticity E in (10.24). This example illustrates that dimensionally reduced models can be derived from the three-dimensional formulation.

Exercise 10.2.5 Neglecting the term P/A in (10.24), determine the approximate value of P at which a column fixed at both ends will buckle. Use one finite element and $p = 4$. Report the relative error. Hint: Refer to Equation (9.27) and the definition $N_5(\xi) = \psi_4(\xi)$ where $\psi_j(\xi)$ ($j = 4, 5, \ldots$) is given by Equation (9.26). The exact value of the critical force is $4\pi^2 EI/\ell^2$.

Exercise 10.2.6 Show that the work done by σ^0_{ij} due to the linear strain terms is exactly canceled by the work done by F_i^0, T_i^0 and δ_i^0 in the sense of Equation (10.9).

Exercise 10.2.7 Consider a 50 mm × 50 mm square plate of thickness 1.0 mm. The material properties are: $E = 6.96 \times 10^4$ MPa; $\nu = 0.365$; $\varrho = 2.71 \times 10^{-9}$ N s^2/mm^4. The plate is fixed on one edge, loaded by a normal traction T_n on the opposite edge and simply supported on the other edges (soft simple support). Determine the first natural frequency corresponding to $T_n = -50$ MPa, $T_n = 0$ and $T_n = 50$ MPa. (Partial answer: For $T_n = -50$ MPa the first natural frequency is 578.6 hertz.)

Exercise 10.2.8 If the multi-span beam described in Exercise 9.1.9 had been pre-stressed by an axial force such that a constant positive initial stress σ^0 acted on the beam, would the rotation at support B be larger or smaller than if the beam had not been pre-stressed? Explain.

10.2.3 Plasticity

The formulation of mathematical models based on the incremental and deformational theories of plasticity is persented in the following. In the incremental theory, as the name implies, a relationship between the increment in the strain tensor and the corresponding increment in the stress tensor is defined. In the deformational theory the strain tensor (rather than the increment in the strain tensor) is related to the stress tensor. It is assumed that the strain components are sufficiently small to justify small strain representation and the plastic deformation is contained, that is, the plastic zone is surrounded by an elastic zone. Uncontained plastic flow, as in metal forming processes, is not within the scope of the following discussion.

The formulation of plastic deformation is based on three fundamental relationships: (a) yield criterion; (b) flow rule; and (c) hardening rule. We will use the von Mises yield criterion[4] and an associative flow rule known as the Prandtl–Reuss flow rule.[5] For a comprehensive discussion we refer to [72].

Notation

The stress deviator tensor is defined by

$$\tilde{\sigma}_{ij} := \sigma_{ij} - \frac{1}{3}\sigma_{kk}\delta_{ij}. \tag{10.26}$$

The second invariant of the stress deviator tensor is denoted by J_2 and is defined by

$$J_2 := \frac{1}{2}\tilde{\sigma}_{ij}\tilde{\sigma}_{ij}$$
$$= \frac{1}{3}\left[(\sigma_{11} - \sigma_{22})^2 + (\sigma_{22} - \sigma_{33})^2 + (\sigma_{33} - \sigma_{11})^2 + 6(\sigma_{12} + \sigma_{23} + \sigma_{31})\right].$$

In a uniaxial test σ_{11} is the only non-zero stress component. Therefore $J_2 = 2\sigma_{11}^2/3$. The equivalent stress, often called the von Mises stress, is defined such that in the special case of uniaxial stress it is equal to the uniaxial stress:

$$\bar{\sigma} = \sqrt{\frac{1}{2}\left[(\sigma_{11} - \sigma_{22})^2 + (\sigma_{22} - \sigma_{33})^2 + (\sigma_{33} - \sigma_{11})^2 + 6(\sigma_{12} + \sigma_{23} + \sigma_{31})\right]}. \tag{10.27}$$

---
[4] Richard von Mises (1883–1953).
[5] Ludwig Prandtl (1875–1953), Endre Reuss (1900–1968).

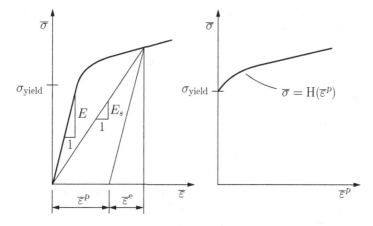

Figure 10.2 Typical uniaxial stress–strain curve.

We will interpret uniaxial stress–strain diagrams as a relationship between the equivalent stress and the equivalent strain. The elastic (resp. plastic) strains will be indicated by the superscript e (resp. p). The three principal strains are denoted by ϵ_1, ϵ_2, ϵ_3. The equivalent elastic strain is defined by

$$\bar{\epsilon}^e := \frac{\sqrt{2}}{2(1+\nu)}\sqrt{(\epsilon_1^e - \epsilon_2^e)^2 + (\epsilon_2^e - \epsilon_3^e)^2 + (\epsilon_3^e - \epsilon_1^e)^2} \qquad (10.28)$$

where ν is Poisson's ratio. The definition of the equivalent plastic strain follows directly from Equation (10.28) by setting $\nu = 1/2$:

$$\bar{\epsilon}^p := \frac{\sqrt{2}}{3}\sqrt{(\epsilon_1^p - \epsilon_2^p)^2 + (\epsilon_2^p - \epsilon_3^p)^2 + (\epsilon_3^p - \epsilon_1^p)^2}. \qquad (10.29)$$

A typical uniaxial stress–strain curve is shown in Figure 10.2

Exercise 10.2.9 Show that the first derivatives of J_2 with respect to σ_x, σ_y, σ_z and τ_{xy} are equal to $\tilde{\sigma}_x$, $\tilde{\sigma}_y$, $\tilde{\sigma}_z$ and $\tilde{\tau}_{xy}$ respectively.

Assumptions

The assumptions on which the formulation of the mathematical problem of plasticity is based are described in the following:

1. **Confined plastic deformation.** The strain components are much smaller than unity on the solution domain and its boundary, and the deformations are small in the sense that equilibrium equations written for the undeformed configuration are essentially the same as the equilibrium equations written for the deformed configuration.

NONLINEAR MODELS

2. **Decomposition of strain.** An increment in the total strain is the sum of the increment in the elastic strain and the increment in the plastic strain:

$$d\epsilon_{ij} = d\epsilon_{ij}^e + d\epsilon_{ij}^p. \tag{10.30}$$

We will not consider thermal strain here.

3. **Yield criterion.** We define

$$F(\sigma_{ij}, \bar{\epsilon}^p) := \bar{\sigma} - H(\bar{\epsilon}^p). \tag{10.31}$$

When $F < 0$ then the material is elastic. Plastic deformation may occur only when $F = 0$. Any stress state for which $F > 0$ is inadmissible. This is known as the consistency condition. Therefore, in plastic deformation

$$dF = 0: \quad \frac{\partial F}{\partial \sigma_{ij}} d\sigma_{ij} - H' d\bar{\epsilon}^p = 0. \tag{10.32}$$

4. **Flow rule.** The Prandtl–Reuss flow rule states that

$$d\epsilon_{ij}^p = \frac{\partial F}{\partial \sigma_{ij}} d\bar{\epsilon}^p. \tag{10.33}$$

Incremental stress–strain relationship

An increment in stress is proportional to the elastic strain:

$$d\sigma_{ij} = C_{ijkl} d\epsilon_{kl}^e = C_{ijkl}(d\epsilon_{kl} - d\epsilon_{kl}^p).$$

Substituting Equation (10.33) we get

$$d\sigma_{ij} = C_{ijkl} d\epsilon_{kl} - C_{ijkl} \frac{\partial F}{\partial \sigma_{kl}} d\bar{\epsilon}^p. \tag{10.34}$$

Using (10.32) we obtain

$$H' d\bar{\epsilon}^p = \frac{\partial F}{\partial \sigma_{ij}} d\sigma_{ij} = \frac{\partial F}{\partial \sigma_{ij}} C_{ijkl} d\epsilon_{kl} - \frac{\partial F}{\partial \sigma_{pq}} C_{pqrs} \frac{\partial F}{\partial \sigma_{rs}} d\bar{\epsilon}^p \tag{10.35}$$

where the dummy indices were suitably renamed. From Equation (10.35) an expression for $d\bar{\epsilon}^p$ is obtained:

$$d\bar{\epsilon}^p = \frac{\dfrac{\partial F}{\partial \sigma_{ij}} C_{ijkl}}{H' + \dfrac{\partial F}{\partial \sigma_{pq}} C_{pqrs} \dfrac{\partial F}{\partial \sigma_{rs}}} d\epsilon_{kl}.$$

On substituting into Equation (10.34) we get the incremental elastic–plastic stress-strain relationship:

$$d\sigma_{ij} = \left(C_{ijkl} - \frac{C_{ijmn}\dfrac{\partial F}{\partial \sigma_{mn}}\dfrac{\partial F}{\partial \sigma_{uv}}C_{uvkl}}{H' + C_{pqrs}\dfrac{\partial F}{\partial \sigma_{pq}}\dfrac{\partial F}{\partial \sigma_{rs}}} \right) d\epsilon_{kl}. \tag{10.36}$$

The bracketed expression in (10.36) is well defined for elastic–perfectly plastic materials (i.e., materials for which $H' = 0$).

The computations involve the following. Given the current stress state σ_{ij}, compute

$$d\sigma_{ij} = C_{ijkl}d\epsilon_{kl}$$

corresponding to an increment in the applied load. In each integration point compute $F(\sigma_{ij} + d\sigma_{ij})$. If $F(\sigma_{ij} + d\sigma_{ij}) \leq 0$ then nothing further needs to be done. If $F(\sigma_{ij} + d\sigma_{ij}) > 0$ then recompute $d\sigma_{ij}$ using Equation (10.36). Repeat the process until $F(\sigma_{ij} + d\sigma_{ij}) \approx 0$. The process is started by a linear analysis.

An alternative algorithm, known as the return mapping algorithm, has been used by several investigators. For details we refer to [72].

The deformation theory of plasticity

In the deformation theory of plasticity it is assumed that the plastic strain tensor is proportional to the stress deviator tensor. Referring to Figure 10.2,

$$\bar{\epsilon}^e + \bar{\epsilon}^p = \frac{\bar{\sigma}}{E_s}$$

where E_s is the secant modulus. Since the elastic part of the strain is related to the stress by Hooke's law

$$\bar{\epsilon}^e = \frac{\bar{\sigma}}{E}$$

we have

$$\bar{\epsilon}^p = \left(\frac{1}{E_s} - \frac{1}{E} \right) \bar{\sigma}. \tag{10.37}$$

In a uniaxial stress state $\tilde{\sigma} = 2\bar{\sigma}/3$ and hence Equation (10.37) can be written as

$$\bar{\epsilon}^p = \frac{3}{2}\left(\frac{1}{E_s} - \frac{1}{E} \right) \tilde{\sigma}.$$

This is generalized to

$$\epsilon^p_{ij} = \frac{3}{2}\left(\frac{1}{E_s} - \frac{1}{E} \right) \tilde{\sigma}_{ij}. \tag{10.38}$$

For example, in planar problems,

$$\begin{Bmatrix} \epsilon_x^p \\ \epsilon_y^p \\ \epsilon_z^p \\ \epsilon_{xy}^p \end{Bmatrix} = \frac{3}{2}\left(\frac{1}{E_s} - \frac{1}{E}\right) \begin{Bmatrix} \tilde{\sigma}_x \\ \tilde{\sigma}_y \\ \tilde{\sigma}_z \\ \tilde{\tau}_{xy} \end{Bmatrix}. \qquad (10.39)$$

Example 10.2.2 In this example an algorithm based on the deformation theory of plasticity is formulated for plane stress. In the deformation theory of plasticity the elastic–plastic compliance matrix is the matrix $[C]$ which establishes the relationship between the total strain and stress:

$$\{\epsilon\} = [C]\{\sigma\}.$$

The elastic–plastic material stiffness matrix $[E_{ep}]$ is the inverse of the elastic–plastic compliance matrix. Using the definition of the stress deviator and the relationship between the plastic strain and deviatoric stress (10.39), we have

$$\begin{Bmatrix} \epsilon_x^p \\ \epsilon_y^p \\ \gamma_{xy}^p \end{Bmatrix} = \frac{3}{2}\left(\frac{1}{E_s} - \frac{1}{E}\right) \begin{bmatrix} 2/3 & -1/3 & 0 \\ -1/3 & 2/3 & 0 \\ 0 & 0 & 2 \end{bmatrix} \begin{Bmatrix} \sigma_x \\ \sigma_y \\ \tau_{xy} \end{Bmatrix}.$$

Using $\{\epsilon\} = \{\epsilon^e\} + \{\epsilon^p\}$, a relationship is obtained between the total strain components and the stress tensor:

$$\begin{Bmatrix} \epsilon_x \\ \epsilon_y \\ \gamma_{xy} \end{Bmatrix} = \left(\frac{1}{E}\begin{bmatrix} 1 & -\nu & 0 \\ -\nu & 1 & 0 \\ 0 & 0 & 2(1+\nu) \end{bmatrix} + \frac{E - E_s}{E_s E}\begin{bmatrix} 1 & -1/2 & 0 \\ -1/2 & 1 & 0 \\ 0 & 0 & 3 \end{bmatrix}\right) \begin{Bmatrix} \sigma_x \\ \sigma_y \\ \tau_{xy} \end{Bmatrix}.$$

The solution is obtained by iteration: for each integration point the equivalent stress and strain are computed. From the stress–strain relationship the secant modulus is computed and the appropriate material stiffness matrix $[E_{ep}]$ is evaluated. The process is continued until the equivalent stress and strain do not deviate from the uniaxial stress–strain curve by more than a pre-set tolerance (usually 1% or less). An interesting benchmark study on the performance of the h- and p-versions of the finite element method is presented in [27].

Exercise 10.2.10 Show that in uniaxial tension or compression, $\bar{\epsilon}^e = \bar{\epsilon}_1^e$, $\bar{\epsilon}^p = \bar{\epsilon}_1^p$ and $\bar{\sigma} = \sigma_1$.

Exercise 10.2.11 Derive Equation (10.28) by specializing the root mean square of the differences of principal strains to the one-dimensional case so that $\bar{\epsilon}^e = \bar{\epsilon}_1^e$.

10.2.4 Mechanical contact

Mathematical models of mechanical contact between solid bodies involve non-linear boundary conditions written in terms of a gap function $g = g(s, t) \geq 0$ where s and t are surface parameters: Whenever $g = 0$, the normal and shearng tractions in corresponding points must

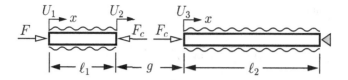

Figure 10.3 Example of mechanical contact: notation

have equal value and opposite sense. When $g > 0$ then the tractions are zero. The condition $g < 0$ is not allowed because it would correspond to penetration.

In many practical problems contacting bodies are lubricated and therefore shearing tractions are negligibly small in comparison with the normal tractions. On the other hand, shearing tractions in dry friction strongly influence wear, see for example [56].

In the following example a procedure for solving elastic contact problems is illustrated in a one-dimensional setting.

Example 10.2.3 Let us consider the problem of contact between two elastic bars. The notation is shown in Fig. 10.3. We assume that the axial stiffness $(AE)_i$ and the spring coefficient c_i ($i = 1, 2$) are constants for each bar and bar 2 is fixed on the right. The gap between the bars is $g = g_0 - U_2 + U_3$ where g_0 is the initial gap.

The goal is to determine the contact force F_c as a function of the applied force F. The differential equations for the bars are:

$$-(AE)_i u_i'' + c_i u_i = 0, \quad i = 1, 2 \tag{10.40}$$

and the corresponding solutions are:

$$u_i(x) = a_i \cosh \lambda_i x + b_i \sinh \lambda_i x \quad \text{where } \lambda_i = \sqrt{c_i/(AE)_i}. \tag{10.41}$$

We associate a local coordinate system with each bar, such that $x = 0$ at the left end. Therefore $U_1 = u_1(0)$, $U_2 = u_1(\ell_1)$ and $U_3 = u_2(0)$.

To the bar on the left we apply the boundary conditions

$$(AE)_1 u_1'(0) = -F, \quad (AE)_1 u_1'(\ell_1) = -F_c$$

and obtain the solution

$$u_1(x) = \frac{F}{(AE)_1 \lambda_1} \left(\frac{\cosh \lambda_1 \ell_1}{\sinh \lambda_1 \ell_1} \cosh \lambda_1 x - \sinh \lambda_1 x \right)$$

$$- \frac{F_c}{(AE)_1 \lambda_1} \frac{\cosh \lambda_1 x}{\sinh \lambda_1 \ell_1}. \tag{10.42}$$

To the bar on the right we apply the boundary conditions

$$(AE)_2 u_2'(0) = -F_c, \quad u_2(\ell_2) = 0$$

and obtain the solution:

$$u_2(x) = \frac{F_c}{(AE)_2 \lambda_2} \left(\frac{\sinh \lambda_2 \ell_2}{\cosh \lambda_2 \ell_2} \cosh \lambda_2 x - \sinh \lambda_2 x \right). \quad (10.43)$$

Given an initial gap $g_0 > 0$, the force needed to close the gap, denoted by F_0, can be computed from Equation (10.42) by setting $u_1(\ell_1) = g_0$ and $F_c = 0$. This yields:

$$F_0 = (AE)_1 \lambda_1 g_0 \sinh \lambda_1 \ell_1. \quad (10.44)$$

For any $F > F_0$ a contact force $F_c > 0$ will develop and the condition $g = 0$ must be satisfied. Let us now assume that $F = \alpha F_0$ was applied where $\alpha > 1$ and hence $g < 0$ which violates the contact condition. From Equations (10.42) and (10.44) we get

$$U_2 \equiv u_1(\ell_1) = \alpha g_0 - \frac{F_c}{(AE)_1 \lambda_1} \frac{\cosh \lambda_1 \ell_1}{\sinh \lambda_1 \ell_1} \quad (10.45)$$

and from Equation (10.43) we get

$$U_3 \equiv u_2(0) = \frac{F_c}{(AE)_2 \lambda_2} \frac{\sinh \lambda_2 \ell_2}{\cosh \lambda_2 \ell_2}. \quad (10.46)$$

Therefore the gap is

$$g = g_0 - U_2 + U_3 = g_0 - \alpha g_0 + F_c Q \quad (10.47)$$

where

$$Q := \frac{1}{(AE)_1 \lambda_1} \frac{\cosh \lambda_1 \ell_1}{\sinh \lambda_1 \ell_1} + \frac{1}{(AE)_2 \lambda_2} \frac{\sinh \lambda_2 \ell_2}{\cosh \lambda_2 \ell_2}. \quad (10.48)$$

Letting $g = 0$ we get

$$F_c = \begin{cases} \dfrac{(\alpha - 1) g_0}{Q} & \text{for } \alpha \geq 1 \\ 0 & \text{for } \alpha < 1. \end{cases} \quad (10.49)$$

In this example it was possible to find F_c in two steps. In the first step the force needed to close the gap, denoted by F_0, was determined. In the second step the contact force F_c was determined for $F = \alpha F_0$ where $\alpha \geq 1$. In two and three dimensions the problem is more complicated because it is necessary to determine the contact surfaces which depend on the contact force, the material properties and the geometric attributes of the contacting bodies.

A challenging aspect of the numerical solution of contact problems is that small errors in the numerical approximation of the contacting surfaces have substantial effects on the computed values of the tractions. The method most commonly used for the numerical treatment of contact conditions is known as the augmented Lagrangian method, see for example [57], [73], [92].

10.3 Chapter summary

In the process of conceptualization one usually starts with linear models bearing in mind the restrictive assumptions incorporated in those models. In order to properly formulate a mathematical model for the representation of some physical reality, it is necessary to consider the effects of those assumptions and, when necessary, remove or modify them. Virtual experimentation is a very useful tool for the assessment of the effects of modeling assumptions on the data of interest. The concept of model hierarchies and the ability to seamlessly pass from linear to nonlinear analyses are the practical prerequisites to estimating and controlling modeling errors. In this chapter the algorithmic aspects of nonlinear formulations were outlined.

Appendix A

Definitions

The properties of norms, normed linear spaces, linear functionals and bilinear forms are listed in the following. The symbols α and β denote real numbers. In Section A.6 the Legendre polynomials are defined and in Section A.8 the Schwarz inequality is derived.

A.1 Norms and seminorms

The norm of a function or vector is a measure of the size of the function or vector. Norms are real non-negative numbers defined on some space X. Norms, denoted by $\|\cdot\|_X$, have the following properties:

1. $\|u\|_X \geq 0$.
2. $\|u\|_X \neq 0$ if $u \neq 0$.
3. $\|\alpha u\|_X = |\alpha| \|u\|_X$.
4. $\|u + v\|_X \leq \|u\|_X + \|v\|_X$. This property is known as the triangle inequality.

As noted in Section 2.4.1, a familiar example of norms is the distance in Euclidean space. In connection with finite element analysis, the commonly used norms are the maximum norm, defined in Equation (2.55); the L_2 norm, defined in Equation (2.56); and the energy norm, defined in Equation (2.53).

Seminorms satisfy properties 1, 3 and 4 of norms but do not satisfy property 2. Instead of property 2 seminorms have the property

$$\|u\|_X = 0 \quad u \in \bar{X} \subset X, \quad u \neq 0.$$

For example, $\|u\|_E$ defined by Equation (2.53) is a seminorm on the space defined by Equation (2.38) when $c = 0$. In this case $\bar{X}$ is the set of constant functions on the interval $(0, \ell)$.

Introduction to Finite Element Analysis: Formulation, Verification and Validation, First Edition.
Barna Szabó and Ivo Babuška. © 2011 John Wiley & Sons, Ltd. Published 2011 by John Wiley & Sons, Ltd.

A.2 Normed linear spaces

A normed linear space X is a family of elements $u, v, \ldots$ which have the following properties:

1. If $u \in X$ and $v \in X$ then $(u + v) \in X$.
2. If $u \in X$ then $\alpha u \in X$.
3. $u + v = v + u$.
4. $u + (v + w) = (u + v) + w$.
5. There is an unique element in X, denoted by 0, such that $u + 0 = u$ for any $u \in X$.
6. Associated with every element $u \in X$ there is an unique element $-u \in X$ such that $u + (-u) = 0$.
7. $\alpha(u + v) = \alpha u + \alpha v$.
8. $(\alpha + \beta) u = \alpha u + \beta u$.
9. $\alpha(\beta u) = (\alpha \beta) u$.
10. $1 \cdot u = u$.
11. $0 \cdot u = 0$.
12. Associated with every $u \in X$ there is a real number $\|u\|_X$, called the norm. The norm has the properties listed in Section A.1.

A.3 Linear functionals

Let X be a normed linear space and $F(v)$ a process which associates with every $v \in X$ a real number $F(v)$. $F(v)$ is called a linear functional or linear form on X if it has the following properties:

1. $F(v_1 + v_2) = F(v_1) + F(v_2)$.
2. $F(\alpha v) = \alpha F(v)$.
3. $|F(v)| \leq C \|v\|_X$ with C independent of v. The smallest possible value of C is called the norm of F.

A.4 Bilinear forms

Let X and Y be normed linear spaces and $B(u, v)$ a process that associates with every $u \in X$ and $v \in Y$ a real number $B(u, v)$. $B(u, v)$ is a bilinear form on $X \times Y$ if it has the following properties:

1. $B(u_1 + u_2, v) = B(u_1, v) + B(u_2, v)$.
2. $B(u, v_1 + v_2) = B(u, v_1) + B(u, v_2)$.
3. $B(\alpha u, v) = \alpha B(u, v)$.

4. $B(u, \alpha v) = \alpha B(u, v)$.

5. $|B(u, v)| \leq C \|u\|_X \|v\|_Y$ with C independent of u and v. The smallest possible value of C is called the norm of B.

The space X is called the trial space and functions $u \in X$ are called trial functions. The space Y is called the test space and functions $v \in Y$ are called test functions. $B(u, v)$ is not necessarily symmetric.

A.5 Convergence

A sequence of functions $u_n \in X$ ($n = 1, 2, \ldots$) converges in the space X to the function $u \in X$ if for every $\epsilon > 0$ there is a number n_ϵ such that for any $n > n_\epsilon$ the following relationship holds:

$$\|u - u_n\|_X < \epsilon. \tag{A.1}$$

A.6 Legendre polynomials

The Legendre polynomials $P_n(x)$ are solutions of the Legendre differential equation for $n = 0, 1, 2, \ldots$:

$$(1 - x^2)y'' - 2xy' + n(n+1)y = 0, \quad -1 \leq x \leq 1. \tag{A.2}$$

The first eight Legendre polynomials are

$$P_0(x) = 1 \tag{A.3}$$

$$P_1(x) = x \tag{A.4}$$

$$P_2(x) = \frac{1}{2}(3x^2 - 1) \tag{A.5}$$

$$P_3(x) = \frac{1}{2}(5x^3 - 3x) \tag{A.6}$$

$$P_4(x) = \frac{1}{8}(35x^4 - 30x^2 + 3) \tag{A.7}$$

$$P_5(x) = \frac{1}{8}(63x^5 - 70x^3 + 15x) \tag{A.8}$$

$$P_6(x) = \frac{1}{16}(231x^6 - 315x^4 + 105x^2 - 5) \tag{A.9}$$

$$P_7(x) = \frac{1}{16}(429x^7 - 693x^5 + 315x^3 - 35x). \tag{A.10}$$

Legendre polynomials can be generated from the recursion formula

$$(n+1)P_{n+1}(x) = (2n+1)x P_n(x) - n P_{n-1}(x), \quad n = 1, 2, \ldots \tag{A.11}$$

and Legendre polynomials satisfy the following relationship:

$$(2n+1)P_n(x) = P'_{n+1}(x) - P'_{n-1}(x), \quad n = 1, 2, \ldots \tag{A.12}$$

where the primes represent differentiation with respect to x. Legendre polynomials satisfy the following orthogonality property:

$$\int_{-1}^{+1} P_i(x) P_j(x) \, dx = \begin{cases} \dfrac{2}{2i+1} & \text{for } i = j \\ 0 & \text{for } i \neq j. \end{cases} \tag{A.13}$$

All roots of Legendre polynomials are located in the interval $-1 < x < +1$.

A.7 Analytic functions

A.7.1 Analytic functions in $\mathbb{R}^2$

A function $f(x, y)$ is analytic on $\bar{\Omega} \in \mathbb{R}^2$ if for any $(x_0, y_0) \in \bar{\Omega}$ there is an $r(x_0, y_0) > 0$ and a $\kappa(x_0, y_0) > 0$ such that

$$f = \sum_{i,j=0}^{\infty} a_{ij}(x-x_0)^i(y-y_0)^j \quad \text{and} \quad \sum_{i,j=0}^{\infty} |a_{ij}|(i+j)! \, r^{i+j} \leq \kappa(x_0, y_0) \tag{A.14}$$

where $r \leq \sqrt{x_0^2 + y_0^2}$.

From this definition it follows that if a solution is analytic then its derivatives are bounded. Specifically, for any s,

$$\left| \frac{\partial^s u}{\partial x^k \partial y^{s-k}} \right| \leq s! \, r^s \kappa(x_0, y_0), \quad k = 0, 1, \ldots, s. \tag{A.15}$$

A.7.2 Analytic curves in $\mathbb{R}^2$

A plane curve or arc is the set of points Γ defined as follows:

$$\Gamma = \{(x, y) \mid x = x(t), \, y = y(t), \, t \in \bar{I} = [-1, 1]\}. \tag{A.16}$$

A plane curve Γ is analytic if $x(t)$ and $y(t)$ are analytic functions of $t \in \bar{I}$ and

$$\left(\frac{dx}{dt}\right)^2 + \left(\frac{dy}{dt}\right)^2 > 0 \quad \text{for all } t \in \bar{I}. \tag{A.17}$$

A.8 The Schwarz inequality for integrals

Definition A.8.1 The function $f(x)$ defined on the interval $a < x < b$ is square integrable if

$$\int_a^b f^2 \, dx < \infty.$$

Theorem A.8.1 Let $f(x)$ and $g(x)$ be square integrable functions defined on the interval $a < x < b$. Then

$$\left| \int_a^b fg \, dx \right| \le \left(\int_a^b f^2 \, dx \right)^{1/2} \left(\int_a^b g^2 \, dx \right)^{1/2}.$$

This is the Schwarz inequality for integrals. To prove this inequality we observe that

$$\int_a^b (f + \lambda g)^2 \, dx \ge 0 \quad \text{for any } \lambda$$

and therefore

$$\int_a^b f^2 \, dx + 2\lambda \int_a^b fg \, dx + \lambda^2 \int_a^b g^2 \, dx \ge 0 \quad \text{for any } \lambda. \tag{A.18}$$

On the left of this inequality is a quadratic expression for λ. To find the roots of this expression we need to compute

$$\lambda = \frac{-\int_a^b fg \, dx \pm \sqrt{\left(\int_a^b fg \, dx \right)^2 - \int_a^b g^2 \, dx \int_a^b f^2 \, dx}}{\int_a^b g^2 \, dx}.$$

Denoting the roots by λ_1 and λ_2, (A.18) can be written as $(\lambda - \lambda_1)(\lambda - \lambda_2) \ge 0$. We now observe that the roots cannot be real and simple because then we could select any λ so that $\lambda_1 < \lambda < \lambda_2$ and we would have $(\lambda - \lambda_1)(\lambda - \lambda_2) < 0$. Therefore the radicand must be less than or equal to zero. This completes the proof.

Appendix B

Numerical quadrature

In the finite element method the terms of the coefficient matrices and right hand side vectors are computed by numerical quadrature. Most commonly Gaussian[1] quadrature is used, while in some cases the Gauss–Lobatto quadrature is used. In one dimension the domain of integration is the standard element I_{st}. An integral expression on the standard element is approximated by a sum

$$\int_{-1}^{+1} f(\xi)\,d\xi \approx \sum_{i=1}^{n} w_i f(\xi_i) + R_n \qquad (B.1)$$

where w_i are the weights, ξ_i are the abscissas and R_n is the error term which depends on the smoothness of f. The abscissas and weights are symmetric with respect to the point $\xi = 0$.

To evaluate an integral on other than the standard domain, the mapping function defined by Equation (2.60) is used for transforming the domain of integration to the standard domain. For example,

$$\int_{x_1}^{x_2} F(x)\,dx = \int_{-1}^{+1} \underbrace{F(Q(\xi))\frac{x_2 - x_1}{2}}_{f(\xi)}\,d\xi \quad \text{where} \quad Q(\xi) = \frac{1-\xi}{2}x_1 + \frac{1+\xi}{2}x_2.$$

---

[1] Johann Carl Friedrich Gauss (1777–1855).

Introduction to Finite Element Analysis: Formulation, Verification and Validation, First Edition.
Barna Szabó and Ivo Babuška. © 2011 John Wiley & Sons, Ltd. Published 2011 by John Wiley & Sons, Ltd.

B.1 Gaussian quadrature

In Gaussian quadrature the abscissa x_i is the ith zero of Legendre polynomial P_n. The weights are computed from[2]

$$w_i = \frac{2}{(1 - x_i^2)[P_n'(x_i)]^2}. \qquad (B.2)$$

The abscissas and weights for Gaussian quadrature are listed in Table B.1 up to $n = 8$. The error term is:

$$R_n = \frac{2^{2n+1}(n!)^4}{(2n+1)[(2n)!]^3} f^{(2n)}(\zeta) \qquad -1 < \zeta < +1$$

where $f^{(2n)}$ is 2nth derivative of f. It can be seen from the error term that if $f(\xi)$ is a polynomial of degree p and Gaussian quadrature is used, then the integral will be exact (up to round-off errors) provided that $n \geq (p+1)/2$. For example, to integrate a polynomial

Table B.1 Abscissas and weights for Gaussian quadrature.

n	$\pm \xi_i$	w_i
2	0.577 350 269 189 626	1.000 000 000 000 000
3	0.000 000 000 000 000	0.888 888 888 888 889
	0.774 596 669 241 483	0.555 555 555 555 556
4	0.339 981 043 584 856	0.652 145 154 862 546
	0.861 136 311 594 053	0.347 854 845 137 454
5	0.000 000 000 000 000	0.568 888 888 888 889
	0.538 469 310 105 683	0.478 628 670 499 366
	0.906 179 845 938 664	0.236 926 885 056 189
6	0.238 619 186 083 197	0.467 913 934 572 691
	0.661 209 386 466 265	0.360 761 573 048 139
	0.932 469 514 203 152	0.171 324 492 379 170
7	0.000 000 000 000 000	0.417 959 183 673 469
	0.405 845 151 377 397	0.381 830 050 505 119
	0.741 531 185 599 394	0.279 705 391 489 277
	0.949 107 912 342 759	0.129 484 966 168 870
8	0.183 434 642 495 650	0.362 683 783 378 362
	0.525 532 409 916 329	0.313 706 645 877 887
	0.796 666 477 413 627	0.222 381 034 453 374
	0.960 289 856 497 536	0.101 228 536 290 376

---

[2] See, for example, Abramowitz, M. and Stegun, I., *Handbook of Mathematical Functions with Formulas, Graphs and Mathematical Tables*, US Department of Commerce, National Bureau of Standards, Applied Mathematics Series 55, 1964. An updated version of this reference entitled NIST Handbook of Mathematical Functions was published by Cambridge University Press in 1910. Editors: F. W. J. Olver, D. W. Lozier, R. F. Boisvert and C. W. Clark. ISBN: 978 0521140638.

of degree 5, $n = 3$ is sufficient. For other than polynomial functions the rate of convergence depends on how well the integrand can be approximated by polynomials. It can be shown that if $f(\xi)$ is a continuous function on I_{st} then the sum in Equation (B.1) converges to the true value of the integral.

The integration procedure can be extended directly to the standard quadrilateral element and the standard hexahedral element. For the standard quadrilateral element

$$\int_{-1}^{+1}\int_{-1}^{+1} f(\xi,\eta)\,d\xi d\eta = \sum_{i=1}^{n_\xi}\sum_{j=1}^{n_\eta} w_i w_j f(\xi_i,\eta_j) \tag{B.3}$$

where n_ξ (resp. n_η) is the number of quadrature points along the ξ- (resp. η-axis). Usually but not necessarily $n_\xi = n_\eta$ is used. Analogously, for the standard hexahedral element,

$$\int_{-1}^{+1}\int_{-1}^{+1}\int_{-1}^{+1} f(\xi,\eta,\zeta)\,d\xi d\eta d\zeta = \sum_{i=1}^{n_\xi}\sum_{j=1}^{n_\eta}\sum_{k=1}^{n_\zeta} w_i w_j w_k f(\xi_i,\eta_j,\zeta_k). \tag{B.4}$$

B.2 Gauss–Lobatto quadrature

In the Gauss–Lobatto quadrature the abscissas are as follows: $x_1 = -1$, $x_n = 1$ and for $i = 2, 3, \ldots, n-1$ the $(i-1)$th zero of $P'_{n-1}(x)$ where $P_{n-1}(x)$ is the $(n-1)$th Legendre

Table B.2 Abscissas and weights for Gauss–Lobatto quadrature.

n	$\pm\xi_i$	w_i
2	1.000 000 000 000 000	1.000 000 000 000 000
3	0.000 000 000 000 000	1.333 333 333 333 333
	1.000 000 000 000 000	0.333 333 333 333 333
4	0.447 213 595 499 958	0.833 333 333 333 333
	1.000 000 000 000 000	0.166 666 666 666 667
5	0.000 000 000 000 000	0.711 111 111 111 111
	0.654 653 670 707 977	0.544 444 444 444 444
	1.000 000 000 000 000	0.100 000 000 000 000
6	0.285 231 516 480 645	0.554 858 377 035 486
	0.765 055 323 929 465	0.378 474 956 297 847
	1.000 000 000 000 000	0.066 666 666 666 667
7	0.000 000 000 000 000	0.487 619 047 619 048
	0.468 848 793 470 714	0.431 745 381 209 863
	0.830 223 896 278 567	0.276 826 047 361 566
	1.000 000 000 000 000	0.047 619 047 619 048
8	0.209 299 217 902 479	0.412 458 794 658 704
	0.591 700 181 433 142	0.341 122 692 483 504
	0.871 740 148 509 607	0.210 704 227 143 506
	1.000 000 000 000 000	0.035 714 285 714 286

polynomial. The weights are

$$w_i = \begin{cases} \dfrac{2}{n(n-1)} & \text{for } i = 1 \text{ and } i = n \\ \dfrac{2}{n(n-1)[P_{n-1}(x_i)]^2} & \text{for } i = 2, 3, \ldots, (n-1). \end{cases} \quad (B.5)$$

The abscissas and weights for Gauss–Lobatto quadrature are listed in Table B.2 up to $n = 8$.
The error term is

$$R_n = \frac{-n(n-1)^3 2^{2n-1}[(n-2)!]^4}{(2n-1)[(2n-2)!]^3} f^{(2n-2)}(\zeta) \quad -1 < \zeta < +1$$

from which it follows that if $f(\xi)$ is a polynomial of degree p and Gauss–Lobatto quadrature is used, then the integral will be exact (up to round-off errors) provided that $n \geq (p+3)/2$.

Appendix C

Properties of the stress tensor

Some of the basic properties of the stress tensor and traction vector are reviewed in the following.

C.1 The traction vector

Let us assume that the state of stress at a point is known and let us determine the components of the traction vector T_x, T_y, T_z acting on the inclined face of the infinitesimal tetrahedral volume element, as shown in Figure C.1. By definition, the traction vector, also known as the stress vector, is

$$\vec{T} = \lim_{\Delta A \to 0} \frac{\Delta \vec{F}}{\Delta A}$$

where $\Delta \vec{F}$ is the differential force, acting on the inclined face of the tetrahedron, the area of which is ΔA. This force is in equilibrium with the resultants of the stresses acting on the other three faces of the tetrahedron.

We first show that the unit normal to the inclined plane, denoted by $\vec{n}$, has the components

$$n_x = \frac{\Delta A_x}{\Delta A}, \quad n_y = \frac{\Delta A_y}{\Delta A}, \quad n_z = \frac{\Delta A_z}{\Delta A} \qquad \text{(C.1)}$$

where ΔA_x (resp. ΔA_y, ΔA_z) is the area of that face of the tetrahedron to which the x-axis (resp. y-, z-axis) is normal, see Figure C.1(a). Consider the cross-product

$$\vec{c} = \vec{a} \times \vec{b} = (\Delta y \vec{e}_y - \Delta x \vec{e}_x) \times (\Delta z \vec{e}_z - \Delta y \vec{e}_y)$$
$$= \Delta y \Delta z \vec{e}_x + \Delta x \Delta z \vec{e}_y + \Delta x \Delta y \vec{e}_z = 2 \Delta A_x \vec{e}_x + 2 \Delta A_y \vec{e}_y + 2 \Delta A_z \vec{e}_z$$

Introduction to Finite Element Analysis: Formulation, Verification and Validation, First Edition.
Barna Szabó and Ivo Babuška. © 2011 John Wiley & Sons, Ltd. Published 2011 by John Wiley & Sons, Ltd.

APPENDIX C

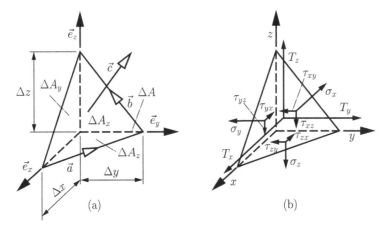

Figure C.1 Notation.

where $\vec{e}_x, \vec{e}_y, \vec{e}_z$ are the unit basis vectors. The vector $\vec{c}$ is normal to the inclined plane and its absolute value is $2\Delta A$. Therefore $\vec{n} = \vec{c}/|\vec{c}| = \vec{c}/2\Delta A$ and the components of the unit normal are as given in Equation (C.1).

The equations of equilibrium are

$$\sum \Delta F_x = 0: \quad T_x \Delta A - \sigma_x \Delta A_x - \tau_{yx} \Delta A_y - \tau_{zx} \Delta A_z = 0$$

$$\sum \Delta F_y = 0: \quad T_y \Delta A - \tau_{xy} \Delta A_x - \sigma_y \Delta A_y - \tau_{zy} \Delta A_z = 0$$

$$\sum \Delta F_z = 0: \quad T_z \Delta A - \tau_{xz} \Delta A_x - \tau_{yz} \Delta A_y - \sigma_z \Delta A_z = 0.$$

On dividing by ΔA, and making use of the fact that the stress tensor is symmetric, we have

$$\begin{Bmatrix} T_x \\ T_y \\ T_z \end{Bmatrix} = \begin{bmatrix} \sigma_x & \tau_{xy} & \tau_{xz} \\ \tau_{yx} & \sigma_y & \tau_{yz} \\ \tau_{zx} & \tau_{zy} & \sigma_z \end{bmatrix} \begin{Bmatrix} n_x \\ n_y \\ n_z \end{Bmatrix}. \tag{C.2}$$

In index notation this can be written as

$$T_i = \sigma_{ij} n_j. \tag{C.3}$$

C.2 Principal stresses

Let us now ask whether it would be possible to find a plane such that the traction vector, acting on that plane, would be normal to the plane (i.e., the shearing components would be zero). This condition is fulfilled when

$$\sigma_{ij} n_j = T n_i \equiv T \delta_{ij} n_j$$

where T is the magnitude of the traction vector. This is a characteristic value problem:

$$(\sigma_{ij} - T\delta_{ij})n_j = 0. \tag{C.4}$$

Since σ_{ij} is symmetric, all characteristic values are real. The characteristic values are called principal stresses and the normalized characteristic vectors are the unit vectors that define the directions of principal stresses. Since the characteristic vectors are mutually orthogonal, at every point there is an orthogonal coordinate system in which the stress state is characterized by normal stresses only. This coordinate system is uniquely defined only when all characteristic values are simple.

It follows from Equation (C.4) that the principal stresses are the roots of the following characteristic equation:

$$T^3 - I_1 T^2 - I_2 T - I_3 = 0 \tag{C.5}$$

where

$$I_1 = \sigma_{kk}, \quad I_2 = \frac{1}{2}(\sigma_{ij}\sigma_{ij} - \sigma_{ii}\sigma_{jj}), \quad I_3 = \det(\sigma_{ij}). \tag{C.6}$$

The principal stresses characterize the state of stress at a point and hence do not depend on the coordinate system in which the stress components are given. Therefore the coefficients I_1, I_2 and I_3 are invariant with respect to rotation of the coordinate system. These coefficients are called the first, second and third stress invariant respectively.

C.3 Transformation of vectors

Consider a Cartesian coordinate system x_i' rotated relative to the x_i system as shown in Figure C.2. Let α_{ij} be the angle between the axis x_i' and the axis x_j and let

$$g_{ij} := \cos \alpha_{ij}. \tag{C.7}$$

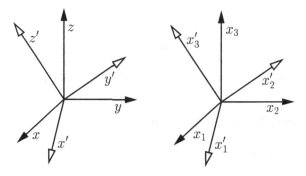

Figure C.2 Coordinate transformation: notation.

In other words, the ith row of g_{ij} is the unit vector in the direction of axis x'_i in the unprimed system. Therefore, if a_i is an arbitrary vector in the unprimed system then the same vector in the primed system is

$$a'_i = g_{ij} a_j. \tag{C.8}$$

Conversely, the jth column of g_{ij} is the unit vector in the direction of axis x_j in the primed system. Therefore, if a'_r is an arbitrary vector in the primed system then the same vector in the unprimed system is

$$a_r = g_{kr} a'_k. \tag{C.9}$$

Given the definition of g_{ij} and the orthogonality of the coordinate systems, g_{ij} multiplied by its transpose must be the identity matrix:

$$g_{ri} g_{rj} = g_{is} g_{js} = \delta_{ij}. \tag{C.10}$$

In other words, g_{ij} is an orthogonal matrix. This can be proven formally as follows:

$$a_i = g_{ri} a'_r$$

$$a'_s = g_{sj} a_j$$

and

$$a'_r = \delta_{rs} a'_s = \delta_{rs} g_{sj} a_j.$$

Therefore,

$$a_i \equiv \delta_{ij} a_j = \underbrace{g_{ri} \delta_{rs} g_{sj}}_{g_{ri} g_{rj}} a_j$$

and, for an arbitrary vector a_j, we have

$$(\delta_{ij} - g_{ri} g_{rj}) a_j = 0.$$

Consequently, the bracketed term must vanish. This completes the proof.

C.4 Transformation of stresses

Referring to the definition of traction vectors given in Section C.1, we have

$$T'_i = \sigma'_{ik} n'_k = \sigma'_{ik} g_{ks} n_s$$

and applying the transformation rule (C.8),

$$T'_i = g_{ir}T_r = g_{ir}\sigma_{rs}n_s.$$

Therefore we have

$$(g_{ks}\sigma'_{ik} - g_{ir}\sigma_{rs})n_s = 0.$$

Since this equation holds for arbitrary n_s, the bracketed term must vanish. Consequently,

$$g_{ks}\sigma'_{ik} = g_{ir}\sigma_{rs},$$

multiplying by g_{js},

$$\underbrace{g_{js}g_{ks}}_{\delta_{jk}}\sigma'_{ik} = g_{ir}g_{js}\sigma_{rs}$$

and using Equation (C.10), we have the transformation rule for stresses:

$$\sigma'_{ij} = g_{ir}g_{js}\sigma_{rs}. \qquad (C.11)$$

Remark C.4.1 Denoting the stress tensor by $[\sigma]$ and g_{ij} by $[g]$, Equation (C.11) is the symmetric matrix triple product

$$[\sigma'] = [g][\sigma][g]^T. \qquad (C.12)$$

Appendix D

Computation of stress intensity factors

The goal of computation in linear elastic fracture mechanics is to determine the stress intensity factors. The contour integral method and the energy release rate method, outlined in the following, are frequently used for this purpose.

D.1 The contour integral method

The contour integral method was described in connection with the Laplace equation in Section 7.4.1. In this section the contour integral method for the computation of the Mode I stress intensity factor in planar elasticity is outlined. The Airy stress function corresponding to the symmetric part of the asymptotic expansion is given by Equation (6.38). The extraction function for the coefficient of the first term, denoted by **w**, corresponds to the first negative eigenvalue $\lambda_1 = -1/2$. Therefore $\varphi(z) = z^{-1/2}$ and $\chi(z) = 3z^{1/2}$ in Equation (6.17) and the Airy stress function is

$$U = C\Re(\bar{z}z^{-1/2} + 3z^{1/2}) \tag{D.1}$$

where C is an arbitrary real number in units of MPa m$^{3/2}$. The stress components $\sigma_x^{(w)}$, $\sigma_y^{(w)}$, $\tau_{xy}^{(w)}$ are determined from Equations 6.20) and (6.21):

$$\sigma_x^{(w)} = -\frac{C}{4}r^{-3/2}(\cos(3\theta/2) + 3\cos(5\theta/2)) \tag{D.2}$$

$$\sigma_y^{(w)} = -\frac{C}{4}r^{-3/2}(7\cos(3\theta/2) - 3\cos(5\theta/2)) \tag{D.3}$$

$$\tau_{xy}^{(w)} = \frac{3C}{4}r^{-3/2}(\sin(3\theta/2) - \sin(7\theta/2)) \tag{D.4}$$

Introduction to Finite Element Analysis: Formulation, Verification and Validation, First Edition.
Barna Szabó and Ivo Babuška. © 2011 John Wiley & Sons, Ltd. Published 2011 by John Wiley & Sons, Ltd.

APPENDIX D

and the traction vector components are computed from

$$T_x^{(w)} = \sigma_x^{(w)} \cos\theta + \tau_{xy}^{(w)} \sin\theta \qquad (D.5)$$

$$T_y^{(w)} = \tau_{xy}^{(w)} \cos\theta + \sigma_y^{(w)} \sin\theta. \qquad (D.6)$$

The components of **w** are determined from Equation (6.23):

$$w_x = C \frac{r^{-1/2}}{2G} \left[\left(\kappa - \frac{3}{2}\right) \cos(\theta/2) + \frac{1}{2} \cos(5\theta/2) \right] \qquad (D.7)$$

$$w_y = -C \frac{r^{-1/2}}{2G} \left[\left(\kappa + \frac{3}{2}\right) \sin(\theta/2) - \frac{1}{2} \sin(5\theta/2) \right]. \qquad (D.8)$$

Referring to Equation (D.10), the path-independent integral evaluated on a circle of radius r is:

$$I_\Gamma(\mathbf{u}, \mathbf{w}) := \int_{-\pi}^{\pi} (T_x^{(u)} w_x + T_y^{(u)} w_y) r \, d\theta - \int_{-\pi}^{\pi} (T_x^{(w)} u_x + T_y^{(w)} u_y) r \, d\theta. \qquad (D.9)$$

The stress intensity factor K_I is defined by convention to be $a_1\sqrt{2\pi}$ where a_1 is the coefficient of the first term in Equation (6.38). Using the orthogonality of the characteristic functions,

$$I_\Gamma(\mathbf{u}, \mathbf{w}) = \frac{a_1 C}{G} F(\kappa) \qquad (D.10)$$

where $F(\kappa)$ is defined by

$$F(\kappa) := G I_\Gamma(\mathbf{u}_1/a_1, \mathbf{w}/C). \qquad (D.11)$$

The function $\mathbf{u}_1$ is the displacement field corresponding to the first term in Equation (6.38). It is determined using Equation (6.23):

$$u_x^{(1)} = \frac{a_1 r^{1/2}}{2G} \left[\left(\kappa - \frac{1}{2}\right) \cos(\theta/2) - \frac{1}{2} \cos(3\theta/2) \right] \qquad (D.12)$$

$$u_y^{(1)} = \frac{a_1 r^{1/2}}{2G} \left[\left(\kappa + \frac{1}{2}\right) \sin(\theta/2) - \frac{1}{2} \sin(3\theta/2) \right]. \qquad (D.13)$$

The corresponding stress field is given by Equations (7.24) to (7.26) with a_1 substituted for $K_I/\sqrt{2\pi}$. Note that $F(\kappa)$ is dimensionless and can be computed explicitly. Values of $F(\kappa)$ are listed in Table D.1.

From Equation (D.10), using $K_I = a_1\sqrt{2\pi}$, we get

$$K_I = \frac{\sqrt{2\pi} G}{C F(\kappa)} I_\Gamma(\mathbf{u}, \mathbf{w}) = \frac{\sqrt{2\pi} G}{F(\kappa)} I_\Gamma(\mathbf{u}, \mathbf{w}/C). \qquad (D.14)$$

Table D.1 Values of $F(\kappa)$.

ν	plane stress	plane strain
0	-4π	-4π
0.1	-11.4240	-11.3097
0.2	-10.4720	-10.0531
0.3	-9.66644	-8.79646
0.4	-8.97598	-7.53982
0.5	-8.37758	-6.28319

Note that since $I_\Gamma(\mathbf{u}, \mathbf{w}/C)$ has the dimension m$^{1/2}$, the dimension of K_I is MPa m$^{1/2}$. Note further that since C is arbitrary, it can be chosen to be 1 MPa m$^{3/2}$. To obtain an approximation for K_I we substitute the finite element solution $\mathbf{u}_{FE}$ for $\mathbf{u}$.

The derivation of the extraction function for K_{II} is analogous.

D.2 The energy release rate

The relationship between the stress intensity factors and the energy release rate $\mathcal{G}$ is derived in the following for a plane elastic body of thickness t_z. We assume that the body force and thermal load are zero. By definition

$$\mathcal{G} := -\frac{\partial \Pi}{\partial a} \tag{D.15}$$

where Π is the potential energy defined by Equation (4.31) and a is the crack length.

D.2.1 Symmetric (Mode I) loading

Consider the state of stress at the crack tip. In the coordinate system centered on the crack tip, as shown in Figure D.1, neglecting terms of higher order,

$$\sigma_y = \frac{K_I}{\sqrt{2\pi x}} \qquad 0 \le x. \tag{D.16}$$

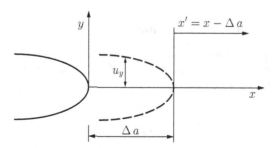

Figure D.1 Notation.

APPENDIX D

The displacement of the crack face is

$$u_y = \frac{(1+\nu)(\kappa+1)}{E} \frac{K_I}{\sqrt{2\pi}} \sqrt{-x} \qquad x \leq 0 \qquad (D.17)$$

where κ is defined by Equation (4.39).

Assume now that the crack length increases by a small amount Δa, as shown in Figure D.1. In this case

$$u_y = \frac{(1+\nu)(\kappa+1)}{E} \frac{K_I}{\sqrt{2\pi}} \sqrt{\Delta a - x} \qquad 0 \leq x \leq \Delta a. \qquad (D.18)$$

The work required to return the crack to its original length, that is, to close the length increment Δa, is

$$\Delta W = 2 \int_0^{\Delta a} \frac{1}{2} \sigma_y u_y \, t_z \, dx = \frac{(1+\nu)(\kappa+1)}{E} \frac{K_I^2 t_z}{2\pi} \underbrace{\int_0^{\Delta a} \sqrt{\frac{\Delta a - x}{x}} \, dx}_{\Delta a \pi / 2}. \qquad (D.19)$$

Hence

$$\Delta W = \frac{(1+\nu)(\kappa+1)}{4E} K_I^2 t_z \Delta a. \qquad (D.20)$$

In order to restore the crack to its initial length, energy equal to ΔW has to be imparted to the elastic body. This is the energy expended in crack growth, called Griffith's surface energy.[1] The potential energy has to decrease by the same amount when the crack increment occurred. Hence

$$\mathcal{G} = -\lim_{\Delta a \to 0} \frac{\Delta \Pi}{\Delta a} = -\frac{\partial \Pi}{\partial a} = \frac{(1+\nu)(\kappa+1)}{4E} K_I^2 t_z. \qquad (D.21)$$

Exercise D.2.1 Explain the reasons for writing

$$\Delta W = 2 \int_0^{\Delta a} \frac{1}{2} \sigma_y u_y \, t_z \, dx$$

in Equation (D.19).

D.2.2 Antisymmetric (Mode II) loading

When the loading is purely antisymmetric then the relationship between the energy release rate and the stress intensity factor is analogous to the symmetric case; however, instead of Equations (D.16) and (D.17) we have

$$\tau_{xy} = \frac{K_{II}}{\sqrt{2\pi x}} \qquad 0 \leq x \qquad (D.22)$$

---

[1] Alan Arnold Griffith (1893–1963).

and

$$u_x = \frac{(1+\nu)(\kappa+1)}{E} \frac{K_{II} t_z}{\sqrt{2\pi}} \sqrt{-x} \qquad x \le 0. \tag{D.23}$$

The derivation of the relationship between $\mathcal{G}$ and K_{II}, under the assumption of perfectly antisymmetric loading, is left to the reader in the following exercise.

Exercise D.2.2 Draw a sketch, analogous to Figure D.1, for the case of perfectly antisymmetric loading and show that

$$\mathcal{G} = \frac{(1+\nu)(\kappa+1)}{4E} K_{II}^2 t_z.$$

D.2.3 Combined (Mode I and Mode II) loading

In view of the fact that the solutions corresponding to Mode I and Mode II loadings are *energy orthogonal*, we have

$$\Pi(\vec{u}_I + \vec{u}_{II}) = \Pi(\vec{u}_I) + \Pi(\vec{u}_{II}) \tag{D.24}$$

where $\vec{u}_I$ and $\vec{u}_{II}$ are the Mode I and Mode II solutions respectively. Therefore in the case of combined loading we have

$$\mathcal{G} = \frac{(1+\nu)(\kappa+1)}{4E} (K_I^2 + K_{II}^2) t_z. \tag{D.25}$$

Consequently, the stress intensity factors are related to $\mathcal{G}$, which is computable, by

$$(K_I^2 + K_{II}^2) = \begin{cases} \dfrac{E\mathcal{G}}{t_z} & \text{for plane stress} \\[1em] \dfrac{E\mathcal{G}}{(1-\nu^2) t_z} & \text{for plane strain.} \end{cases} \tag{D.26}$$

D.2.4 Computation by the stiffness derivative method

In the following the vector of coefficients of the basis functions computed by the finite element method is denoted by x which is a function of the crack length a. Let us assume that we have computed $x = x(a)$ for a problem of linear elastic fracture mechanics. The potential energy is

$$\Pi(a) = \frac{1}{2} x^T K x - x^T r$$

where $K = K(a)$ is the stiffness matrix and $r = r(a)$ is the load vector. Following crack extension, the potential energy is

$$\Pi(a + \Delta a) = \frac{1}{2}(x^T + \Delta x^T)(K + \Delta K)(x + \Delta x) - (x^T + \Delta x^T)(r + \Delta r)$$

$$= \Pi(a) + \Delta x^T \underbrace{(Kx - r)}_{\text{this is zero}} + \frac{1}{2}\Delta x^T K \Delta x + \frac{1}{2}x^T \Delta K x$$

$$+ x^T \Delta K \Delta x + \frac{1}{2}\Delta x^T \Delta K \Delta x - x^T \Delta r - \Delta x^T \Delta r.$$

Therefore

$$\mathcal{G} = -\lim_{\Delta a \to 0} \frac{\Pi(a + \Delta a) - \Pi(a)}{\Delta a} = -\frac{1}{2}x^T \frac{\partial K}{\partial a} x + x^T \frac{\partial r}{\partial a}. \quad (D.27)$$

In finite element computations, $\partial K / \partial a$ and $\partial r / \partial a$ are approximated by finite differences:

$$\frac{\partial K}{\partial a} \approx \frac{K(a + \Delta a) - K(a - \Delta a)}{2\Delta a}$$

$$\frac{\partial r}{\partial a} \approx \frac{r(a + \Delta a) - r(a - \Delta a)}{2\Delta a}.$$

This involves recomputation of the stiffness matrices for only those elements which have a vertex on the crack tip. In most cases $\partial r / \partial a$ is either zero or negligibly small.

Appendix E

Saint-Venant's principle

Saint-Venant's principle is often invoked in the formulation of mathematical models to justify the idea that the precise distribution of tractions on the boundary is unimportant when the region of primary interest is far from where the tractions are applied, provided that the tractions are statically equivalent. Similarly, in heat conduction the precise functional form of fluxes prescribed on the boundary is unimportant far from where the fluxes are applied, provided that the heat flow rate corresponding to the fluxes is the same.

In this appendix a precise statement of Saint-Venant's principle is presented for the Laplace equation on the half space in two dimensions. The results are applicable to elasticity; a presentation based on elasticity would be more complicated.

E.1 Green's function for the Laplace equation

Green's function for the Laplace equation is defined as follows:

$$G(x, y, x_0, y_0) := \frac{1}{2\pi} \ln \sqrt{(x - x_0)^2 + (y - y_0)^2}. \tag{E.1}$$

Note that $\Delta G = 0$ everywhere except at the point (x_0, y_0). Introducing the transformation

$$x = x_0 + r \cos\theta, \quad y = y_0 + r \sin\theta,$$

we have

$$G(r, \theta, x_0, y_0) = \frac{1}{2\pi} \ln r.$$

APPENDIX E

This is the solution corresponding to a concentrated unit flux at the point (x_0, y_0). This can be seen by evaluating the contour integral on a circle of radius ϱ centered on the point (x_0, y_0):

$$\oint \frac{\partial G}{\partial r} ds = \int_0^{2\pi} \left(\frac{\partial G}{\partial r}\right)_{r=\varrho} \varrho\, d\theta = 1.$$

Note that the value of this integral is independent of ϱ.

E.2 Model problem

Saint-Venant's principle is illustrated on the basis of the following model problem:

$$\Delta u = 0 \text{ on } \Omega := \{x, y \mid -\infty < x < \infty,\ y < 0\}. \tag{E.2}$$

The boundary condition is

$$\frac{\partial u}{\partial y} = \begin{cases} q(x) & \text{on } \{x, y \mid |x| < d,\ y = 0\} \\ 0 & \text{on } \{x, y \mid |x| \geq d,\ y = 0\}. \end{cases} \tag{E.3}$$

The notation is shown in Figure E.1. It is assumed that $q(x)$ satisfies the condition

$$\int_{-\infty}^{\infty} q(x)\, dx = 0. \tag{E.4}$$

It is further assumed that $u(x, y) \to 0$ as $\sqrt{x^2 + y^2} \to \infty$. This assumption is made in order to ensure that $u(x, y)$ is unique. Otherwise $u(x, y)$ would be known up to an arbitrary constant only.

We denote the point $(x_0, -y_0)$ by P_0 and obtain estimates for u and the derivatives of u at the point P_0 in the following.

Remark E.2.1 We can associate the physical meaning of $q(x)$ as the difference between any two flux distributions on $\{x, y \mid |x| < d,\ y = 0\}$ that have the same heat flow rate. Specifically,

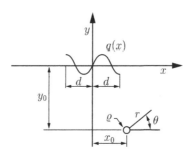

Figure E.1 Model problem: notation.

let

$$q_1(x) = -k\frac{\partial u_1}{\partial y} \quad \text{and} \quad q_2(x) = -k\frac{\partial u_2}{\partial y}.$$

Setting $k = 1$ and defining $u(x) = u_2(x) - u_1(x)$ and $q(x) = q_1(x) - q_2(x)$, we get Equation (E.3).

Estimation of the bound on $|u(P_0)|$

We define

$$H(x, y, x_0, y_0) := G(x, y, x_0, y_0) + G(x, y, x_0, -y_0) \tag{E.5}$$

and

$$\omega_\varrho := \{x, y \mid (x - x_0)^2 + (y + y_0)^2 < \varrho^2\}. \tag{E.6}$$

The function H is symmetric with respect to the x-axis. This can be verified by showing that $H(x, y, x_0, y_0) = H(x, -y, x_0, y_0)$. Note that H is a harmonic function on $\Omega - \omega_\varrho$. Therefore

$$\int_{\Omega-\omega_\varrho} \Delta u H \, dxdy = \int_{-\infty}^{\infty} (\nabla u \vec{n} \, H)_{y=0} \, dx + \int_{\partial \omega_\varrho} \nabla u \vec{n} \, H \, ds$$

$$- \int_{-\infty}^{\infty} (u \nabla H \vec{n})_{y=0} \, dx - \int_{\partial \omega_\varrho} u \nabla H \vec{n} \, ds$$

$$+ \int_{\Omega-\omega_\varrho} u \Delta H \, dxdy = 0 \tag{E.7}$$

where $\partial \omega_\varrho$ is the perimeter of the circle of radius ϱ centered on the point P_0 and ds is the positive (clockwise) differential increment along $\partial \omega_\varrho$. Noting that $\Delta u = \Delta H = 0$ on $\Omega - \omega_\varrho$ and $\nabla u \vec{n} = q(x)$ and, due to the symmetry of H with respect to the x-axis, $\nabla H \vec{n} = 0$ on $(-\infty < x < \infty, y = 0)$, we have

$$\int_{-\infty}^{\infty} q(x) \, (H)_{y=0} \, dx + \int_{\partial \omega_\varrho} \nabla u \vec{n} \, H \, ds - \int_{\partial \omega_\varrho} u \nabla H \vec{n} \, ds = 0. \tag{E.8}$$

We define

$$F(x) := \int_{-\infty}^{x} q(t) \, dt. \tag{E.9}$$

340   APPENDIX E

Note that $F(-d) = F(d) = 0$. The first term in Equation (E.8) is

$$\int_{-\infty}^{\infty} q(x) (H)_{y=0} \, dx = \frac{1}{2\pi} \int_{-d}^{d} \frac{dF}{dx} \ln\left((x-x_0)^2 + y_0^2\right) dx$$

$$= -\frac{1}{\pi} \int_{-d}^{d} F(x) \frac{x - x_0}{(x - x_0)^2 + y_0^2} \, dx \quad \text{(E.10)}$$

where we have integrated by parts and used $F(-d) = F(d) = 0$.

The second term in Equation (E.8) is

$$\int_{\partial \omega_\varrho} \nabla u \vec{n} \, H \, ds = -\frac{1}{4\pi} \int_0^{2\pi} \left(\frac{\partial u}{\partial r}\right)_{r=\varrho} \ln(\varrho^4 - 4y_0 \varrho^3 \sin\theta + 4y_0^2 \varrho^2) \varrho \, d\theta \quad \text{(E.11)}$$

where we used the transformation $x = x_0 + \varrho \cos\theta$, $y = -y_0 + \varrho \sin\theta$. Noting that u is a harmonic function on Ω and therefore its derivatives are bounded, we have

$$\lim_{\varrho \to 0} \int_{\partial \omega_\varrho} \nabla u \vec{n} \, H \, ds = 0. \quad \text{(E.12)}$$

The third term in Equation (E.8) is

$$\int_{\partial \omega_\varrho} u \nabla H \vec{n} \, ds = -\int_0^{2\pi} u \left(\frac{\partial H}{\partial r}\right)_{r=\varrho} \varrho \, d\theta$$

$$= -\frac{1}{4\pi} \int_0^{2\pi} u \frac{4\varrho^3 - 12y_0 \varrho^2 \sin\theta + 8y_0^2 \varrho}{\varrho^4 - 4y_0 \varrho^3 \sin\theta + 4y_0^2 \varrho^2} \varrho \, d\theta. \quad \text{(E.13)}$$

Therefore

$$\lim_{\varrho \to 0} \int_{\partial \omega_\varrho} u \nabla H \vec{n} \, ds = -u(P_0). \quad \text{(E.14)}$$

From Equations (E.8), (E.10), (E.12) and (E.14) we have

$$u(P_0) = \frac{1}{\pi} \int_{-d}^{d} F(x) \frac{x - x_0}{(x - x_0)^2 + y_0^2} \, dx \quad \text{(E.15)}$$

and therefore

$$|u(P_0)| \leq \frac{1}{\pi} \max |F| \int_{-d}^{d} \frac{|x - x_0|}{(x - x_0)^2 + y_0^2} \, dx. \quad \text{(E.16)}$$

Recalling the mean value theorem, there exists a $\xi \in (-d, d)$ such that

$$|u(P_0)| \leq \frac{1}{\pi} \max |F| \frac{2d|\xi - x_0|}{(\xi - x_0)^2 + y_0^2}. \quad \text{(E.17)}$$

To obtain an upper bound for $|u(P_0)|$ we select a ξ for the numerator such that the numerator is maximum and one for the denominator such that the denominator is minimum. We distinguish between two cases:

Case (a) $x_0 \in (-d, d)$. For the numerator

$$\max_{\xi} |\xi - x_0| \leq 2d, \quad \xi \in (-d, d)$$

and for the denominator

$$\min_{\xi}((\xi - x_0)^2 + y_0^2) \geq y_0^2, \quad \xi \in (-d, d).$$

Therefore

$$|u(P_0)| \leq \frac{1}{\pi} \max |F| \frac{4d^2}{y_0^2}. \tag{E.18}$$

Case (b) $x_0 \notin (-d, d)$. For the numerator

$$\max_{\xi} |\xi - x_0| \leq d + |x_0|, \quad \xi \in (-d, d)$$

and for the denominator

$$\min_{\xi}((\xi - x_0)^2 + y_0^2) \geq (d - |x_0|)^2 + y_0^2, \quad \xi \in (-d, d).$$

Therefore

$$|u(P_0)| \leq \frac{1}{\pi} \max |F| \frac{2d(d + |x_0|)}{(d - |x_0|)^2 + y_0^2}. \tag{E.19}$$

Estimates (E.18) and (E.19) are based on the assumption that F is bounded. If q is also bounded then $\max |F| \leq 2d \max |q|$.

Remark E.2.2 If $u(x)$ is prescribed on the boundary satisfying the conditions

$$u(x, 0) = \begin{cases} \tilde{u}(x) & \text{on } \{x, y \mid |x| < d, \ y = 0\} \\ 0 & \text{on } \{x, y \mid |x| \geq d, \ y = 0\} \end{cases}$$

and

$$\int_{-d}^{d} \tilde{u}(x) \, dx = 0$$

then the procedure for estimating $|u(P_0)|$ is analogous; however, instead of H defined in Equation (E.5) the function $\bar{H}$, defined by

$$\bar{H}(x, y, x_0, y_0) := G(x, y, x_0, y_0) - G(x, y, x_0, -y_0), \qquad (E.20)$$

has to be used. This is because $\bar{H}(x, 0) = 0$ and hence the first term on the right of Equation (E.7) is zero.

Estimation of the bound on the first derivatives of u in point P_0

The first derivatives of G defined by Equation (E.1) with respect to x_0 and y_0 are denoted as follows:

$$G_x(x, y, x_0, y_0) = -\frac{1}{2\pi} \frac{x - x_0}{(x - x_0)^2 + (y - y_0)^2} \qquad (E.21)$$

$$G_y(x, y, x_0, y_0) = -\frac{1}{2\pi} \frac{y - y_0}{(x - x_0)^2 + (y - y_0)^2}. \qquad (E.22)$$

We define

$$H_x(x, y, x_0, y_0) = G_x(x, y, x_0, y_0) + G_x(x, y, x_0, -y_0) \qquad (E.23)$$

$$H_y(x, y, x_0, y_0) = G_y(x, y, x_0, y_0) - G_y(x, y, x_0, -y_0). \qquad (E.24)$$

Note that both H_x and H_y are symmetric with respect to the x-axis and therefore

$$\left(\frac{\partial H_x}{\partial y}\right)_{y=0} = 0, \quad \left(\frac{\partial H_y}{\partial y}\right)_{y=0} = 0. \qquad (E.25)$$

This can be easily seen by verifying that $H_x(x, y, x_0, y_0) = H_x(x, -y, x_0, y_0)$ and analogously for H_y.

On replacing H with H_x in Equation (E.7) an equation analogous to Equation (E.8) is obtained:

$$\int_{-\infty}^{\infty} q\,(H_x)_{y=0}\,dx + \int_{\partial\omega_\varrho} \nabla u \vec{n}\, H_x\, ds - \int_{\partial\omega_\varrho} u \nabla H_x \vec{n}\, ds = 0. \qquad (E.26)$$

The first term in Equation (E.26) is

$$\int_{-\infty}^{\infty} q\,(H_x)_{y=0}\,dx = -\frac{1}{\pi}\int_{-d}^{d} \frac{dF}{dx} \frac{x - x_0}{(x - x_0)^2 + y_0^2}\, dx$$

$$= \frac{1}{\pi}\int_{-d}^{d} F(x) \frac{-(x - x_0)^2 + y_0^2}{((x - x_0)^2 + y_0^2)^2}\, dx \qquad (E.27)$$

where we have used Equation (E.9) and integrated by parts.

APPENDIX E

Once again using the transformation $x = x_0 + \varrho \cos \theta$, $y = -y_0 + r \sin \theta$ and recalling that the positive normal to $\partial \omega_\varrho$ points toward the center of the circle, the second term in Equation (E.26) is

$$\int_{\partial \omega_\varrho} \nabla u \vec{n} \, H_x \, ds =$$

$$\frac{1}{2\pi} \int_0^{2\pi} \left(\frac{\partial u}{\partial x} \cos \theta + \frac{\partial u}{\partial y} \sin \theta \right) \left(\frac{\varrho \cos \theta}{\varrho^2 - 4 y_0 \varrho \sin \theta + 4 y_0^2} + \frac{1}{\varrho} \cos \theta \right) \varrho \, d\theta.$$

Since u has bounded derivatives, we have

$$\lim_{\varrho \to 0} \int_{\partial \omega_\varrho} \nabla u \vec{n} \, H_x \, ds = \frac{1}{2} \left(\frac{\partial u}{\partial x} \right)_{P_0}. \tag{E.28}$$

The third term in Equation (E.26) is

$$\int_{\partial \omega_\varrho} u \nabla H_x \vec{n} \, ds = - \int_0^{2\pi} u \left(\frac{\partial H_x}{\partial r} \right)_{r=\varrho} \varrho \, d\theta$$

$$= \frac{1}{2\pi} \int_0^{2\pi} u \left(-\frac{\cos \theta}{\varrho} + O(\varrho) \right) \varrho \, d\theta. \tag{E.29}$$

Expanding u in a Taylor series we obtain

$$u = u(P_0) + \left(\frac{\partial u}{\partial x} \right)_{P_0} \varrho \cos \theta + \left(\frac{\partial u}{\partial y} \right)_{P_0} \varrho \sin \theta + O(\varrho^2). \tag{E.30}$$

On substituting the Taylor series expansion for u into Equation (E.29) and letting $\varrho \to 0$ we get

$$\lim_{\varrho \to 0} \int_{\partial \omega_\varrho} u \nabla H_x \vec{n} \, ds = -\frac{1}{2} \left(\frac{\partial u}{\partial x} \right)_{P_0}. \tag{E.31}$$

From Equations (E.26), (E.27). (E.28), (E.31) we have

$$\left(\frac{\partial u}{\partial x} \right)_{P_0} = -\frac{1}{\pi} \int_{-d}^{d} F(x) \frac{-(x - x_0)^2 + y_0^2}{((x - x_0)^2 + y_0^2)^2} \, dx \tag{E.32}$$

therefore, assuming $F(x)$ is bounded,

$$\left| \left(\frac{\partial u}{\partial x} \right)_{P_0} \right| \leq \frac{1}{\pi} \max |F| \int_{-d}^{d} \frac{1}{((x - x_0)^2 + y_0^2)} \, dx. \tag{E.33}$$

Therefore

$$\left|\left(\frac{\partial u}{\partial x}\right)_{P_0}\right| \leq \frac{2d}{\pi} \max |F| \begin{cases} \dfrac{1}{y_0^2} & \text{when } x_0 \in (-d, d) \\ \dfrac{1}{(d - |x_0|)^2 + y_0^2} & \text{when } x_0 \notin (-d, d). \end{cases}$$

It can be seen that the size of $\partial u/\partial x$ decreases at least in proportion to d/y_0^2.

Estimation of the size of the derivative $\partial u/\partial y$ is analogous, but in Equation (E.26) the function H_x has to be replaced by H_y.

Appendix F

Solutions for selected exercises

Exercise 2.1.2

$$-AEu'' + cu = 0, \quad u(0) = \hat{u}_0, \quad AEu'(\ell) = k_\ell(d_\ell - u(\ell)).$$

The solution is

$$u = a_1 \cosh \lambda x + a_2 \sinh \lambda x$$

where $\lambda := \sqrt{c/AE}$. From the boundary conditions we have $a_1 = \hat{u}_0$ and

$$a_2 = \frac{d_\ell - \hat{u}_0(\beta \sinh \lambda \ell + \cosh \lambda \ell)}{\beta \cosh \lambda \ell + \sinh \lambda \ell} \quad \text{where } \beta := \frac{\sqrt{AEc}}{k_\ell}.$$

Exercise 2.1.4 On substituting $AE = 2.804 \times 10^8$ kN, $k = 325.0$ kN/m^3, $\ell = 17.5$ m into Equation (2.22) we get $F_0/u_0 \approx 42.75$ kN/mm.

Exercise 2.2.2

$$\pi(u_2) = \frac{1}{2} \int_0^\ell [\kappa(a_1\varphi_1'(x) + a_2\varphi_2'(x))^2 + c(a_1\varphi_1(x) + a_2\varphi_2(x))^2] \, dx.$$

Letting $\partial\pi/\partial a_1 = 0$, $\partial\pi/\partial a_2 = 0$ and recognizing that the coefficients

$$k_{ij} := \int_0^\ell (\kappa\varphi_i'(x)\varphi_j'(x) + c\varphi_i(x)\varphi_j(x)) \, dx$$

are symmetric ($k_{ij} = k_{ji}$), the algebraic equations of Example 2.2.1 are obtained.

Introduction to Finite Element Analysis: Formulation, Verification and Validation, First Edition.
Barna Szabó and Ivo Babuška.   © 2011 John Wiley & Sons, Ltd. Published 2011 by John Wiley & Sons, Ltd.

Exercise 2.2.4 Partial solution: to compute the second row of matrix $[B]$ we write

$$x(\ell - x)^2 = b_{21}x(\ell - x) + b_{22}x^2(\ell - x) + b_{23}x^3(\ell - x).$$

On factoring $x(\ell - x)$ we find

$$\ell - x = b_{21} + b_{22}x + b_{23}x^2$$

and therefore $b_{21} = \ell$, $b_{22} = -1$, $b_{23} = 0$.

Exercise 2.3.2 Definitions:

$$E(I) = \left\{ u \mid \int_0^\ell \left(\kappa(u')^2 + cu^2 \right) dx \leq C < \infty \right\}$$

$$F(v) = \int_0^\ell fv \, dx$$

$$\tilde{E}(I) = E^0(I) = \{u \mid u \in E(I), \, u(\ell) = 0\}.$$

Problem statement: "Find $u \in E^0(I)$ such that $B(u, v) = F(v)$ for all $v \in E^0(I)$."

Exercise 2.3.4 Definitions:

$$E(I) = \left\{ u \mid \int_0^\ell \left(\kappa(u')^2 + cu^2 \right) dx \leq C < \infty \right\}$$

$$B(u, v) = \int_0^\ell (\kappa u'v' + cuv) \, dx + \beta_\ell u(\ell)v(\ell)$$

$$F(v) = \int_0^\ell fv \, dx + \hat{q}_0 v(0) + \beta_\ell U_\ell v(\ell).$$

Problem statement: "Find $u \in E(I)$ such that $B(u, v) = F(v)$ for all $v \in E(I)$."

Exercise 2.5.2 $N_i(-1) = 0$ for $i \geq 3$ because the interval of integration is zero. $N_i(1) = 0$ for $i \geq 3$ due to the orthogonality of Legendre polynomials:

$$\int_{-1}^1 P_{i-2}(t) \, dt = \int_{-1}^1 P_{i-2}(t) P_0(t) \, dt = 0 \text{ when } i \geq 3.$$

Exercise 2.5.6 For constant c_k

$$m_{11}^{(k)} = \frac{c_k \ell_k}{2} \int_{-1}^1 \left(\frac{1}{2}\xi(\xi - 1) \right)^2 d\xi = \frac{2c_k \ell_k}{15}$$

$$m_{13}^{(k)} = \frac{c_k \ell_k}{2} \int_{-1}^1 \frac{1}{2}\xi(\xi - 1)(1 - \xi^2) \, d\xi = \frac{c_k \ell_k}{15}.$$

Exercise 2.5.8

$$r_5^{(k)} = \int_{x_k}^{x_{k+1}} f(x) N_5(Q_k^{-1}(x))\, dx = \frac{f_k \ell_k}{2} \int_{-1}^{1} \sin \frac{1+\xi}{2} \pi\, N_5(\xi)\, d\xi$$

where the shape function $N_5(\xi)$ is defined in Figure 2.9. Using the weights and abscissas in Table B.1 we need to evaluate

$$r_5^{(k)} \approx \frac{f_k \ell_k}{2} \sum_{i=1}^{n} w_i \sin \frac{1+\xi_i}{2} \pi\, N_5(\xi_i)$$

for $n = 3, 4, 5$. The answer for $n = 3$ is $r_5^{(k)} \approx 0.034\,138\,75\, f_k \ell_k$.

Exercise 2.5.10 Using Equation (2.62) we have

$$\frac{dN_i}{d\xi} = \sqrt{\frac{2i-3}{2}} P_{i-2}(\xi), \quad i = 3, 4, \ldots, p+1.$$

Therefore for $i \geq 3$

$$r_i^{(k)} = \sqrt{\frac{2i-3}{2}} \int_{-1}^{+1} \left(\frac{1-\xi}{2} \tau(x_k) + \frac{1+\xi}{2} \tau(x_{k+1}) \right) P_{i-2}\, d\xi$$

and

$$r_i^{(k)} = \begin{cases} (\tau(x_{k+1}) - \tau(x_k))/\sqrt{6} & \text{for } i = 3 \\ 0 & \text{for } i > 3. \end{cases}$$

Exercise 2.5.12

$$[C] = \begin{bmatrix} \dfrac{2AE}{\ell} + \dfrac{c\ell}{6} + k_0 & -\dfrac{2AE}{\ell} + \dfrac{c\ell}{12} & 0 \\[6pt] -\dfrac{2AE}{\ell} + \dfrac{c\ell}{12} & \dfrac{4AE}{\ell} + \dfrac{c\ell}{3} & -\dfrac{2AE}{\ell} + \dfrac{c\ell}{12} \\[6pt] 0 & -\dfrac{2AE}{\ell} + \dfrac{c\ell}{12} & \dfrac{2AE}{\ell} + \dfrac{c\ell}{6} \end{bmatrix} \quad \{r\} = \begin{Bmatrix} k_0 d_0 \\ 0 \\ 0 \end{Bmatrix}.$$

Exercise 2.5.14 Assume that force F_k is acting on node k and no traction load, thermal load or spring load is acting of elements $k-1$ and k. Let us consider the following possibilities: (a) F_k is assigned to element $k-1$. In this case $b_2^{(k-1)} r_2^{(k-1)} = b_2^{(k-1)} F_k$ and $b_1^{(k)} r_1^{(k)} = 0$ where the lower indices refer to the standard element-level numbering and the upper indices refer to the global numbering. (b) F_k is assigned to element k. In this case $b_2^{(k-1)} r_2^{(k-1)} = 0$ and $b_1^{(k)} r_1^{(k)} = b_1^{(k)} F_k$. In order to enforce continuity on the test function we assign $b_2^{(k-1)} = b_1^{(k)} = b_k$ and sum the load vector terms multiplied by b_k. Therefore, following the assembly we will have $b_k F_k$ in either case.

Exercise 2.5.16

$$\begin{bmatrix} 29/5 & 0 & -4\sqrt{21}/35 \\ 0 & 15/7 & 0 \\ -4\sqrt{21}/35 & 0 & 23/15 \end{bmatrix} \begin{Bmatrix} a_3 \\ a_4 \\ a_5 \end{Bmatrix} = \begin{Bmatrix} 33.068 \\ 5.376 \\ 0 \end{Bmatrix}.$$

Exercise 2.5.18 Denoting $e = u_{EX} - u_{FE}$ and $e_k = e(x_k)$, selecting $v = \varphi_k(x)$ as defined in Figure 2.10, and applying the Galerkin orthogonality we get

$$\frac{(AE)_{k-1}}{\ell_{k-1}} \int_{x_{k-1}}^{x_k} e' \, dx - \frac{(AE)_k}{\ell_k} \int_{x_k}^{x_{k+1}} e' \, dx = 0 \quad \text{for } k = 2, 3, \ldots, M(\Delta)$$

and hence

$$-\frac{(AE)_{k-1}}{\ell_{k-1}} e_{k-1} + \left(\frac{(AE)_{k-1}}{\ell_{k-1}} + \frac{(AE)_k}{\ell_k} \right) e_k - \frac{(AE)_k}{\ell_k} e_{k+1} = 0.$$

For $k = 1$ and $k = M(\Delta + 1)$ we have $e_1 - e_2 = 0$ and $-e_{M(\Delta)} + e_{M(\Delta)+1} = 0$ respectively. Therefore we have $M(\Delta) + 1$ tridiagonal homogeneous equations. Assume that the essential boundary condition is prescribed at x_1. In that case $e_1 = 0$ and hence $e_k = 0$ for $k = 1, 2, \ldots, M(\Delta) + 1$. The same would be true if the essential boundary condition were prescribed at $x = x_{M(\Delta)+1}$.

Exercise 2.5.20 Let $T_\Delta = T - T_0 = 1$ K and solve the problem for $u(\ell)$. The gap closes at $u(\ell) = \Delta/2$. Since $u(\ell)$ is a linear function of $T - T_0$, we find $T_c = T_0 + \Delta/(2u(\ell))$ where $\Delta/(2u(\ell))$ is in K (kelvin) units because it is the multiplier of 1 K. Reaction force by the direct method:

$$F_0(T) = AE \left(u'_{FE}(0) - \alpha(T - T_0) \right) \quad \text{for } T_0 < T \le T_c$$
$$F_0(T) = AE \left(u'_c(0) - \alpha(T_c - T_0) \right) - AE\alpha(T - T_c) \quad \text{for } T > T_c$$

where $u'_c(0)$ is $u'_{FE}(0)$ evaluated at $T = T_c$.

Reaction force by the indirect method: the differential equation is

$$-\left(AE(u' - \alpha T_\Delta) \right)' + cu = 0.$$

Multiplying by v and integrating by parts we get

$$\int_0^\ell \left(AE(u' - \alpha T_\Delta)v' \, dx + cuv \right) dx = F_\ell v(\ell) - F_0 v(0)$$

where

$$F_\ell = AE(u' - \alpha T_\Delta)_{x=\ell}, \quad F_0 = AE(u' - \alpha T_\Delta)_{x=0}.$$

APPENDIX F

Selecting $v = 1 - x/\ell$ we have

$$F_0 = -\int_0^\ell (AEu'v' + cuv)\,dx - AE\alpha T_\Delta.$$

Therefore

$$F_0(T) = -\int_0^\ell (AEu'_{FE}v' + cu_{FE}v)\,dx - AE\alpha(T-T_0) \quad \text{for } T_0 < T \leq T_c$$

$$F_0(T) = -\int_0^\ell (AEu'_c v' + cu_c v)\,dx - AE\alpha(T-T_0) \quad \text{for } T > T_c$$

where u_c is u_{FE} evaluated at $T = T_c$.

Exercise 2.5.22 From the definition of nodal forces given by Equation (2.91) we have

$$f_1^{(k)} + f_2^{(k)} = \sum_{j=1}^{p_k+1} \left(k_{1,j}^{(k)} + k_{2,j}^{(k)}\right) a_j^{(k)} - (\bar{r}_1 + \bar{r}_2).$$

Referring to Equation (2.71),

$$k_{1j}^{(k)} + k_{2j}^{(k)} = \frac{2}{\ell_k}\int_{-1}^{+1} \kappa(Q_k(\xi)) \frac{d(N_1+N_2)}{d\xi}\frac{dN_j}{d\xi}\,d\xi = 0.$$

This is because $N_1 + N_2 = 1$. By definition

$$\bar{r}_1 + \bar{r}_2 = \frac{\ell_k}{2}\int_{-1}^{+1} f(Q_k^{-1}(\xi))(N_1+N_2)\,d\xi + F_0(N_1(\xi_0)+N_2(\xi_0)) + \tilde{r}_1 + \tilde{r}_2.$$

Exercise 2.5.24 By definition the nodal forces for element 2 are

$$\left\{\begin{matrix}f_1^{(2)}\\ f_2^{(2)}\end{matrix}\right\} = \begin{bmatrix} \dfrac{2AE}{\ell}+\dfrac{c\ell}{6} & -\dfrac{2AE}{\ell}+\dfrac{c\ell}{12} \\ -\dfrac{2AE}{\ell}+\dfrac{c\ell}{12} & \dfrac{2AE}{\ell}+\dfrac{c\ell}{6} \end{bmatrix}\left\{\begin{matrix}a_1^{(2)}\\ a_2^{(2)}\end{matrix}\right\}$$

where $a_1^{(2)} = a_2$ is computed by the finite element solution and $a_2^{(2)} = \hat{u}_\ell$ is the prescribed boundary condition. The nodal force $f_2^{(2)}$ is the estimated reaction force at $x = \ell$.

Exercise 2.5.26 The exact solution is $u_{EX} = x - x^\alpha$. Therefore $u'_{EX}(0) = 1$. Let $h = 1/M(\Delta)$.

(a) The direct method:

$$u'_{FE}(0) = \frac{u_{FE}(h) - u_{FE}(0)}{h} = \frac{u_{EX}(h)}{h} = 1 - h^{\alpha-1}.$$

The condition that the relative error must not exceed 1% can be stated as

$$\left|\frac{u'_{EX}(0) - u'_{FE}(0)}{u'_{EX}(0)}\right| = h^{\alpha-1} \leq 0.01.$$

Therefore $h \leq 10^{-2/(\alpha-1)}$ from which we have $h \leq 10^{-40}$ or $M(\Delta) > 10^{40}$.

(b) The nodal force method: By definition, the nodal forces for element 1 are

$$\begin{Bmatrix} F_1^{(1)} \\ F_2^{(1)} \end{Bmatrix} = \frac{1}{h}\begin{bmatrix} 1 & -1 \\ -1 & 1 \end{bmatrix}\begin{Bmatrix} u_{FE}(x_1) \\ u_{FE}(x_2) \end{Bmatrix} - \begin{Bmatrix} r_1^{(1)} \\ r_2^{(1)} \end{Bmatrix}$$

where

$$r_1^{(1)} = \alpha(\alpha-1)\int_0^h x^{\alpha-2}\left(1 - \frac{x}{h}\right)dx = h^{\alpha-1}.$$

Using $u_{FE}(x_1) = u_{FE}(0) = 0$ and $u_{FE}(x_2) = u_{FE}(h) = u_{EX}(h) = h - h^{\alpha}$, we find $F_1^{(1)} = -1$. Referring to the definition of nodal forces, for this problem $F_1^{(1)} = -u'_{FE}(x_1)$ and we find that the nodal force method yields $u'_{FE}(0) = 1$, hence the relative error is zero independent of h.

Exercise 3.4.6 We can select, for example, $T_x = \pm 1$ on $x = \pm a$, $T_z = \pm 1$ on $z = \pm c$ and $T_y = \pm y$ on $y = \pm b$ with the other traction components zero.

Exercise 3.4.10 Let the angle between the positive x-axis and the unit normal be α. Then the unit normal is

$$\vec{n} = \cos\alpha \vec{e}_x + \sin\alpha \vec{e}_y \equiv n_x \vec{e}_x + n_y \vec{e}_y.$$

The tangent vector $\vec{t}$ is rotated 90° counterclockwise relative to the normal:

$$\vec{t} = \cos(\alpha + \pi/2)\vec{e}_x + \sin(\alpha + \pi/2)\vec{e}_y$$
$$\equiv -\sin\alpha \vec{e}_x + \cos\alpha \vec{e}_y \equiv -n_y \vec{e}_x + n_x \vec{e}_y.$$

By definition

$$\vec{T} = T_n \vec{n} + T_t \vec{t} = \underbrace{(T_n n_x - T_t n_y)}_{T_x} \vec{e}_x + \underbrace{(T_n n_y + T_t n_x)}_{T_y} \vec{e}_y$$

which was to be shown.

Exercise 6.3.6

1. Considering the symmetric case and $\alpha = 3\pi/2$, Equation (6.33) takes the form

$$\frac{\sin\lambda\alpha}{\lambda\alpha} - \frac{2}{3\pi} = 0.$$

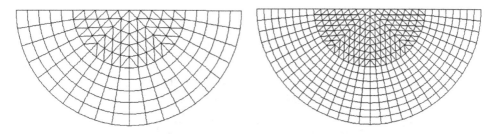

Figure F.1 Exercise 6.4.2: examples of h-refinement, consisting of 200 and 512 elements.

Referring to Figure 6.5 (point D) we see that the lowest root lies between $\pi/2$ and π. Using the root finding routine in Mathematica, or some other program, we find $\lambda\alpha = 2.56581916$, hence $\lambda = \lambda_1 = 0.544483737$.

2. From Equation (6.28)

$$\frac{a_3}{a_1} \equiv Q_1 = -\frac{\cos[(\lambda_1 - 1)3\pi/4]}{\cos[(\lambda_1 + 1)3\pi/4]} = 0.543075579.$$

Therefore the symmetric terms in Equation (6.25) can be written in the form of Equation (6.40). The equivalence of Equation (6.40) and Equation (6.41) is demonstrated by substituting $z = r(\cos\theta + i\sin\theta)$ and $\bar{z} = r(\cos\theta - i\sin\theta)$ into Equation (6.41).

Exercise 6.4.2 Partial solution: having computed finite element solutions corresponding to a sequence of meshes, similar to those shown in Figure F.1, we have information such as that shown in Table F.1.

Using Theorem 2.6.3 and Equation (2.102) we have

$$\|e_{200}\|_E^2 = \pi_{200} - \pi \approx k^2/N_{200}^{2\beta}$$
$$\|e_{512}\|_E^2 = \pi_{512} - \pi \approx k^2/N_{512}^{2\beta}$$

where $\pi = -0.375722076321\, p_0^2 R^2 t/E$. We find $\beta \approx 0.51$.

Table F.1 Data for Exercise 6.4.2.

NEL	N	$\pi E/p_0^2 R^2 t$
200	1079	-0.375627708
512	2687	-0.375685015

Exercise 8.3.6

$$\sigma_x^{(d)} = \frac{1}{d}\int_a^{a+d} \sigma_x(r,\pi/2)\,dr = \frac{\sigma_\infty}{d}\int_a^{a+d}\left(1+\frac{1}{2}\frac{a^2}{r^2}+\frac{3}{2}\frac{a^4}{r^4}\right)dr$$

$$= \sigma_\infty\left(1+\frac{1}{2}\frac{1}{1+d/a}+\frac{3}{2}\frac{1+d/a+d^2/(3a^2)}{(1+d/a)^3}\right)$$

$$= \sigma_\infty\left(1+2\frac{1}{1+d/a}-\frac{3}{2}\frac{d/a+2d^2/(3a^2)}{(1+d/a)^3}\right)$$

$$= \sigma_\infty\left(1+2\frac{1}{1+d/a}+O(d/a)\right).$$

Letting $\alpha = d$, $\varrho = a$ in Equation (8.14) and neglecting terms of order d/a we get

$$K_e = \frac{\sigma_x^{(d)}}{\sigma_\infty} \approx \bar{q}(K_t - 1) + 1$$

where $K_t = 3$ was used.
The relative error for $d/a = 0.2$ is -4.46%.

Exercise 9.1.10 The solution is analogous to the solution of Exercise 9.1.1. Letting

$$\mathbf{w}(x,y) \equiv \begin{Bmatrix} w_x \\ w_y \end{Bmatrix} = \begin{Bmatrix} -w'(x)y \\ w(x) \end{Bmatrix}, \quad \mathbf{v}(x,y) \equiv \begin{Bmatrix} v_x \\ v_y \end{Bmatrix} = \begin{Bmatrix} -v'(x)y \\ v(x) \end{Bmatrix}$$

and assuming that $\varrho = \varrho(x)$, we get

$$\int_0^\ell EIw''v''\,dx + \int_0^\ell c_s wv\,dx - \omega^2\int_0^\ell \varrho(x)(Iw'v' + Awv)\,dx = 0.$$

The extraction function for M_0 is the unit rotation at $x = 0$, which is $v = -N_2(\xi)$. The sign is negative because a positive moment would cause negative rotation. Therefore

$$M_0 = B(w,v) = k_{12}\delta - k_{24}\frac{3\delta}{\ell} = 12\frac{EI\delta}{\ell^2}.$$

Exercise 9.2.4 We have to interpret the potential

$$P =: F(\mathbf{u}) = \int_\Omega F_i u_i\,dV + \int_{\partial\Omega_T} \hat{T}_i u_i\,dS$$

in Equation (4.16) in terms of the dimensionally reduced problem of the Reissner–Mindlin plate. Letting $F_1 = F_2 = 0$, $F_3 = q/d$ and $u_1 = u_2 = 0$, $u_3 = w$, we have

$$\int_{\Omega_{3D}} F_i u_i\,dV = \int_{\Omega_{2D}}\int_{-d/2}^{d/2} \frac{q}{d}w\,dz\,dx\,dy = \int_{\Omega_{2D}} qw\,dx\,dy.$$

Bibliography

[1] Actis R, Szabó BA and Schwab C. Hierarchic models for laminated plates and shells. *Comput. Methods Appl. Mech. Eng.* **172** (1999) 79–107.

[2] Ainsworth M and Oden JT. *A Posteriori Error Estimation in Fiinite Element Analysis*. John Wiley & Sons, Inc., New York, 2000.

[3] ASTM E 1049-85 (Reapproved 1997). Standard Practices for Cycle Counting in Fatigue Analysis. In: *Annual Book of ASTM Standards*, Vol. 03.01 Philadelphia, 1999, pp. 710–718.

[4] Babuška I, d'Harcourt JM and Schwab C. Optimal shear correction factors in hierarchical plate modelling. *Math. Modelling Sci. Comput.* **1** (1993) 1–30.

[4a] Babuška I, Szabó BA and Katz IN. The p-version of the finite element method. SIAM J. Numer. Anal. **18** (1981) 515–545.

[5] Babuška I and Miller A. The post-processing approach in the finite element method - Part 1: Calculation of displacements, stresses and other higher derivatives of the displacements. *Int. J. Numer. Meth. Eng.* **20** (1984) 1085–1109.

[6] Babuška I and Oden JT. Verification and validation in computational engineering and science: basic concepts. *Comput. Methods Appl. Mech. Eng.* **193** (2004) 4057–4066.

[7] Babuška I and Oden JT. Reliability of computer predictions: can they be trusted? *Int. J. Numer. Anal. Modeling* **3** (2006) 255–272.

[8] Babuška I and Osborn J. Eigenvalue problems. In: *Handbook of Numerical Analysis*, P. G. Ciarlet and J. L. Lions, editors. Elsevier, North-Holland, Amsterdam, Vol. II, 1991, 642–787.

[9] Babuška I and Rheinboldt WC. A posteriori error estimates for adaptive finite element computations. *SIAM J. Numer. Anal.* **15** (1978) 736–754.

[10] Babuška I and Scapolla T. Benchmark computation and performance evaluation for a rhombic plate bending problem. *Int. J. Numer. Meth. Eng.* **28** (1989) 155–179.

[11] Babuška I and Silva RS. Numerical treatment of engineering problems with uncertainties. The fuzzy set approach and its application to the heat exchanger problem. *Int. J. Numer. Meth. Eng.* Published in electronic form in 2010. To appear in 2011.

[12] Babuška I and Strouboulis T. *The Finite Element Method and its Reliability*. Oxford University Press, Oxford, 2001.

[13] Babuška I and Suri M. The p- and hp-versions of the finite element method: an overview. *Comput. Methods: Appl. Mech. Eng.* **80** (1990) 5–26.

[14] Babuška I and Szabó B. On the generalized plane strain problem in thermoelasticity. *Comput. Methods Appl. Mech. Eng.* **195** (2006) 5390–5402.

[15] Babuška I and Tempone R. Static frame challenge problem: summary. *Comput. Methods Appl. Mech. Eng.* **197** (2008) 2572–2577.

[16] Babuška I, Whiteman JR and Strouboulis T. *Finite Elements: An Introduction to the Method and Error Estimation.* Oxford University Press, Oxford, 2010.

[17] Bathias C and Paris PC. *Gigacycle Fatigue in Mechanical Practice.* Marcel Dekker, New York, 2005.

[18] Brezzi F and Fortin M. *Mixed and Hybrid Finite Element Methods.* Springer-Verlag, New York, 1991.

[19] Chen Q and Babuška I. Approximate optimal points for polynomial interpolation of real functions in an interval and in a triangle. *Comput. Methods Appl. Mech. Eng.* **128** (1995) 405–417.

[20] Ciarlet PG. *The Finite Element Method for Elliptic Problems.* SIAM, Philadelphia, 2002.

[21] Clough RW and Tocher JL. Finite element stiffness matrices for analysis of plate bending. In: *Proceedings of Conference on Matrix Methods in Structural Mechanics,* Report AFFDL-TR-66-80, Wright–Patterson Air Force Base, Ohio (1966) 515–545.

[22] Corum JM, Bolt SE, Greenstreet WL and Gwaltney RC. Theoretical and experimental stress analysis of ORNL thin-shell cylinder-to-cylinder model No. 1. Oak Ridge National Laboratory Report ORNL 4553, Oak Ridge, TN, (October 1972).

[23] Costabel M, Dauge M and Suri M. Numerical approximation of a singularly perturbed contact problem. *Comput. Methods Appl. Mech. Eng.* **157** (1998) 349–363.

[24] Cowper GR. The shear coefficient in Timoshenko's beam theory. *J. App. Mech.* **33** (June 1966) 335–340.

[24a] Demkowicz L. *Computing with hp-adaptive finite elements. Vol. 1: One and two-dimensional elliptic and Maxwell problems.* Chapman & Hall/CRC Boca Raton 2007.

[25] Demmel JW. *Applied Numerical Linear Algebra.* SIAM, Philadelphia, 1997.

[26] Düster A. High order finite elements for three-dimensional thin-walled nonlinear continua. Dissertation. Technische Universität München, 2002.

[27] Düster A and Rank E. The p-version of the finite element method compared to an adaptive h-version for the deformation theory of plasticity. *Comput. Methods Appl. Mech. Eng.* **190** (2001) 1925–1935.

[28] Ferson S, Oberkampf WL and Ginzburg L. Model validation and predictive capability for the thermal challenge problem. *Comput. Methods Appl. Mech. Eng.* **197** (2008) 2408–2430.

[29] Fung YC. *Foundations of Solid Mechanics.* Prentice Hall, Englewood Cliffs, NJ, 1965.

[30] Girkmann K. *Flächentragwerke,* 4th edition. Springer-Verlag, Vienna, 1956.

[31] Golub GH and Van Loan CF. *Matrix Computations.* The Johns Hopkins University Press, Baltimore, MD, 1989.

[32] Gui B and Babuška I. The *h-, p-* and *hp-*versions of the FEM in 1 dimension. Parts I, II, and III. *Num. Math.* **80** (1986) 577–683.

[33] Guide for Verification and Validation in Computational Solid Mechanics. ASME V&V 10-2006, American Society of Mechanical Engineers, New York (2006).

[34] Gwaltney RC, Corum JM, Bolt SE and Bryson JW. Experimental stress analysis of cylinder-to-cylinder shell models and comparisons with theoretical predictions. *J. Pressure Vessel Technol.* (1976) 283–290.

[35] Hu S and Pagano NJ. On the use of a plane strain model to solve generalized plane strain problems. *J. Appl. Mech.* **64** (1997) 236–238.

[36] Kanninen MF and Popelar CH. *Advanced Fracture Mechanics.* Oxford University Press, New York, 1985.

[37] Karniadakis GE and Sherwin SJ. *Spectral/hp Element Methods for CFD.* Oxford University Press, Oxford, 1999.

[38] Királyfalvi G. and Szabó B. Quasi-regional mapping for the p-version of the finite element method. *Finite Elem. Anal. Des.* **27** (1997) 85–97.

[38a] Kleindorfer GB, O'Neill R and Ganeshan R. Validation in Simulation: Various Positions in the Philosophy of Science. *Management Science.* **44** (1998) 1087–1099.

[39] Kozlov VA, Mazia VG and Rossmann J. *Spectral Problems Associated with Corner Singularities of Solutions of Elliptic Equations.* American Mathematical Society, Providence, RI, 2000.

[40] Kuhn P and Hardrath HF. An engineering method for estimating notch-size effect in fatigue tests of steel. NACA Technical Note 2805 (1952).

[41] Lazzeri L and Mariani U. Application of damage tolerance principles to the design of helicopters. *Int. J. Fatigue* **31** (2009) 1039–1045.

[42] Mikhlin SG. *The Numerical Performance of Variational Methods.* Wolters-Noordhoff, Groningen, 1971.

[43] Military Handbook MIL-HDBK-5H: Metallic Materials and Elements for Aerospace Vehicle Structures. US Department of Defense, December 1, 1998.

[44] Morley LSD. *Skew Plates and Structures.* Macmillan, New York, 1963.

[45] Muskhelishvili NI. *Some Basic Problems of the Mathematical Theory of Elasticity*, 3rd edition Published in Russian in 1949. English translation by J. R. M. Radok. Noordhoff, Groningen, 1953.

[46] Naghdi PM. Foundations of elastic shell theory. In: *Progress in Solid Mechanics*, Vol. 4. North-Holland, Amsterdam, 1963.

[46a] Nazarov SA and Plamenevsky BA. *Elliptic Problems in Domains with Piecewise Smooth Boundaries.* Walter de Gruyter & Co. Berlin 1994.

[47] Nervi S and Szabó BA. On the estimation of residual stresses by the crack compliance method. *Comput. Methods Appl. Mech. Eng.* **196** (2007) 3577–3584.

[48] Nervi S, Szabó BA and Young KA. Prediction of distortion of airframe components made from aluminum plates. *AIAA J.* **47** (2009) 1635–1641.

[49] Neuber H. *Kerbspannungslehre.* Springer-Verlag, Berlin, 1937.

[50] Novozhilov VV. *Thin Shell Theory.* Noordhoff, Groningen, 1964.

[51] Oberkampf WL and Roy CJ. *Verification and Validation in Scientific Computing.* Cambridge University Press, Cambridge, 2010.

[52] Oberkampf WL and Trucano TG. Verification and validation benchmarks. Sandia Report SAND2007-0853 (February 2007).

[53] Oreskes N, Shrader-Frechette K and Belitz K. Verification, validation and confirmation of numerical models in the earth sciences. *Nature* **263** (1994) 641–646.

[54] Osgood CC. *Fatigue Design.* John Wiley & Sons, Inc., New York, 1970.

[55] Overgaard V and Hutchinson JW. Effect of T-stress on mode I crack growth resistance in a ductile solid. *Int. J. Solids Struct.* **31** (1994) 823–833.

[56] Páczelt I and Mróz Z. Optimal shape of contact interfaces due to wear in the steady relative motion. *Int. J. Solids Struct.* **44** (2006) 1–30.

[57] Páczelt I, Szabó BA and Szabó T. Solution of contact problem using the hp-version of the finite element method. *Comput. Math. App.* **38** (1999) 49–69.

[58] Papadakis PJ and Babuška I. A numerical procedure for the determination of certain quantities related to the stress intensity factors in two-dimensional elasticity. *Comput. Methods Appl. Mech. Eng.* **122** (1995) 69–92.

[59] Paris PC, Gomez MP and Anderson WE. A rational analytic theory of fatigue. *Trend Eng.* **13** (1961) 9–14.

[60] Peterson RE. *Stress Concentration Factors*. John Wiley & Sons, Inc., New York, 1974.

[61] Petroski H. *To Engineer is Human*. Vintage, New York, 1992.

[62] Petroski H. *Design Paradigms: Case Histories of Error and Judgment in Engineering*. Cambridge University Press, New York, 1994.

[63] Pitkäranta J, Matache A-M and Schwab C. Fourier mode analysis of layers in shallow shell deformation. *Comput. Methods Appl. Mech. Eng.* **190** (2001) 2943–2975.

[64] Polyanin AD. *Linear Partial Differential Equations for Engineers and Scientists*. Chapman & Hall/CRC Press, Boca Raton, FL, 2002.

[65] Rank E and Babuška I. An expert system for the optimal mesh design in the hp-version of the finite element method. *Int. J. Numer. Meth. Eng.* **24** (1987) 2087–2106.

[66] Reddy JN. *An Introduction to Nonlinear Finite Element Analysis*. Oxford University Press, Oxford, 2004.

[67] Rektorys K. *Variational Methods in Mathematics, Science and Engineering*, 2nd edition, Reidel, Dordrecht 1980.

[68] Roache PJ. *Verification and Validation in Computational Science and Engineering*. Hermosa, Socorro, NM, 1998, pp. 403–412.

[69] Schijve J. *Fatigue of Structures and Material*, 2nd edition. Springer-Verlag, Berlin, 2009.

[70] Schwab C. *p- and hp-Finite Element Methods: Theory and Applications in Solid and Fluid Mechanics*. Clarendon Press, Oxford, 1998.

[71] Schwab C, Suri M and Xenophontos C. The hp finite element method for problems in mechanics with boundary layers. *Comput. Methods Appl. Mech. Eng.* **157** (1998) 311–333.

[72] Simo JC and Hughes TJR. *Computational Inelasticity*. Springer-Verlag, New York, 1998.

[73] Simo JC and Laursen TA. An augmented Lagrangian treatment of contact problems involving friction. *Comput. Struct.* **42** (1992) 97–116.

[74] Sivia DS. *Data Analysis: A Bayesian Tutorial*. Oxford University Press, Oxford, 1996.

[75] Sokolnikoff IS. *Mathematical Theory of Elasticity*, 2nd edition. McGraw-Hill, New York, 1956.

[76] Suresh S. *Fatigue of Materials*, 2nd edition. Cambridge University Press, Cambridge, 1998.

[77] Szabó B and Actis R. On the importance and uses of feedback information in FEA. *Appl. Numer. Math.* **52** (2005) 219–234.

[78] Szabó B and Actis R. On the role of hierarchic spaces and models in verification and validation. *Comput. Methods Appl. Mech. Eng.* **198** (2009) 1273–1280.

[79] Szabó BA and Babuška I. Computation of the amplitude of stress singular terms for cracks and reentrant corners. In: *Fracture Mechanics: Nineteenth Symposium ASTM STP 969*, T. A. Cruse, editor, American Society for Testing and Materials, Philadelphia, 1988, pp. 101–124.

[80] Szabó B and Babuška I. *Finite Element Analysis*. John Wiley & Sons, Inc., New York, 1991.

[81] Szabó B, Babuška I, Pitkäranta J and Nervi S. The problem of verification with reference to the Girkmann problem. *Eng. Comput.* **26** (2010) 171–183. doi: 10.1007/s00366-009-0155-0.

[82] Szabó B, Düster A and Rank E. The p-version of the Finite Element Method. In: *Encyclopedia of Computational Mechanics. Volume 1: Fundamentals*, E. Stein, R. de Borst and T. J. R. Hughes, editors. John Wiley & Sons, Ltd, Chichester, 2004, Chapter 5.

[83] Szabó BA and Muntges DE. Procedures for the verification and validation of working models for structural shells. *J. Appl. Mech.* **72** (2005) 907–915.

[84] Szabó BA and Sahrmann GJ. Hierarchic plate and shell models based on p-extension. *Int. J. Numer. Meth. Eng.* **26** (1988) 1855–1881.

[85] Szabó BA and Yosibash Z. Numerical analysis of singularities in two dimension. Part 2: Computation of generalized flux/stress intensity factors. *Int. J. Numer. Meth. Eng.* **39** (1996) 409–434.

[86] Taylor D. *The Theory of Critical Distance: A New Perspective in Fracture Mechanics*. Elsevier, Amsterdam, 2007.

[87] Timoshenko S and Goodier JN. *Theory of Elasticity*, 2nd edition. McGraw-Hill, New York, 1951.

[88] Timoshenko S and Woinowsky-Krieger S. *Theory of Plates and Shells*, 2nd edition. McGraw-Hill, New York, 1959.

[89] Vasilopoulos D. On the determination of higher order terms of singular elastic stress fields near corners. *Numer. Math.* **53** (1998) 51–95.

[90] Vigilante VN, Underwood JH and Crayon D. Use of instrumented bolt and constant displacement bolt-loaded specimen to measure in-situ hydrogen crack growth in high-strength steels. In: *Fatigue and In: Fracture Mechanics: 30th Volume, ASTM STP 1360*, P. C. Paris and K. L. Jerina, editors, American Society for Testing and Material, West Conshohocken, PA, 2000, pp. 377–387.

[91] Walker K. The effect of stress ratio during crack propagation and fatigue for 2024-T3 and 7075-T6 aluminum. *ASTM STP 462*. American Society for Testing and Materials, Philadelphia, 1970.

[92] Wriggers P. *Computational Contact Mechanic*, 2nd edition. Springer-Verlag, Berlin and Heidelberg, 2006.

[93] Yosibash Z and Szabó BA. Generalized stress intensity factors in linear elastostatics. *Int. J. Fract.* **72** (1995) 223–240.

[94] Yosibash Z and Szabó BA. Numerical analysis of singularities in two dimension. Part 1: Computation of eigenpairs. *Int. J. Numer. Meth. Eng.* **38** (1995) 2055–2082.

[95] Young WC and Budynas RG. *Roark's Formulas for Stress and Strain*, 7th edition. McGraw-Hill, New York, 2002.

[96] Zienkiewicz OC and Zhu JZ. The superconvergent patch recovery and a posteriori error estimates. Part 1: The recovery technique. *Int. J. Numer. Meth. Eng.* **33** (1992) 1331–1364.

[97] Zienkiewicz OC and Zhu JZ. The superconvergent patch recovery and a posteriori error estimates. Part 2: Error estimates and adaptivity. *Int. J. Numer. Meth. Eng.* **33** (1992) 1365–1382.

[98] Zienkiewicz OC and Zhu JZ. The superconvergent patch recovery (SPR) and adaptive finite element refinement. *Comput. Meth. Appl. Mech. Eng.* **101** (1992) 207–224.

Index

a priori estimate, 72
adaptive methods, 211
admissible data, 168, 173
admissible functions, 35, 118, 120, 141
Airy equation, 26
Airy stress function, 176
analytic arc, 168, 194
analytic curves, 318
analytic function, 167, 318
antisymmetry, 21, 29, 175
assembly, 55
asymptotic consistency, 277, 285
augmented Lagrangian method, 312

bar force, 18
basis function, 29, 32, 39
basis vectors
    curvilinear, 283
Bayesian analysis, 7
beam models, 261
    Bernoulli–Euler beam, 269
    Timoshenko beam, 265
bending moment, 274
biharmonic equation, 176, 281
bilinear form, 36, 314
boundary condition, 83
    antisymmetric, 22
    antisymmetry, 85, 99, 126
    convection, 83
    Dirichlet, 29, 35, 93, 110

    displacement, 20, 98
    essential, 29, 35, 58
    flux, 83
    force, 21
    homogeneous, 20, 35, 93, 134
    kinematic, 20, 98
    mixed, 29
    naural, 29
    Neumann, 29, 37, 69, 70, 93, 110, 111
    periodic, 22, 35, 85, 99
    radiation, 297
    Robin, 29, 37, 93, 110
    spring, 21
    symmetric, 22, 85
    symmetry, 99
    temperature, 83
    traction, 98
    Winkler spring, 98
boundary layer, 206, 288, 290

CAD tools, 3
calibration, 4, 24
characteristic function, 135
characteristic value, 135
coefficient
    of permeability, 92
    of convective heat transfer, 84
    of dynamic viscosity, 105
    of thermal conductivity, 82, 89, 216

Introduction to Finite Element Analysis: Formulation, Verification and Validation, First Edition.
Barna Szabó and Ivo Babuška. © 2011 John Wiley & Sons, Ltd. Published 2011 by John Wiley & Sons, Ltd.

coefficient (*Cont.*)
   of thermal expansion, 95
   spring, 98
coefficient matrix, 57
complex potentials, 176
concentrated force, 53, 186
conceptualization, 2, 24, 237, 241, 313
conservation law, 81
consistency condition, 308
constant
   Stefan–Boltzmann, 298
constraint
   rigid body, 120, 125, 139
contact, 312
continuity, 48, 69
   C^1, 282
   exact and minimal, 48, 282
contour integral, 161
contour integral method, 231, 233
contour plot, 216
convergence, 70, 143, 225, 317
   algebraic, 74, 193, 198
   asymptotic, 193, 198
   exponential, 193, 198
   monotonic, 128
   pre-asymptotic, 198
   radius of, 174
coordinates
   Eulerian, 299
   Lagrangian, 299
   triangular, 149
crack, 168
cycle ratio, *see* stress ratio, 240
cyclic frequency, 135

d'Alembert's principle, 96
damage accumulation, 237
Darcy's law, 92
data of interest, 23, 31
degrees of freedom, 39, 59
design rules, 12
differential volume, 159
differentiation, 161
dimensional reduction, 86
director functions, 261, 284
Dirichlet, *see* boundary condition, 93

discretization, 2, 7
displacement formulation, *see* formulation, 143
display grid, 216
divergence theorem, 80, 109, 116
domain, 3, 39
drivers of damage accumulation, 238

effectivity index, 74, 199, 209
eigenfunction, 135
   Mode I, 178
   Mode II, 178
eigenpair, 135
eigenvalue, 135, 178
element
   constrained linear strain triangle (CLST), 291
   eight-node quadrilateral, 148
   four-node quadrilateral, 148
   Hsieh–Clough–Tocher (HCT) triangle, 282, 291
   nine-node quadrilateral, 149
   six-node triangle, 149
   three-node triangle, 149
emissivity, 298
endurance limit, 241
energy norm, *see* norm, 41
energy release rate, 233, 332
energy space, *see* space, 34
equations of motion, 97
equations of static equilibrium, 97
equilibrium
   dynamic, 96
equivalent stress, 306
error estimation, 73, 199, 211
error indicator, 212
errors
   conceptual, 8
   modeling, 2, 5, 296
   of approximation, 41
   of discretization, 2, 7
   of idealization, 2
   pollution, 222, 223, 225
   programming, 8
exact solution, 43
existence, *see* generalized solution, 112
extension operator, 212

INDEX 361

extraction, 2, 8, 212
extraction function, 63, 75, 221, 230

factor of safety, 11
fatigue limit, 241
fatigue strength, 241
feedback information, 222
feedback methods, 211
field functions, 261
fillets, 139, 200
finite element, 39
   hexahedral, 157
   pentahedral, 157
   standard quadrilateral, 145
   standard triangular, 145
   tetrahedral, 157
finite element mesh, 39
finite element model, 9
finite element solution, 43
   uniqueness of, 143
finite element space, *see* space, 196
flux, 92
   heat, 81
flux intensity factor, 213, 228
forcing function, 185
   inadmissible, 186
formulation
   displacement, 143
   mixed, 141, 143
Fourier's law, 81
fracture mechanics, 232, 250
free surface, 92
functional
   linear, 314
   quadratic, 68

Galerkin orthogonality, 43
generalized formulation, 34
generalized solution, 111
   existence of, 112
   uniqueness of, 67, 112, 119, 143
Girkmann's problem, 227
grading factor, 48
Green's function, 76

h-extension, 48, 193
h-method, 194

h-version, 194
Hartford Civic Center Arena, 9
hierarchic finite element spaces, *see*
   hierarchic spaces, 193
hierarchic spaces, 48, 128, 193, 222
Hooke's law, 18, 95, 101
   generalized, 96
hp-extension, 48, 193

incompressibility
   condition of, 104
index notation, 79
inequality
   Schwarz, 34, 70, 319
   triangle, 13, 315
initial condition, 83, 84, 99, 134
input data, 168
integration, 159
isoparametric mapping, *see* mapping, 152

Jacobian
   determinant, 159
   matrix, 159, 216
      inverse of, 161

Kármán vortices, 9
Kronecker delta, 80

L-shaped domain, 197, 226
Lamé constants, 95
Laplace equation, 173
Legendre polynomial, 45, 152, 317, 322
linear functional, 36
linear independence, 32
linear space, 70
load spectrum, 243
locking, 143, 287
   shear, 268

mapping
   by blending functions, 154
   high-order elements, 156
   improper, 160
   inverse, 44, 155, 215
   isoparametric, 152, 153, 156
   linear isoparametric, 152
   proper, 159

mapping (*Cont.*)
  quadratic isoparametric, 152
  subparametric, 153
  superparametric, 153
mapping function, 44
margin of safety, 12
material stiffness matrix, 119
Mathematica, 130
mathematical model, 2, 14
  hierarchic models, 106, 261
  working models, 4
matrix
  Gram, 50
  mass, 51
  orthogonal, 328
  stiffness, 49
mechanical contact, 310
membrane force, 274
mesh
  geometric, 48, 189, 197, 198
  irregular, 193
  quasiuniform, 48, 194
  radical, 48, 198
  regular, 193
metal fatigue, 238
mixed formulation, *see* formulation, 141
Mode I eigenfunctions, 178
mode of vibration, 135
model, *see* mathematical model, 2
modulus of elasticity, 95
modulus of rigidity, *see* shear modulus, 96
Mohr circle, 276
Monte Carlo method, 3

natural frequency, 135
Navier equations, 97
Neumann, *see* boundary condition, 93
Newton–Raphson method, 25, 155, 215
nodal force, 64, 192, 217
node point, 39, 44
  irregular, 193
nominal stress, 242
norm, 41, 313
  L_2 norm, 42
  energy, 41
  maximum, 42, 190

notch sensitivity, 246
  index, 247
numbering of basis functions, 55

p-extension, 48, 193
p-method, 194
p-version, 194
Paris ΔK region, 250
Paris' law, 250
path-independent integral, 229
periodic boundary condition, *see* boundary condition, 22
piecewise analytic function, 168
piezometric head, 92
plane strain, 101
  generalized, 101, 252
plane stress, 101, 277
plasticity, 306
plate constant, 278
plates
  Kirchhoff plate, 280
  Reissner–Mindlin plate, 276
Poisson's ratio, 95
pollution error
  seeerrors, 222
potential energy, 32, 68, 111, 127
  sign of, 68
potential flow, 81
potential function, 92
prediction, 27, 28
pressure, 105
principal direction, 84, 327
principal moments, 276
principal stress, 327
principle of minimum potential energy, 68
  application to beams, 263
  application to plates, 278
principle of virtual work, 117
problem
  in Category A, 170, 196
  in Category B, 172, 196
  in Category C, 173, 201
process zone, 238
product space, 147, 150, 195
  anisotropic, 287
projection, 76

proportional limit, 23
pull-back polynomials, 154

quadrature, 160, 321
    Gauss–Lobatto, 321, 324
    Gaussian, 130, 321
quarter-point element, 49

radiation, 84
rate of convergence, 74, 222
Rayleigh quotient, 136, 138
Rayleigh–Ritz method, 32, 41
region of primary interest, 222, 335
region of secondary interest, 222
regularity of functions, 167
reliability of estimators, 74
representative volume element, 238
resolvent set, 304
Reynolds number, 105
Richardson extrapolation, 211
rigid body
    displacement, 67, 120
    rotation, 120, 156, 301
rigid body constraint, *see* constraint, 139
Ritz method, *see* Rayleigh–Ritz method, 32
Robin, *see* boundary condition, 93
robustness
    of error estimators, 209

S–N curve, 240
saddle point, 142
Saint-Venant's principle, 187, 337
secant modulus, 309
seminorm, 67, 315
separation of variables, 134
serendipity space
    see trunk space, 146
shape function vector, 162
shape functions, 45
    three-dimensional, 157
    hierarchic, 46
    Lagrange, 45, 147
shear correction factor, 264, 266, 277, 279
shear force, 274
shear modulus, 96

shell
    hierarchic models, 284
    hyperboloidal, 287
    structural, 283
shell model
    Naghdi, 285
    Novozhilov–Koiter, 286
singular point, 168, 225
    geometric, 168
    neighborhood of, 172
singularity
    degree of, 173, 175, 201
    strong, 193, 290
    weak, 193
Sleipner A offshore platform, 10
smoothness of functions, 167
solvability, 143
solver
    direct, 61
    iterative, 61
source function, 92
space
    energy, 34, 118
    Euclidean, 41, 79
    finite element, 7, 39, 47, 113, 145, 193, 196
span, 32, 33
spectrum, 135, 304
    continuous, 304
    point, 304
SRQ, *see* system response quantity, 2
standard element, 44, 157
stationary problems, 83
steady state problems, 83
Steklov method, 183
stiffness matrix, 50
    global unconstrained, 57
Stokes equations, 105
strain
    Almansi, 299
    engineering shear, 94
    equivalent elastic, 307
    Green, 300
    infinitesimal, 93
    linear, 301
    mechanical, 18, 95
    normal, 93

strain (*Cont.*)
　shear, 94
　thermal, 18, 95
　total, 18, 95
　volumetric, 104
strain energy, 34
stress
　residual, 3, 6, 96
　resultant, 265
　RMS measure of, 75
　transformation of, 327
stress intensity factor, 228, 233, 332
stress invariants, 327
stress ratio, 240
stress stiffening, 303, 304
stress vector, 325
StressCheck, 77, 130
strong form, 79
superconvergence, 76
surface integral, 161
symmetry, 21, 29, 137, 175
　axial, 89, 102
　line of, 21
　mirror image, 21
system response quantity, 2

T-stress, 180, 233
Tacoma Narrows Bridge, 9
Taylor series, 171
temperature
　absolute, 84, 297
test function, 34, 35, 118, 141
test space, 35, 317
thermal load, 54
thin-solid models, 286
traction, 97
traction force, 18

traction load, *see* traction force, 18
traction vector, 325
trial function, 29, 34, 35, 39
trial space, 35, 317
trunk space, 146, 150
　anisotropic, 287, 290
twisting moment, 274

uncertainty, 106
　aleatory, 3, 15, 237
　epistemic, 3, 15, 28, 237
　quantification (UQ), 295
　statistical, 3
uniqueness, *see* generalized solution, 67
units of physical data, 17

Vaiont Dam, 10
validation, 5, 14, 27, 28, 200, 256
　criteria, 6
　metric, 6
vector
　transformation of, 328
verification, 8, 15, 20, 200, 222
virtual displacement, 116
virtual experimentation, 4, 290, 297, 313
virtual work, 116
viscosity, 105
von Mises stress, 306

Wöhler curve, *see* S–N curve, 240
weak form, 79
Winkler spring, *see* boundary condition, 98

Zienkiewicz–Zhu estimator, 210
ZZ estimator, *see* Zienkiewicz–Zhu estimator, 210